High-Level Data Fusion

For a listing of recent titles in the
Artech House Electronic Warfare Library,
turn to the back of this book.

High-Level Data Fusion

Subrata Das

ARTECH
HOUSE

BOSTON | LONDON
artechhouse.com

Library of Congress Cataloging-in-Publication Data
A catalog record for this book is available from the Library of Congress.

British Library Cataloguing in Publication Data
A catalogue record for this book is available from the British Library.

ISBN-13: 978-1-59693-281-4

Cover design by Igor Valdman

685 Canton Street
Norwood, MA 02062

10 9 8 7 6 5 4 3 2 1

For my family…

Who always wanted me to succeed and whom I have always drawn inspiration from.

Table of Contents

Preface

High-level data fusion (or, simply, high-level fusion) is the study of relationships among objects and events of interest within a dynamic environment. The study is supported by analyses of data produced by the sensors placed within the environment. By dynamic we mean the state of the environment, and hence relationships among its objects and events, changes due to both natural/internal events and external events by players (also called actions) within the environment. Therefore, assessing the impact of a current situation is also part of the study.

A battlefield operational area along with its friendly, adversarial, and neutral objects and entities constitute, for example, a battlefield dynamic environment. In the cyber domain, a dynamic environment includes networks, computers, firewalls, and human operators. A dynamic space environment is comprised of satellites, ground stations, and human operators. Such dynamic environments are associated with one or more fusion centers, such as intelligence cells and control rooms, to fuse data collected by the sensors. Other examples of fusion centers include power plant control rooms, city traffic centers, patient monitoring rooms, business transaction processing units, and disease surveillance centers.

Objects in a dynamic environment are something physical that can be perceived, such as aircrafts, soldiers, vehicles, satellites, tanks, missiles, bridges, houses, and explosives. Events are observed physical reality represented in space and time involving objects. An example event would be a vehicle crossing the bridge in the eastern part of the city around noontime today. An interdiction by a hostile force or a bio-attack is a high-level event that changes its respective environment. An interdiction event, for example, changes the environment by placing units involved at a certain location and formation. Relationships are high-level abstract concepts conceived in mind via appropriate spatiotemporal arrangements and aggregation of interrelated objects and events. For example, a defensive formation defines certain spatial relationships among the units involved. The study of relationships in data fusion spans across past, current, and future, both spatially and temporally, and reaches to a higher order by recursively considering relationships of relationships.

High-level fusion thus provides estimation of hostile entity states for enhanced awareness of situations, and threats they pose. This book provides a computational account of automated high-level data fusion to include issues such as representation of relationships among objects and events in terms of models, reasoning with models for inferencing situations and threats, handling of associated uncertainty in inferencing, and adaptation of models over time. In the sense of the revised Joint Directors of Laboratories (JDL) model for fusion processes (Steinberg et al., 1998), this book covers Level 2 fusion (a.k.a. situation assessment) and limited Level 3 fusion (a.k.a. threat or impact assessment), but not JDL Level 4 (a.k.a. process refinement or collection management).

A threat assessment task inevitably involves analyses of Courses of Action (COAs) planning of friendly and adversarial forces. But this book does not cover detailed COA generation techniques, though very high-level COAs as hypotheses will be evaluated using the Expected Utility Theory (EUT) and symbolic argumentation-based decision making. The target tracking aspect of Level 1 fusion (a.k.a. object refinement) has been presented for the sake of completeness. This book also covers those aspects of fusion that fall between Levels 1 and 2 (explained in the next chapter why we term as Level 1½). In the rest of the book, the phrase "high-level fusion" would mean aspects of fusion at levels higher than Level 1.

Our view is necessarily a closer interaction between Level 1 and high-level fusion processes. Not only an assessment of situation requires results from Level 1 processes of estimating an entity's position, velocity, attributes, and identity, but results from high-level fusion processes can also be fed back to the Level 1 process for updating states of tracked entities in the environment, and for providing prior information about new entities. In addition to a systematic account of situations and threat assessments, the book goes beyond the JDL model via an associated generation of appropriate response recommendations for decision making under uncertainty. From this point of view, we discuss EUT and argumentation to evaluate hypotheses of alternative actions.

Our approach to high-level fusion is model-based in that inferences for situation and threat assessments in the context of a problem are made on a symbolic and/or subsymbolic representation of the problem, called a model of the problem. The emphasis on building temporal models reflects the fact that we are not only dealing with situation and threat snapshots but also their evolution. Moreover, models are not necessarily static prebuilt and monolithic but will be adapted over time via learning from events as they occur. Finally, we introduce intelligent agent and distributed fusion technologies with a view to carry out high-level fusion collaboratively in Network Centric Warfare (NCW) environments to generate a common picture.

So how do we build these models? By observing various battlefield situations as they unfold and by interacting with peers and with battlefield artifacts, such as sensors and information processing systems, intelligence analysts form internal, mental models of things that they are observing and with which they are interacting. These models provide predictive and explanatory power for understanding a specific situation at hand, for which there are no algorithmic solutions. This implies that one needs to capture the mental model of an analyst to "automate" the situation understanding process, which is critical in the presence of huge volumes of data continuously generated by many battlefield sensors. In the consideration of capturing an intelligence analyst's mental model, one must consider following four things, as shown in Figure P-1:

1) the analyst's mental model;
2) the knowledge engineer's conceptualization of that mental model;
3) the target system that captures the analyst's mental model for prediction and explanation of situations; and
4) the computational model of the target fusion system.

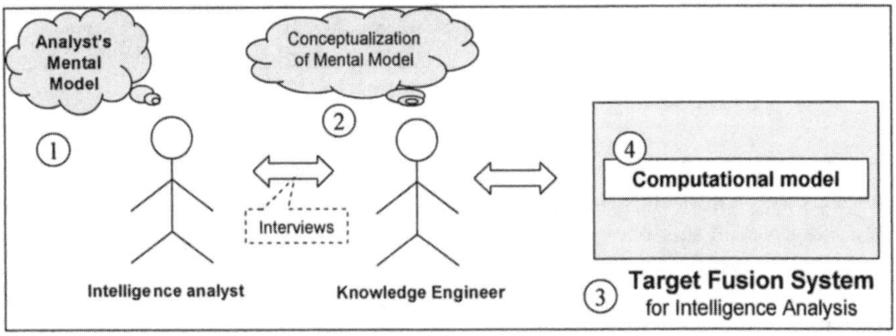

Figure P-1: Steps in building a model-based system

A computational model is invented to provide an accurate, consistent, and complete representation of the target system. The knowledge engineer helps to transform an analyst's mental model to the computational model of a target system. However, this transformation process via knowledge acquisition is a serious bottleneck (Das, 2007) in the development of a knowledge intensive system, and, more generally, artificial intelligence (AI) systems. Computational representations are often complex in structures and semantics, and do not always naturally lend themselves to easy transformation from mental models. Fortunately, there are other ways one can build models for fusion systems without always needing knowledge engineers.

Computational models (or, simply, models) for fusion systems to be presented in this book are in three categories as shown and explained in Figure P-2: physics-based, data-based (or knowledge-lean), and knowledge-based (or knowledge-rich). For example, a fusion system built on the knowledge-based models is unable to detect any unusual enemy activities or movements that have not been modeled explicitly. So the system should be hybridized with data-based models, for example, spatiotemporal clustering to indicate that there is something going on. The physics-based approach, for example, models vehicle dynamics to build trackers. Irrespective of an approach to modeling a fusion system, representation and propagation of uncertainty in both data and knowledge are a common problem that we address extensively in this book.

Physics-based	Data-based	Knowledge-based
• Linear/nonlinear equations specify system behavior (transition and observation) models • *Technologies*: Kalman filtering, extended and unscented Kalman filtering, particle filtering	• System input/output behavior is observed, and machine learning techniques extract system behavior models • *Technologies*: clustering, neural network, and various linear, nonlinear, and symbolic approaches to learning	• Domain experts provide knowledge of system behavior, and knowledge engineers develop computational models using an underlying ontology • *Technologies*: logic-based expert systems, fuzzy logic, Bayesian networks

Figure P-2: Approaches to modeling fusion systems

Most fusion systems will be "hybrid" in nature in that a combination of the above three approaches, namely physics-, data-, and knowledge-based. Most of the examples in this book are drawn from the battlefield information processing domain, though fusion techniques have been applied in numerous other domains, including life status estimation, bioterrorism, space situational awareness, and cyber security. Most of the techniques presented here have roots in control theory, artificial intelligence (AI), and machine learning, and thus there is an abundance of relevant literature for further studies. The chapters in this book are arranged mostly by fusion topic areas, not by technologies.

Chapter Organization and Content

Chapter 1 details the high-level fusion concept with examples drawn from various application domains. It provides a brief account of some well-known models and architectures of fusion, including the JDL model. The chapter also systematically introduces various forms of intelligence that are input to a fusion process. A generic fusion architecture is presented at the end of the chapter associating its modules with the book's content, thus guiding readers with a bird's-eye-view of the book and interrelationships among its chapters. Chapter 1 should be an easy-to-read chapter for everyone and it's a must read one from the perspective of understanding the basics of fusion in general.

Chapter 2 presents a background on mathematical preliminaries, including graph theory, probability and statistics, mathematical logic, and algorithmic complexity. Readers who have already been exposed to these concepts may omit Chapter 2.

Chapter 3 provides a detailed account of uncertainty and various well-established formal approaches to handling uncertainty, including the probability theory, fuzzy logic, and the theory of belief functions. The fact is that handling uncertainty is at the core of any fusion process. This chapter is a must read for everybody who wants to delve into the area of building a fusion system grounded on a well-founded theory, rather than an ad hoc one.

Chapter 4 is devoted to the core techniques of Level 1 fusion considering all three cases combining sensors and targets, namely single target single sensor, single target multiple sensors, and multiple targets and multiple sensors. Various filtering and data association techniques for handling these cases are presented, namely Kalman Filter (KF), Probabilistic Data Association (PDA), Joint Probabilistic Data Association (JPDA), and Multiple-Hypothesis Tracking (MHT). Though Level 1 fusion is not a major focus of this book, we do provide enough background material for readers to understand interactions between Level 1 and Level 2 fusion processes.

Chapter 5 describes various techniques for classifying Level 1 produced tracks into different target categories. In this process, we establish relationships between an individual target and objects in its environment. We make use of some of the inference techniques presented in Chapter 3. Therefore, readers need to understand Chapter 3. Chapter 5 also introduces the spatiotemporal clustering concept to aggregate units that are moving together to create, for example, an ambush-like situation in urban environments.

Chapter 6 explains in detail the Bayesian Belief Network (BN) technology for model-based high-level fusion. The BN technology avoids maintenance of large joint distributions via certain independence assumptions based on which application BN models are built. The chapter presents various evidence propagation algorithms. There is not always an intuitive explanation of how evidence is propagated up and down the arrows in a BN model via abductive and deductive inferencing. This is largely due to the conditional independence assumption and, as a consequence, separation among variables. To understand the evidence propagation behavior and also to identify sources of inferencing inefficiency, readers are therefore encouraged to go through in as much detail as they can the theory underlying the BN technology and propagation algorithms.

Chapter 7 presents temporal approaches to model-based fusion. We consider Hidden Markov Model (HMM) framework and its extensions Hierarchical Hidden Markov Model (HHMM) and the Dynamic Bayesian Network (DBN) technology. Inference algorithms for these models are also provided. Readers need to understand the BN technology to understand its temporal extension.

Chapter 8 provides inference algorithms for handling nonlinear models. We introduce Extended Kalman Filter (EKF), Unscented Kalman Filter (UKF), and various versions of Particle Filtering (PF) for handling nonlinear transition and observations functions. PF is especially effective for handling hybrid DBNs containing both continuous and categorical variables.

Chapter 9 describes the use of the Influence Diagram (ID) and symbolic argumentation technologies to make decisions. The BN and rule-based formalisms for hypotheses evaluation do not explicitly incorporate the concepts of action and utility that are ubiquitous in decision-making contexts. By incorporating the concepts of action and utility, BNs are converted to IDs, subsuming the functionality of the normative theory of decision making under EUT and decision trees. Decision making based on IDs means choosing the most suitable action (or hypothesis) from a set of alternative actions, which yields the highest utility. We also present an inference algorithm for IDs. Symbolic argumentation allows one to express arguments for and against decision hypotheses with weights from a variety of dictionaries, including the probability dictionary. Arguments are aggregated to rank the considered set of hypotheses to choose the most plausible one.

Chapter 10 provides a basis for inductive reasoning via learning of fusion models from observational data, and therefore supplements abductive and deductive reasonings for high-level fusion. We present techniques for learning HMMs, rules and decision trees, and BNs, and thus readers need to understand these modeling techniques.

Chapter 11 introduces intelligent agent technology, especially cognitive agent architectures and their usefulness in battlefield information processing. The agent technology as a fusion problem-solving paradigm is useful for several reasons. First, sensor devices are proliferating by numbers and producing enormous amounts of information from distributed NCW environments. So, there is a need for "autonomous" local processing at the sensor level to address the bandwidth problem. Second, the information overloading problem needs to be addressed by rapidly filtering out large amounts of irrelevant information similar to the way a "human operator" would do if there were enough time. Third, situation and threat assessments and decision making in a "collaborative" manner need to be supported by taking into account numerous relevant factors. Chapter 2 is a recommended reading before reading this agent chapter.

Chapter 12 takes high-level fusion processes to distributed NCW environments. We show how KF-based tracking can be performed in a distributed manner, and how BN models can be distributed to perform distributed high-level fusion in NCW environments. One therefore needs to know the basics of KF and the BN technology before reading this chapter.

The scope of high-level fusion is broad and interdisciplinary in nature, and likely to touch a breadth of topic areas. It's not an aim of this book to cover each and every aspect of high-level fusion. The book provides a computational account and stays away from areas such as human-in-the-loop fusion and visualization. Moreover, the symbolic thrust of the book naturally excludes from our consideration other potential techniques, such as neural networks. Notable omissions are also case-based reasoning and blackboard approaches to high-level fusion, though technologies presented in the book can provide foundations of such alternative approaches. I have made my best effort to make this book informative, readable, and free from mistakes, and I would welcome any criticism or suggestions for improvement.

Tutorials Source

Much of the materials in this book are based on the slides of the two series of tutorials that I have delivered over the past few years: one series delivered at the International Conference on Information Fusion every year during the period 2004-08, and the other series was organized by a technology training corporation at various locations in the USA during 2006-07. Conference organizers, institutions, and government and commercial organizations interested in hosting such tutorials based on the content of this book may contact the author directly (subrata@skdas.com).

Acknowledgements

My sincere thanks go to my wife, Janique, my son, Sébastien, and my daughter, Kabita, for their love, patience, and inspiration throughout the preparation of this book. Thanks also to all of my colleagues at Charles River Analytics, especially Scott Harrison, Brad Rosenberg, John Chamberlain, Dan Gutchess, Roger Schane, Scott Neil O'Reilly, Greg Zacharias, Karen Harper, Magnus Snorasson, Ted Fichtl, Jonathan Pfautz, Yvonne Fuller, and Joe Gorman, and my ex-colleagues Josh Introne, Dave Lawless, John Langton, Partha Kanjilal, Paul Gonsalves, and Rich Reposa, with whom I have had many technical discussions on various aspects of this book. Many thanks to Artech House, especially Mark Walsh, the series editor Ed Waltz, the copyeditor, and the anonymous reviewer for their help in producing the book from the beginning.

There are many academic and fusion practitioners in government and industry from around the world with whom I have had many technical discussions and arguments that helped me to understand and appreciate better the fusion area. I would like to acknowledge Jim Llinas (State University of New York at Buffalo); Alan Steinberg (Independent Consultant); Avi Pfeffer (Harvard University); Belur V. Dasarathy (Independent Consultant and Editor-in-Chief of the Journal of Information Fusion); John Salerno and Mike Hinman (Air Force Research Laboratory, Rome); Wendy Martinez (Office of Naval Research); Norman Coleman (US Army ARDEC); Katia Sycara (Carnegie-Mellon University); Mica Endsley (SA Technologies); Dan Corbett (DARPA); Erik Blasch (Air Force Research Lab, WPAFB); Mitch Kokar (Northeastern University); Joe Karakowski (US Army RDECOM, I2WD); Chris Bowman (Data Fusion & Neural Networks); Alper Caglayan (Milcord); Hans Miller (Independent Consultant); Pierre Valin (Defence R&D Canada Valcartier); Gregor Pavlin (Thales Research & Technology Netherlands/D-CIS Lab); Walt Truszkowski (NASA Goddard Space Flight Center); and Ivan Kadar (Interlink Systems Sciences).

Finally, I thank my parents, brothers, sisters, and other family members back in one of several thousands of small villages in India for patiently accepting my absence and showing their encouragement and support through many phone calls.

Subrata Das
Charles River Analytics, Inc.
Cambridge, MA

July 2008

Chapter 1

Models, Architectures, and Data

The objective of this chapter is to provide readers with a background in data fusion in general, with an emphasis on the high-level fusion aspect for it being the major focus of the book. The chapter surveys and compares a number of models of fusion processes, including the well-known Joint Directors of Laboratories (JDL) model. Both structured and unstructured data in the form of intelligence messages generated from battlefield areas of operation drive fusion processes. We present the six basic intelligence categories into which these messages belong to. Finally, we present an all-encompassing generic fusion system architecture to provide readers a bird's-eye-view of the chapters and contents of the book. The architecture indicates interactions between various fusion modules and specific technologies that they are based on.

1.1 WHAT IS HIGH-LEVEL FUSION?

Earlier at the start of the Preface we described the high-level fusion concept in a rather abstract manner, and considered Situation Assessment (SA) and Threat Assessment (TA) as its parts. Here we take a more concrete approach by first formulating a set of representative fusion-related questions that are usually posed by analysts, working in a variety of application domains, to fulfill their intelligence requirements:

- *Conventional Warfare*: Will the enemy force interdict from the north or from the south? What is the composition of the defensive forces?
- *Strategic*: What is the capability and capacity of a rogue nation in terms of weapon stockpile and missiles?
- *Urban Warfare*: What is the formation of adversarial entities? Is an ambush situation likely? What is the likelihood of an IED (Improvised Explosive Devices) attack within a certain region at a certain time?
- *Asymmetric Warfare*: Is the suspected site an NBRC (Nuclear, Biological, Radiological, Chemical) site?

- *Air Operation Center*: Is it a time-sensitive target? What is the class of the target?
- *Maritime Domain*: Is the approaching vessel threatening?
- *Missile Defense*: Is the suspected site a reload, logistic, or launch site?
- *Bioterrorism*: Has there been an anthrax attack? At what stage (preclinical, prediagnostic, and diagnostic) is the attack?
- *Cyber Warfare*: Has the system been compromised? What is the nature of the attack?
- *Space Surveillance*: What is the current health and status of a satellite? Is a satellite under threat?
- *Psychological Warfare*: How does the shortage supply of electricity anger the local population?
- *Social Network Analyses*: What's the relationship between the leaders of the two terrorist organizations? How do they influence each other?

The underlying generic problem in the majority of the above cases is one of aggregation of a group of interrelated objects and events to produce an aggregate object (e.g., missile reload site) or event (e.g., an anthrax attack). The term "situation" refers to an estimated aggregate object or event and the phrase "situation assessment" refers to the estimation of an aggregate object or event. The state of the aggregate is therefore represented as a relation among its constituent elements.

Consider, for example, an interdiction event by a hostile force. If such an event occurs then its constituent spatiotemporal events, such as gun shot and reconnaissance, are likely to occur in space and time. Moreover, certain objects are likely to be spatially arranged in a certain manner within the area of operation. All of these constituent events and objects together define a relationship for the interdiction event. Such an event is a high-level abstraction in the sense that it cannot be observed directly but its constituent elementary objects and events can be observed.

There are cases when a relationship defines an aggregate object rather than an event. Consider, for example, a missile reload site which is composed of certain spatially arranged objects, including reload equipment and security measures. This example is a typical physical relationship relating objects, but a variety of other relation types need to be considered (Steinberg et al., 1998), including organizational (e.g., relationship between the leaders) and informational (e.g., state of a compromised system). An estimation problem is spatiotemporal in nature since events occur in space and time, and both current and future locations of objects change over time.

A related concept to a situation is threat. For example, an underlying implicit requirement for assessing the situation of an adversarial force is to determine the threat it poses. Threat is a relative concept between two opposing forces. A squad-size adversarial unit, for example, is not so much of a threat when a battalion-size friendly force is present that can counter the threat very easily. But the nature or amount of threat is different if the adversarial unit is known to be of battalion size.

Determining the "amount" of threat under a certain situation therefore requires some form an abstract wargaming, which is not within the scope of this book. Assessment of situation sometimes, but not always, requires evaluation of a set of competing hypotheses (e.g., enemy interdiction is at the north or south), and from this perspective we have a chapter (Chapter 9) on decision making via hypotheses evaluation.

1.2 FUSION MODELS

This section surveys a set of well-known data fusion models ranging from a variety of perspectives, including a purely computational and a human information processing. A fusion system can be developed based on any of these models. Computational technologies covered in this book help to realize modules and layers of the chosen model. A correspondence between levels and layers of the data fusion models described in this section is presented at the end of the section.

1.2.1 JDL Model

The most influential data fusion model to date is from the Joint Director's of Laboratories (JDL) and is shown in Figure 1-1. The so-called JDL functional model (White, 1988) was intended for communication among data fusion practitioners, rather than a complete architecture detailing various processes and their interactions.

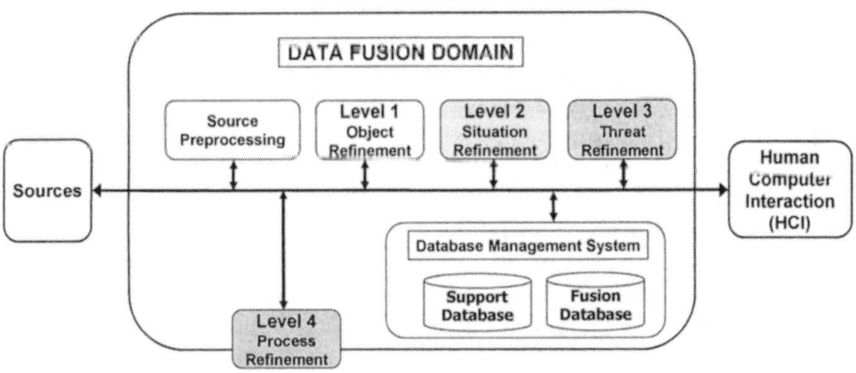

Figure 1-1: JDL data fusion model (White, 1988)

Sources on the left of the figure include local and remote sensors accessible to the data fusion system, information from the reference system, and human input. Source preprocessing involves processing of individual sensor data to extract information, improve signal-to-noise ratio, and preparation of data such as spatiotemporal alignment for subsequent fusion processing. The JDL model has the following four functional levels of fusion:

Level 1: Object Refinement

Combines sensor data to obtain the most reliable and accurate estimate of an entity's position, velocity, attribute, and identity.

Level 2: Situation Refinement

Develops a description of current relationships among entities and events in the context of their environment.

Level 3: Threat Refinement

Projects the current situation into the future to draw inferences about enemy threats, friend and foe vulnerabilities, and opportunities for operations.

Level 4: Process Refinement

Monitors the overall data fusion process to assess and improve real-time system performance (it has been placed on the edge of the data fusion domain due to its meta-level monitoring characteristics).

The Human Computer Interaction (HCI) block provides an interface to allow a human to interact with the fusion system. The Database Management System block provides management of data for fusion (sensor data, environmental information, models, estimations, etc).

Steinberg et al. (1998) revised and expanded the JDL model to broaden the functional model and related taxonomy beyond the original military focus. They introduced a Level 0 to the model for estimation and prediction of signal/object observable states on the basis of pixel/signal-level data association and characterization. They also suggested renaming and re-interpretation of Level 2 and Level 3 to focus on understanding the external world beyond military situation and threat focus.

The distinction between Level 2 and Level 3 is often artificial. Our models for Level 2 fusion are temporal in many cases, and thus both the current situation and its projection to the future come from a single temporal model. We therefore adopt the definition of Level 2 fusion along the line of Steinberg et al. (1998) as the "estimation and prediction of relations among entities, to include force structure and cross force relations, communications and perceptual influences, physical context, etc." As per their terminology, we call Level 2 fusion Situation Assessment (SA). Moreover, drawing inferences about enemy threats, friend and foe vulnerabilities, and opportunities for operations requires generations of Courses of Action (COAs). In this respect, this book does not focus on detailed COA generations via planning, scheduling, resource management, and abstract wargaming. Rather we take the hypotheses evaluation approach where COAs are overall actions and their suitabilities need to be evaluated via some arguments of pros and cons and expected utility measures.

Llinas et al. (2004) discuss issues and functions considered to be important to any further generalization of the current fusion model. Their remarks and assertions include quality control, reliability, and consistency in data fusion; need for coprocessing of abductive, inductive, and deductive inferencing processes; and the case of distributed data fusion. Blasch and Plano (2002, 2003) add Level 5 user refinement into the JDL model to support a user's trust, workload, attention, and situation awareness.

1.2.2 DIKW Hierarchy and Abstraction of Knowledge

The traditional DIKW (Data Information Knowledge and Wisdom) hierarchy (Figure 1-2) organizes data, information, knowledge, and wisdom in layers with an increasing level of abstraction and addition of knowledge, starting from the bottommost data layer. The hierarchy bears some resemblance to the JDL data fusion model in the sense that both start from raw transactional data to yield knowledge at an increasing level of abstraction.

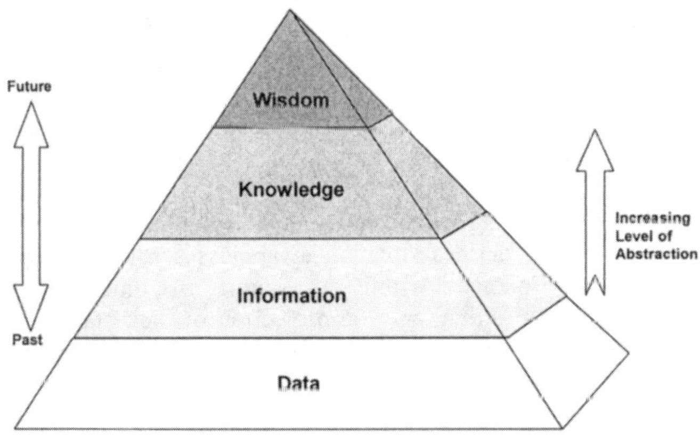

Figure 1-2: The knowledge pyramid

Data Layer

Data are transactional, physical, and isolated records of activity (e.g., Signal Intelligence (SIGINT) reports, fac ts, or figures obtained from experiments or surveys). Data are the most basic level and by themselves have little purpose and meaning.

Information Layer

Information is semantic interpretation of data or represents relationships between data with meaning and purpose (e.g., tank unit at a certain battlefield location, presence of an enemy unit in a defensive posture).

Knowledge Layer

Knowledge is general awareness or possession of information, facts, ideas, truths, or principles. Knowledge is generally personal and subjective.

Wisdom Layer

Wisdom is knowledge of what is true or right coupled with just judgment as to action. Wisdom is the knowledge and experience needed to make the right decisions and judgments in actions.

Thus "data" are the basic unit of "information," which in turn is the basic unit of "knowledge," which in turn is the basic unit of "wisdom."

1.2.3 Assessment Versus Awareness

While the JDL model is a functional model for the data fusion process, it does not model it from a human perspective. A human operator/analyst becomes "aware" of certain situations (or achieves situation awareness) based on either the results from his or her own cognitive processes or the results produced by an automated SA tool built on some processing model such as the JDL model. Such an automated SA tool does not have to mimic the complex human cognitive processes (in fact the JDL model and many other computational models do not). Endsley (1988) formally defines situation awareness as "the perception of the elements in the environment within a volume of time and space, the comprehension of their meaning, and the projection of their state in the future." The three levels of situation awareness, namely perception, comprehension, and projection, as shown in Figure 1-3, parallel the corresponding levels in the JDL model. The levels in the JDL model can be viewed as processes producing results to help a human operator became aware of the situation at hand.

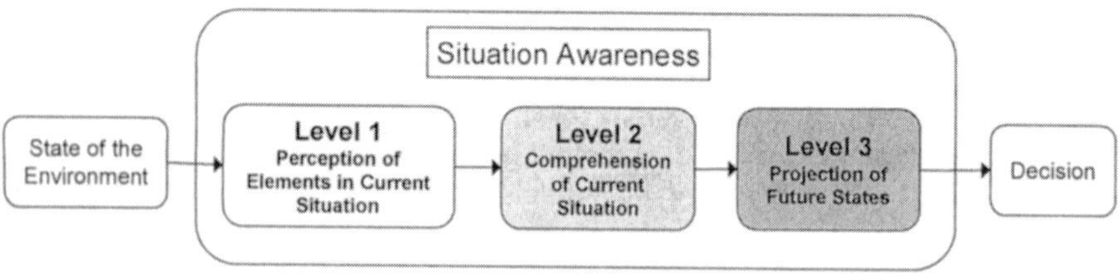

Figure 1-3: Situation awareness model

Level 1: Perception of Elements in Current Situation

A human operator needs to perceive the status, attributes, and dynamics of relevant elements in the environment, including the current and historical movements of both friendly and hostile units, their types, terrain, and weather. The information overloading problem can easily rise in a dynamic, high-tempo, operational environment.

Level 2: Comprehension of Current Situation

A human operator needs to comprehend the situation based on a synthesis of perceived elements along with the mission at hand. The level provides a holistic picture of the environment relating the disjointed perceived elements. For example, upon observing some probing activities by an adversarial force in an area within an urban environment, the operator quickly determines the seriousness of the problem in terms of a coordinated attack by combining other contextual knowledge, which may include high-value targets in the area, previous activities, and capabilities. The degree of comprehension by a human operator varies according to the operator's level of experience. A novice operator may be capable of achieving the same Level 1 situation awareness as a more experienced operator, but may not be able to integrate various perceived elements along with the mission in order to comprehend the situation.

Level 3: Projection of Future States

A human operator needs to understand what will happen in the near future based on both understanding of dynamics of the perceived elements and comprehension of the current situation. This level provides the knowledge and time necessary to decide on a suitable course of action to deal with threatening situations.

1.2.4 OODA Loop

One of the first C4I (Command, Control, Communications, Computers, and Intelligence) architectures is the OODA (Observe-Orient-Decide-Act) Loop (2001), shown in Figure 1-4.

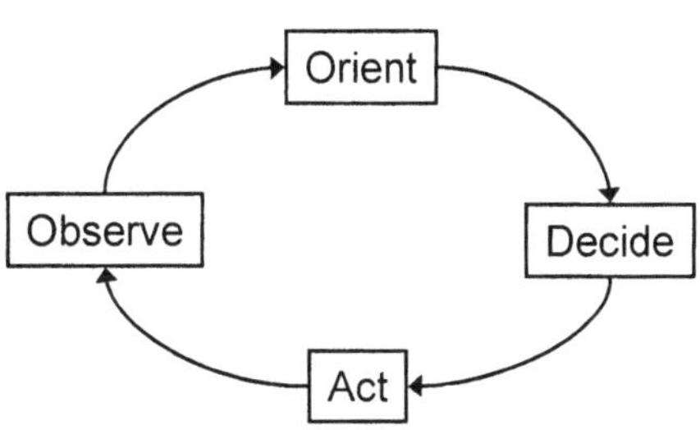

Figure 1-4: Boyd's OODA loop

The OODA architecture was developed by Col. John Boyd, USAF (Ret) during the Korean War while referring to the ability possessed by fighter pilots that allowed them to succeed in combat. Observations in OODA refer to scanning the environment and gathering information from it, orientation is the use of the information to form a mental image of the circumstances, decision is considering options and selecting a subsequent course of action, and action refers to carrying out the conceived decision.

1.2.5 Rasmussen Information Processing Hierarchy

Rasmussen's three-tier model of human information processing (Rasmussen, 1983, 1986) is shown in Figure 1-5. The arch in Rasmussen's SRK (skill, rule, knowledge) model represents the flow of information through the human decision-maker. The left side of the arch corresponds to stimulus processing, and the right side corresponds to motor processing. Processing is divided into three broad categories, corresponding to activities at three different levels of complexity.

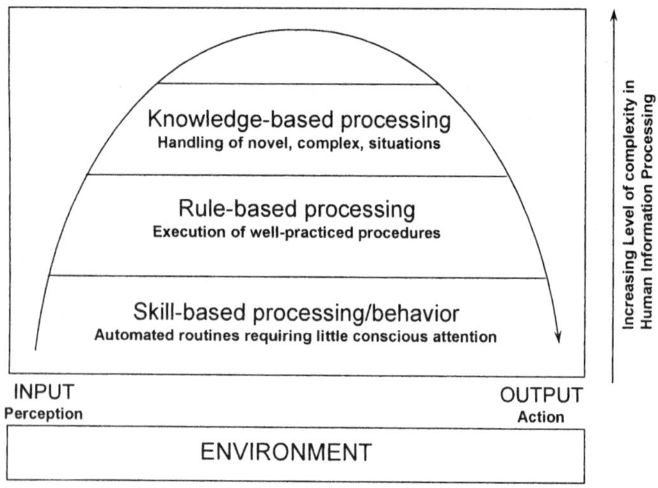

Figure 1-5: Rasmussen hierarchy of human information processing

Skill-Based Processing

At the lowest level is skill-based sensorimotor behavior, such as perceptual feature extraction and hand-eye coordination. This level represents the most automated, largely unconscious level of skilled performance (e.g., identification of a tank by looking at raw sensor reports).

Rule-Based Processing

At the next level is rule-based behavior, exemplified by procedural skills for well-practiced tasks such as the identification of an enemy unit composition based on its numbers and relative locations.

Knowledge-Based Processing

Knowledge-based behavior represents the most complex cognitive processing used to handle novel, complex, situations where no routines or rules are available for handling situations. Examples of this type of processing include interpretation of unusual behavior, and generating COAs based on enemy unit size and behavior.

The Generic Error Modeling System (GEMS) (Reason, 1990) is an extension of Rasmussen's approach, which describes the competencies needed by workers to perform their roles in complex systems. GEMS describes three major categories of errors: skill-based slips and lapses, rule-based mistakes, and knowledge-based mistakes. See Das and Grecu (2000) for an instantiation of the information processing hierarchy to implement an agent to amplify human perception and cognition.

1.2.6 Correspondence among Models

Table 1-1 shows a rough correspondence (not comparison) among levels and layers of various models presented in the last few sections.

JDL Data Fusion Model	Level 1 = Object Assessment	Level 2 = Situation Assessment	Level 3 = Threat Assessment	Action (not explicit in the JDL model)	Level 4 = Process Refinement
Knowledge Pyramid	Data	Information and Knowledge	Knowledge	Wisdom	n/a
Situation Awareness	Perception	Comprehension	Projection	Decision	n/a
OODA Loop	Observe	Orient	Decide	Act	n/a
Information Processing Hierarchy	Skill/Rule-based processing	Rule/Knowledge-based processing	Rule/Knowledge-based processing	n/a	n/a

Table 1-1: Comparison among fusion models

The correspondence in the table by levels and layers along the columns is not as distinct as one would like it to be. Nonetheless, the table can provide some guidance in identifying components of one's own fusion architecture.

1.3 SENSORS AND INTELLIGENCE

Intelligence messages generated from battlefield areas of operation drive fusion processes. Broadly, there are six basic intelligence categories (USAF, 1998; ODNI, 2008):

- Signals Intelligence (SIGINT)
- Imagery Intelligence (IMINT)
- Measurement and Signature Intelligence (MASINT)
- Human Intelligence (HUMINT)
- Open-Source Intelligence (OSINT)
- Geospatial Intelligence (GEOINT)

In addition, there is also Scientific and Technical (S&T) Intelligence resulting from the analysis of foreign scientific and technical information. The above six categories are explained in detail in the following subsections along with relevant examples.

1.3.1 Signals Intelligence (SIGINT)

SIGINT is derived from intercepted electromagnetic emissions. SIGINT subcategories include Electronic Intelligence (ELINT) and Communications Intelligence (COMINT).

ELINT is derived from the processing and analysis of electromagnetic radiation emitted from noncommunication emitters (most commonly radars) other than nuclear detonations or radioactive sources. An emitter is usually tied closely to a specific military function such as an early warning for air defense. Friendly assets scan the electromagnetic spectrum "listening" for these noncommunications emitters and produce a Tactical Report (TACREP) for each intercept. As shown on the left side of Figure 1-6, a report typically consists of an estimate of the emitter category, location, a radius representing error probable for emitter location, type, an elliptical error probable for emitter location, and various electronic attributes, such as frequency and pulse duration. Signal analyses can provide insights about an emitter's type which can then be fed into a high-level fusion process.

COMINT is derived from the processing and analysis of intercepted communications from emitters. The communications that are sometimes encrypted may take a variety of forms such as voice, e-mail, fax, Morse code, and radio-teletype. As shown on the right of Figure 1-6, a typical COMINT report consists of emitter frequency, its latitude and longitude, line of bearing, etc.

ELINT	
TACELINT ID:	14198
Detect Day & Time:	104888756
Emitter Category:	---
Location Latitude:	4736.45
Location Longitude:	-12228.34
Radius:	---
Orient in Bearing:	---
Semi-major/minor:	---
Bearing Error:	409
Radio Frequency:	6757.5
Operating Mode:	---
Pulse Duration:	17.6

COMINT	
COMINT ID:	50450
Delta Time:	2246860
Ground Op Details:	---
Emitter Frequency:	40.0
Location Latitude:	4736.45
Location Longitude:	-12228.34
Frequency Bandwidth:	---
Line of Bearing:	88.0
Signature String:	---

Figure 1-6: Example SIGINT

1.3.2 Imagery Intelligence (IMINT)

IMINT is derived from visible (optical), infrared, radar (e.g., Synthetic Aperture Radar (SAR) and Moving Target Indicator (MTI)), laser, laser radar (LADAR), and multispectral sensors. Each sensor has a unique capability, with distinct advantages and disadvantages. Some work in all weather conditions, some work at night, and some produce high-quality images with detectable signatures. Figure 1-7, for example, is a LADAR reflectance image taken during a tower-test data collection exercise. The targets in the image have been segmented via an automatic target segmentation technique (Stevens and Snorrason, 2003). These targets need to be recognized as an array of missile launchers, tanks, and armored personnel carriers by their extracted features.

Figure 1-7: Segmented LADAR reflectance image

Neither automatic target segmentation nor recognition is within the scope of this book, but a model for situation and threat assessments will incorporate extracted features and recognized targets as input evidence to the model.

1.3.3 Measurement and Signature Intelligence (MASINT)

MASINT is obtained by "quantitative and qualitative analysis of data (metric, angle, spatial, wave length, time dependent, modulation, plasma, and hydromagnetic)" derived from sensing instruments other than those generally used for SIGINT or IMINT (www.fas.org). MASINT therefore works with collected signals, but is more of an analysis discipline. However, there are unique MASINT sensors, typically working in different regions or domains of the electromagnetic spectrum, such as infrared or magnetic fields. MASINT includes an array of intelligence such as Radar Intelligence (RADINT), Nuclear Intelligence (NUCINT), Laser Intelligence (LASINT), and Chemical and Biological Intelligence (CBINT). RADINT, for example, is a specialized form of ELINT, which categorizes and locates as active or passive collection of energy reflected from a target.

Figure 1-8 is an example of a RADINT record from the domain of vessel traffic tracking (Anzano, 1999). The record provides information about a target, including vessel name, position, reporting sensor type and number, and speed.

Vessel Name:	UNK-4410
Universal Time:	120996230119
Sensor Type/No.:	Radar/3
Track ID:	409
Track True Course:	91.0 deg
Vessel Speed:	17.6
Latitude/Longitude:	4736.45, -12228.34
Vessel Size:	252 knots

Figure 1-8: Example RADINT

The assumption here is that this observation has already been assigned to a vessel track. The vessel's name has not been determined (referenced as unknown), but its speed has been derived. A tracker for maritime domain awareness will make use of position, speed, and other contextual information.

1.3.4 Human Intelligence (HUMINT)

HUMINT is the gathering of information through human contact. Information of interest might include target name, size, location, time, movement, and intent. HUMINT typically includes structured text (e.g., tables, lists), annotated imagery, and free-form prose text (e.g., sentences, paragraphs). HUMINT provides insights into adversary actions, capability and capacity, plans and intentions, deliberations and decisions, and research and development goals and strategies. A couple of fictitious semistructured (combination of table and text) and unstructured HUMINT examples are shown in Figure 1-9. These HUMINTs provide insights into adversary capability and capacity.

Semistructured HUMINT	Unstructured HUMINT
The following equipment is located at the suspected terrorist training site: 4 – Armed Vehicles 9 – Surface-to-Air Missile 20 – Armed Men with AK-47	The suspected member of a terrorist organization entered into the building around 11 am on Jan 9, 2005. There were 4 armed men standing in front of the building.

Figure 1-9: Fictitious examples of HUMINT

Text parsing is required to extract evidence from unstructured HUMINTs to feed into an automated high-level fusion system. We will describe HMM-based text processing techniques later in the chapter on modeling time for SA.

1.3.5 Open Source Intelligence (OSINT)

OSINT is publicly available information appearing in print or electronic form including radio, television, newspapers, journals, the Internet, commercial databases, videos, graphics, and drawings. Some example OSINTs are shown in Figure 1-10.

Example OSINT
"Militants from the southern region threatened on Monday to blow up oil facilities ..." "The authorities have released six suspected terrorists belonging to ..." "The country has purchased air defense radar systems from ..." "Members of the terrorist organization met the leader of the country ..."

Figure 1-10: Example OSINT

OSINT complements the other intelligence disciplines and can be used to fill gaps and enhance accuracy and confidence in classified information and actionable intelligence. For example, the classified information about a country's capability and capacity will be updated with the OSINT about its purchase of air defense radar systems. Release of suspected terrorists is an actionable intelligence which calls for a constant surveillance on their whereabouts.

1.3.6 Geospatial Intelligence (GEOINT)

GEOINT is the analysis and visual representation, along with their geographical references, of security-related activities on the earth. It is produced through an integration of imagery, imagery intelligence, and geospatial information.

1.3.7 Intelligent Data Format

Intelligence messages are exchanged in command and control systems via special NATOEX, NATO STANAG (Standardization Agreement) (http://www.nato.int/

docu/standard.htm), and national USMTF (United States Message Text Format) formats. Figure 1-11 shows an example Ground Moving Target Indicator (GMTI) record, generated by surveillance radar, in the NATOEX format.

NATOEX Header Segment

Message Indicator:	---	Sensor Platform Time Stamp:	45556040952770000
Radar Mode:	---	Date Time Stamp:	---
Scan Flags:	---	Sensor Platform ID:	---
Resolution:	---	Sensor Platform Position:	---
Label:	---	Sensor Platform Velocity:	---
Sequence Number:	---	Sensor Platform Heading:	---
Target/Imagery Packet Count:	7	Topocentric Origin:	---
Scan Area (X, Y, Z in 3 Formats):	---	Protocol Version Number:	---
Scan Number:	---	Byte Count:	---
Service Request Number:	---	Sending Platform ID:	---

NATOEX MTI Segment

Target Location (X, Y, Z):	(24094.27, -12609.24, -7.84)
Target Radial Velocity:	5.0445
Target Radar Cross Section:	6.0
Target Classification:	0
Radial Velocity Quality:	---
Count:	1
Truth Tag (Count, Appl., Entity):	(1, 9, 1962)
Errors (Range and Cross-Range):	(200, 9957)

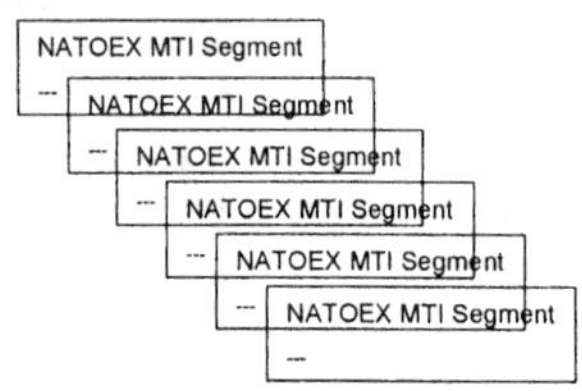

Figure 1-11: GMTI record in NATOEX format

The GMTI data in NATOEX format consists of variable length packets, where each packet has a fixed-length Header Segment plus zero or more Moving Target Indicator (MTI) Segments. Each MTI Segment specifies information such as location, speed, and heading for one target. Such information is primarily used for tracking targets as presented in Chapter 4 on target tracking.

1.4 GENERIC FUSION ARCHITECTURE AND BOOK SCOPE

Figure 1-12 shows a generic system architecture for fusion instantiated in the military domain as a guide to illustrate the scope of this book. Individual architecture modules are appropriately labeled by JDL fusion levels (0-5).

A battlefield environment is shown at the bottom right of the figure. A set of sensors and humans, managed by the Sensors and Management Module, are placed within the environment. Sensors in the environment include sensors for detecting units and their movements, weather and terrain mobility sensors, and so forth. The sensed information in the form of various INTs is then passed to the Target Tracking module.

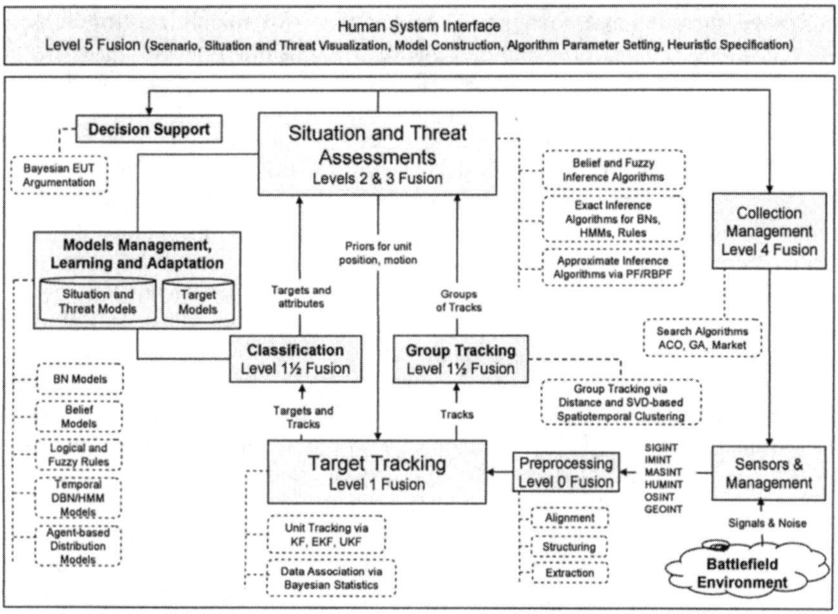

Figure 1-12: Generic fusion architecture and the scope of this book

The Target Tracking (Level 1 fusion) module then carries out tracking of units, including gating, data association, and detection and maintenance of tracks, via Kalman Filtering (Chapter 4) and its extensions (Chapter 8). The generated tracks are then passed on to the Classification and Group Tracking (Level 1½ fusion) modules. The book mostly covers kinematic tracking but one could easily extend the tracking concept into other domains, including the tracking of a disease outbreak in the bioterrorism domain and the tracking of a hacker's activity in the cyber domain.

The Classification module identifies unit target types based on incoming tracks and other contextual information (Chapter 5). The Classification module makes use of a library of preconstructed generic models of targets stored in a Target Model database. The Group Tracking module tracks groups of units by applying spatiotemporal clustering algorithms on incoming tracked information to determine groups of objects with correlated movements (Chapter 5).

Both the Classification and Group Tracking modules identify some form of relationships and hence part of high-level fusion. The Classification module establishes relationships between target attributes and objects in surrounding contexts. On the other hand, the Group Tracking module identifies spatiotemporal relationships without any contextual information. Only the positional attribute of each tracked unit is made use of to extract such clusters. No other attribute or contextual information is necessary in the algorithms presented in the chapter.

Moreover, the relationship among the tracked units within an identified cluster is merely based on their correlated movements. Establishing relationships for target classification and aggregation as described here is certainly a high-level fusion task, but does not generate full complex relationships among targets. For this reason, I depart slightly from the standard JDL terminology and term the classification and aggregation approaches in this book as Level 1½ fusion.

The Situation and Threat Assessment (Levels 2 and 3 fusion) module constructs and runs a set of models based on the situation at hand to identify units individually, relationships among identified units, and threats they pose. The models are instantiated and/or constructed using preconstructed generic components, based on known enemy doctrine and behaviors, stored in a Situation and Threat Model database. The Situation and Threat Assessment module accepts, filters, and processes information (unit position and velocity, and their aggregation) from the Classification and the Group Tracking modules, creates new instances of models as appropriate to assess situations at hand, and posts evidence into these models based on information received and based on information about weather and terrain. The module then predicts or infers higher-level significant events or situations based on various exact Inference Algorithms for BNs and DBNs (Chapters 6 and 7), and approximate Inference Algorithms such as Particle Filtering (Chapter 8). The assessment results are then passed to both the Collection Management and the Decision Support modules.

The Human System Interface (Level 5 fusion) module helps to interact between the analyst and the rest of the systems in terms of accepting input from users and visualizing results of situation and threat assessments. The interface provides a facility for constructing and/or instantiating from libraries, and for setting various parameters and domain-specific heuristics for the inference algorithms.

The Collection Management (Level 4 fusion) module identifies information needs based on assessed situations and mission requirements. The module then generates collection plans to satisfy information needs by taking into account assets' capabilities/availability as well as environmental and operational factors. The plan is then interpreted accordingly for a redeployment of sensor assets to satisfy the information needs. This module may specify, for example, that a UAV should be rerouted to provide more sensor information about enemy troop movements. (Discussion on the collection management aspect is not within the scope of this book.) The Sensors and Management module then carries out actual redeployment of sensor assets.

Finally, the Models Management, Learning, and Adaptation module manages models of targets, situations, and threats to support target classification, and situation and threat assessments tasks. The Situation and Threat Model databases include Bayesian Belief Network (BN) models (Chapter 6), belief models (Chapters 3 and 5), logical and fuzzy rules (Chapters 3 and 5), Dynamic Bayesian Network (DBN) models and Hidden Markov Models (HMMs) (Chapter 7), agent-based models (Chapter 11), and distributed models (Chapter 12). The Target

Model database includes some of these types of models as well as Naïve Bayesian Classifier (NBC) models (Chapter 5). The module learns BN and HMM models and rules from fused data via an array of machine learning techniques, and then continues to adapt the learned models over time (Chapter 10).

The Decision Support module provides recommendations based on the assessed situations and threats. The module makes use of Influence Diagrams (IDs), which generalizes BNs with the concepts of action and utility, and symbolic argumentation technology (Chapter 9).

1.5 FURTHER READING

Steinberg et al. (1998) and Llinas et al. (2004) are good to read to understand deficiencies in the current JDL model and needed enhancements. Endsley's book itself provides a good account to the situational awareness models. Salerno et al. (2004) develop a framework for situation awareness combining aspects of JDL and Endsley's models (and knowledge discovery). An Internet search with relevant keywords will point to a number of websites discussing various intelligence categories in detail.

Chapter 2

Mathematical Preliminaries

This chapter provides the background mathematical notations and concepts needed to understand the rest of the book. The topics discussed in this chapter include graphs and trees, basic probability theory and statistics, mathematical logic, and a concise introduction to the theory of algorithmic complexity (we will use this theory to analyze expected runtime performance of evidence propagation algorithms). But first, we explain our conventions for symbol usage.

2.1 USAGE OF SYMBOLS

In general, the conventions in Table 2-1 are used for symbols representing terms, random variables, clauses, sets, and so on:

Item	Convention	Example
▪ Variables for random variable states ▪ Variables in a logical language	▪ Italicized, lowercase letters	▪ $x, y, z, \ldots, x_1, y_1, z_1, \ldots$
▪ Probabilistic random variables ▪ Vector variables ▪ Graph nodes	▪ Italicized, uppercase letters ▪ Italicized string starting with an uppercase letter	▪ $X, Y, Z, N, U, V, W, \ldots, X_1, Y_1, Z_1, N_1, U_1, V_1, W_1, \ldots$ ▪ Weather, Mobility, ...
▪ Logical predicate symbols	▪ Italicized uppercase letters ▪ Italicized string starting with an uppercase letter	▪ $P, Q, R, \ldots, P_1, Q_1, R_1, \ldots$ ▪ Weather, Mobility, ...
▪ Random variable states	▪ Italicized, lowercase letters	▪ $a, b, c, \ldots, a_1, b_1, c_1, \ldots$ ▪ medium, yes, john,

▪ Constant symbols in a logical language	▪ Italicized string starting with a lowercase letter	canceled, ...
▪ Function symbols	▪ Italicized, lowercase letters ▪ Italicized string starting with a lowercase letter	▪ $f, g, h, \ldots, f_1, g_1, h_1, \ldots$ ▪ *succ*
▪ Atoms ▪ Literals	▪ Italicized, uppercase letters	▪ $A, B, L, M, \ldots, A_1, B_1, L_1, M_1, \ldots$
▪ Formulae ▪ Goals ▪ Clauses	▪ Italicized, uppercase letters	▪ $F, G, H, F_1, G_1, H_1, \ldots$
▪ Sets, matrices, and clusters	▪ Bold or italicized, Times-Roman typeface, uppercase letters	▪ $\mathbf{A}, \mathbf{B}, \mathbf{S}, \ldots, A, B, S, \ldots$

Table 2-1: Conventions used for symbols

New terminology is italicized on its first use. Figures and tables are sequentially numbered by including chapter numbers. A vector with n attributes will be represented by an $n \times 1$ matrix.

2.2 GRAPHS AND TREES

The fundamental modeling tools known as graphs and trees are introduced in this section. Graphs and trees are most often presented graphically (hence their names) as in the examples below; we give their formal nongraphical definitions here, as this background is useful for issues such as theoretical discussions and algorithm development.

A *simple graph* $\mathcal{G}$ is a pair $\langle \mathbf{V}, \mathbf{E} \rangle$, where $\mathbf{V}$ is a nonempty set of elements called *vertices* (or *nodes*), and $\mathbf{E}$ is a set of unordered pairs of distinct elements of $\mathbf{V}$ called *edges*. Edges are denoted as $N_i N_j$, where $N_i, N_j \in \mathbf{V}$. The definition of a *directed graph* (or *digraph*) is given in the same way as a graph except that the set $\mathbf{E}$ is a set of ordered pairs ($N_i N_j \neq N_j N_i$) of elements of $\mathbf{V}$ called *directed edges*. A simple graph is said to have been *obtained* from a directed graph by removing the direction of each of its edges.

A *path* (of length m) in a simple graph $\langle \mathbf{V}, \mathbf{E} \rangle$ is a finite sequence of edges of the form

$$N_0 N_1, N_1 N_2, \ldots, N_{m-1} N_m$$

where $N_i N_{i+1} \in \mathbf{E}$ for each term in the sequence, and where the $N_i \in \mathbf{V}$ are distinct vertices (except possibly $N_0 = N_m$). Informally, a path consists of a sequence of "hops" along edges of the graph to distinct vertices. The sequence above can be written in abbreviated form as

$$N_0 - N_1 - N_2 - \ldots - N_{m-1} - N_m$$

Therefore, if $N_0 - N_1 - \ldots - N_m$ is a path in a simple graph then $N_m - N_{m-1} - \ldots - N_0$ is also a path in the graph. Similarly, a *path* (of length *m*) in a directed graph $\langle \mathbf{V}, \mathbf{E} \rangle$ is a finite sequence of directed edges of the form

$$N_0 \to N_1 \to N_2 \to \ldots \to N_{m-1} \to N_m$$

where each ordered pair $N_i N_{i+1} \in \mathbf{E}$ in the sequence is understood to be a directed edge of the digraph.

Two nodes are *connected* if there is a path between them. A *cycle* (of length $m+1$) in a simple graph $\langle \mathbf{V}, \mathbf{E} \rangle$ is a finite sequence of edges of the form $N_0 - N_1 - N_2 - \ldots - N_m - N_0$, where $m \geq 1$. A *cycle* (of length $m+1$) in a directed graph $\langle \mathbf{V}, \mathbf{E} \rangle$ is a finite sequence of edges of the form $N_0 \to N_1 \to N_2 \to \ldots \to N_m \to N_0$. Thus, if $N_0 = N_m$ in a path then we return to our starting point, and the path is called a *cycle* (or directed cycle in the case of a digraph).

Example

The graph on the left side of Figure 2-1 represents a simple graph $\langle \mathbf{V}, \mathbf{E} \rangle$, where $\mathbf{V} = \{N_1, N_2, N_3, N_4, N_5\}$ and $\mathbf{E} = \{N_1 N_2, N_1 N_3, N_2 N_4, N_3 N_4, N_3 N_5\}$. Since each edge is an unordered pair of elements, $N_i N_j = N_j N_i$, for all *i*, *j*. An example path of length 3 in this simple graph is $N_1 - N_2 - N_4 - N_3$. An example path of length 4 in this simple graph is $N_1 - N_2 - N_4 - N_3 - N_1$.

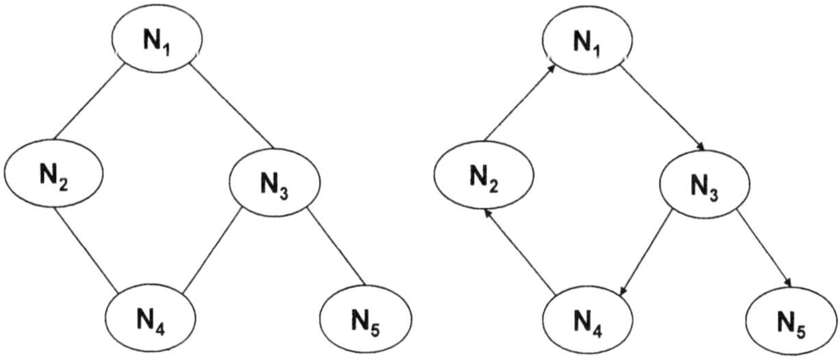

Figure 2-1: Simple (*left*) and directed (*right*) graphs

The graph on the right side of Figure 2-1 represents a directed version of the graph, where $\mathbf{E} = \{N_2N_1, N_4N_2, N_3N_4, N_1N_3, N_3N_5\}$. An example path of length 3 in this directed graph is $N_1 \to N_3 \to N_4 \to N_2$. An example path of length 4 in this directed graph is $N_1 \to N_3 \to N_4 \to N_2 \to N_1$.

A simple graph is said to be *acyclic* if it has no cycles. A directed graph is said to be *acyclic* (or a *directed acyclic graph* or simply a DAG) if it has no cycles. Neither of the simple and directed graphs in Figure 2-1 is acyclic. Examples of simple and directed acyclic graphs are shown in Figure 2-2.

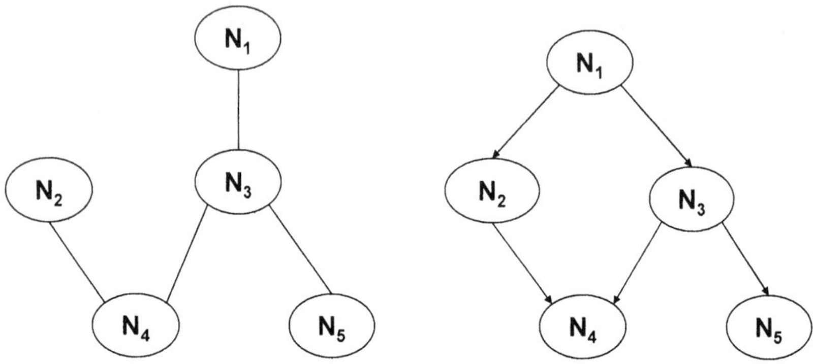

Figure 2-2: Simple (*left*) and directed (*right*) acyclic graphs

A simple graph is a *polytree* if and only if any two vertices of the graph are connected by exactly one path. A directed graph is a *polytree* if and only if its underlying simple graph is a polytree. Example polytrees are shown in Figure 2-3.

Suppose $N_0 \to N_1 \to N_2 \to \ldots \to N_{m-1} \to N_m$ is a path of a directed graph. The vertices occurring in this path are described in genealogical terms as follows:

- N_{i+1} is a *child* of N_i, for $0 \le i < m$
- N_{i-1} is a *parent* of N_i, for $0 < i \le m$
- N_i is an *ancestor* of N_j, for $0 \le i < j \le m$
- N_j is a *descendant* of N_i, for $0 \le i < j \le m$

A *leaf* of a directed polytree is a node without any children in any path of the tree containing the node. A *root* of a directed tree is a node without any parent in any path of the tree containing the node.

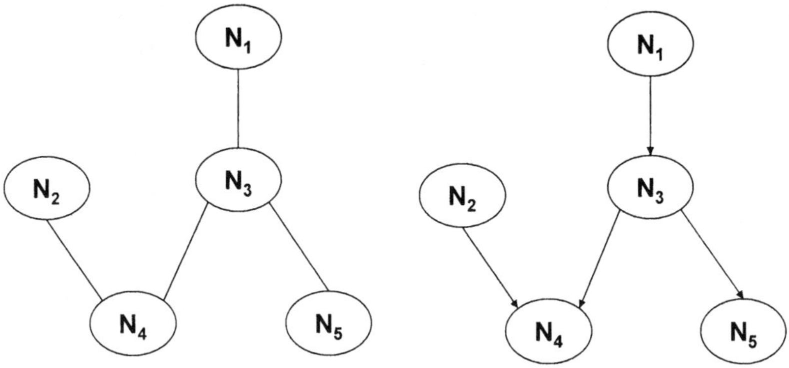

Figure 2-3: Simple (*left*) and directed (*right*) polytrees

Example

The graph on the left side of Figure 2-3 is a simple polytree and the graph on the right side of Figure 2-3 is a directed polytree. Consider the directed polytree in the figure: N_3 is the only child of N_1; N_4 and N_5 are the children of N_3; N_2 and N_3 are the parents of N_4; and N_3 is a parent of N_4 and N_5. N_3, N_4, and N_5 are the descendants of N_1; N_1 and N_3 are the ancestors of N_5; N_1 and N_2 are the root nodes; and N_4 and N_5 are the leaf nodes.

A directed graph is a *tree* if it is a polytree with only one root node. The *level* of a vertex in such a tree is the number of edges in the path between the vertex and the root. The *depth* of such tree is the maximum level of the vertices in the tree. The level of the root in a tree is 0. Example trees are shown in Figure 2-4.

The root and leaf nodes of a simple polytree are not well defined as its edges are undirected. For example, the node N_4 in the simple polytree of Figure 2-3 can be taken as a root node as well as a leaf node. To resolve this kind of ambiguity, we can designate a set of such vertices as root nodes and convert a simple polytree into a *rooted tree*. For example, the nodes N_1 and N_2 in the simple polytree of Figure 2-3 can be designated as roots. Unless otherwise stated, each reference in this book to the term "tree" will be implicitly regarded as "rooted tree."

Example

The graph on the left side of Figure 2-4 is a simple tree and on the right is a directed tree. Vertex N_1 is the root node of the directed tree, whereas the vertex N_1 has been designated as the root of the simple tree. The levels of the vertices N_1, N_2, N_3, N_4, and N_5 are 0, 1, 1, 2, and 2 respectively. Therefore the depth of each of these two trees is 2.

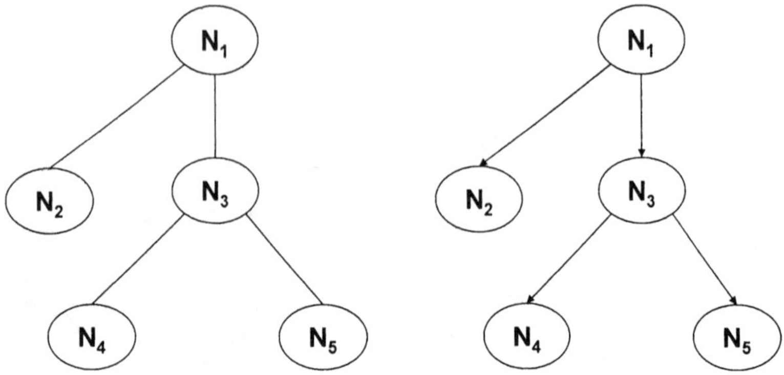

Figure 2-4: Simple (*left*) and directed (*right*) trees

2.3 PROBABILITY AND STATISTICS

Probabilities are defined in terms of likely outcomes of random experiments. A repetitive process, observation, or operation that determines the results of any one of a number of possible outcomes is called a *random experiment.* An *event* is an outcome of a random experiment. The set of all possible outcomes of an experiment is called the *sample space* or *event space.*

Example

Random experiments and outcomes include: tossing a coin a hundred times to determine the number of heads, rolling a pair of dice a couple of hundred times to determine the number of times the sum of the upturned faces is 7, observing the weather throughout the month of March to determine the number of sunny mornings, and sensing daily temperatures over a month to determine the number of hot days. Therefore, tossing a head, rolling a six and a three, and a sunny morning are example events. The sets $\{head, tail\}$, $\{(1,1),(1,2),...,(6,6)\}$, $\{sunny, rain, snow\}$, and $\{t : t \in [0^\circ \text{C}, 100^\circ \text{C}]\}$ are, respectively, examples of sample spaces for these experiments.

A *probability* provides a quantitative description of the likely occurrence of a particular event. The probability of an event x, denoted as $p(x)$, is conventionally expressed on a scale from 0 to 1, inclusive.

Example

In the single die experiment, the probability of rolling a six is 1/6. There are 36 possible combinations of numbers when two dice are rolled. The sample points for the two events x and y consisting of sums of 7 and 10 are, respectively, $x = \{(1, 6), (2, 5), (3, 4), (4, 3), (5, 2), (6, 1)\}$ and $y = \{(4, 6), (5, 5), (6, 4)\}$. Hence, we have $p(x) = 6/36$, $p(y) = 3/36$.

As defined above, an event consists of a single outcome in the sample space. Let us generalize this definition by calling it an *elementary* (or *simple event* or *atomic event*), and by defining a *compound event* as an event that consists of multiple simple events. In general, an event is either a simple event or a compound event. Set theory can be used to represent various relationships among events. In general, if x and y are two events (which may be either simple or compound) in the sample space S then:

- $x \cup y$ means either x or y occurs (or both occur).
- $x \cap y$ or xy means both x and y occur.
- $x \subseteq y$ means if x occurs then so does y.
- $\overline{x}$ means event x does not occur (or equivalently, the complement of x occurs).
- Φ represents an impossible event.
- S is an event that is *certain* to occur.

Two events x and y are said to be *mutually exclusive* if $x \cap y = \Phi$. (The occurrence of both x and y is impossible, and therefore the two events are mutually exclusive.) On the other hand, two events x and y are said to be *independent* if $p(x \cap y) = p(x) \times p(y)$. As a result, when dealing with independent events x and y in an event space, the sets x and y must have a point (event) in common if both x and y have nonzero probabilities. Mutually exclusive, nonimpossible events x and y cannot be independent as $x \cap y = \Phi$, so that $p(x \cap y) = 0$, but $p(x) \times p(y) \neq 0$.

Example

Suppose in the two-dice experiment we want to find the probability that the first die shows even and the second die shows odd. We consider the event x as the set of all sample points with the first element even and event y as the set of all sample points with the second element odd. Therefore, x is $\{(2,1), (2,2), \ldots, (6,6)\}$ and y is $\{(1,1), (2,1), \ldots, (6,5)\}$. Each of these two events has 18 points and the two sets have nine points in common. Hence, $p(x) = 18/36$, $p(y) = 18/36$, and $p(x \cap y) = 9/36$. Therefore, $p(x \cap y) = p(x) \times p(y)$ holds. So by definition, x and y are independent.

There are three approaches that provide guidelines on how to assign probability values to events:

- The classical approach
- The relative frequency approach
- The axiomatic approach

In the *classical approach*, the probability of an event x in a finite sample space S is defined as follows:

$$p(x) = \frac{n(x)}{n(S)}$$

where $n(X)$ is the cardinality of the (finite) set X. Since $x \subseteq S$, $0 \le p(x) \le 1$ and $p(S) = 1$.

In the *relative frequency* approach, the probability of an event x is defined as the ratio of the number (say, n) of outcomes or occurrences of x to the total number (say, N) of trials in a random experiment. The choice of N depends on the particular experiment, but if an experiment is repeated at least N times without changing the experimental conditions, then the relative frequency of any particular event will (in theory) eventually settle down to some value. The probability of the event can then be defined as the limiting value of the relative frequency:

$$p(x) = \lim_{n \to \infty} \frac{n}{N}$$

where n is the number of occurrences of x and N is total number of trials. For example, if a die is rolled many times then the relative frequency of the event "six" will settle down to a value of approximately 1/6.

In the *axiomatic* approach, the concept of probability is axiomatized as follows:

- $p(x) \ge 0$, where x is an arbitrary event.
- $p(S) = 1$, where S is a certain event (i.e., the whole event space).
- $p(x \cup y) = p(x) + p(y)$, where x and y are mutually exclusive events.

Note that while the axiomatic approach merely provides guidance on how to assign values to probabilities, the classical and relative frequency approaches specify what values to assign.

A *Bayesian* or *subjective probability* describes an individual's personal judgment about how likely a particular event is to occur. It is not based on any precise computation, but is an assessment by a subject matter expert based on his or her experience (that is, it's a "guesstimate").

Now we turn to formally defining random variables and probability distributions, the concepts central to the development of probabilistic models for decision making. A *random variable* is a function defined over an event space (that is, the domain of a random variable consists of random events from the sample space) and its value is determined by the outcome of an event. A *discrete random variable* is a random variable whose range is finite or denumerable. The elements in the range (i.e., possible values) of a random variable are called its *states*.

Example

Consider the process of rolling a pair of dice, whose sample space is $\{(1,1),(1,2),...,(6,6)\}$. Consider the random variable *Dice* defined over this sample space, where its values are determined by the sum of the upturned faces, that is, $Dice(i,j)=i+j$, for each sample point (i,j). For example, $Dice(2,3)$ is equal to 5. Therefore, *Dice* is discrete, with a range of $\{2,3,4,...,12\}$. Consider another random variable *Weather* defined over the sample space of the morning weather conditions in a particular month, where the current weather determines its value on a particular morning. The possible values of the discrete random variable *Weather* might be $\{sunny, rain, snow\}$. The domain $\{t: t \in [0^\circ \mathrm{C}, 100^\circ \mathrm{C}]\}$ of the random variable *Temperature* is continuous, and the range could be kept the same as the domain. If the range is considered as, for example, $\{hot, warm, normal, cold, freezing\}$ then it becomes a discrete random variable.

2.3.1 Probability Distributions

The *probability distribution* of a random variable is a function whose domain is the range of the random variable, and whose range is a set of values associated with the probabilities of the elements of the domain. The probability distribution of a discrete random variable is called a *discrete probability distribution*. The probability distribution of a continuous random variable is called a *continuous probability distribution*.

The probability distribution of a random variable is described through its *probability mass function* (or *probability density function*, or PDF). A PDF f of a discrete random variable X with states $\{x_1,...,x_n\}$ is defined as follows: $f(x_i)$ is the probability that X will assume the value x_i. The expected value or expectation of variable X is defined by

$$\mu = E[X] = \sum_i x_i p(X = x_i)$$

Examples

Consider the random variable *Dice* defined as the sum of the upturned faces of two dice, and therefore having the range $\{2,3,4,...,12\}$. Now, $\{(i,j) \in \{(1,1),(1,2),...,(6,6)\}: i+j=5\}$ is equal to $\{(1,4),(2,3),(3,2),(4,1)\}$. Therefore, $p(Dice=5)=4/36$. Similarly, we have the following:

$p(Dice=2)=1/36$, $p(Dice=3)=2/36$, $p(Dice=4)=3/36$,
$p(Dice=5)=4/36$, $p(Dice=6)=5/36$, $p(Dice=7)=6/36$,
$p(Dice=8)=5/36$, $p(Dice=9)=4/36$, $p(Dice=10)=3/36$,
$p(Dice=11)=2/36$, $p(Dice=12)=1/36$.

Also, the sum of all the probabilities is equal to 1, that is, $\sum_{x\in\{2,3,\ldots,12\}} p(Dice=x)=1$. The expected value of the sum of the upturned faces is therefore

$$E[Dice]=2\times\frac{1}{36}+3\times\frac{2}{36}+4\times\frac{3}{36}+5\times\frac{4}{36}+6\times\frac{5}{36}+7\times\frac{6}{36}+ \\ 8\times\frac{5}{36}+9\times\frac{4}{36}+10\times\frac{3}{36}+11\times\frac{2}{36}+12\times\frac{1}{36} \\ =6.7$$

Consider the random variable *Weather* with range $\{sunny, rain, snow\}$. Define

$$p(Weather=sunny)=0.55$$
$$p(Weather=rain)=0.15$$
$$p(Weather=snow)=0.30$$

Figure 2-5 represents the graphs of the two PDFs associated with the random variables *Dice* and *Weather*. Such a graphical depiction of a probability mass function is called a *probability histogram*.

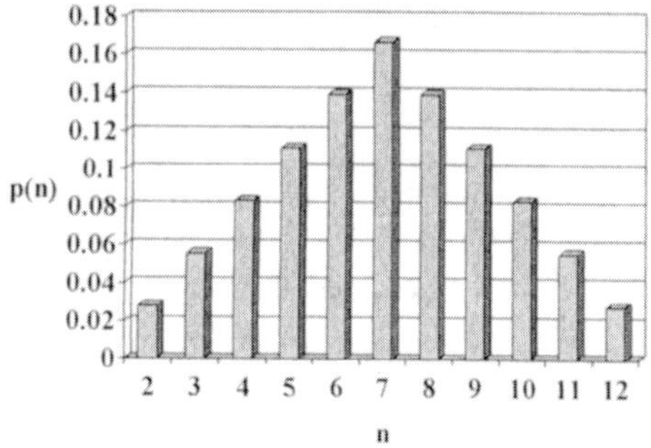

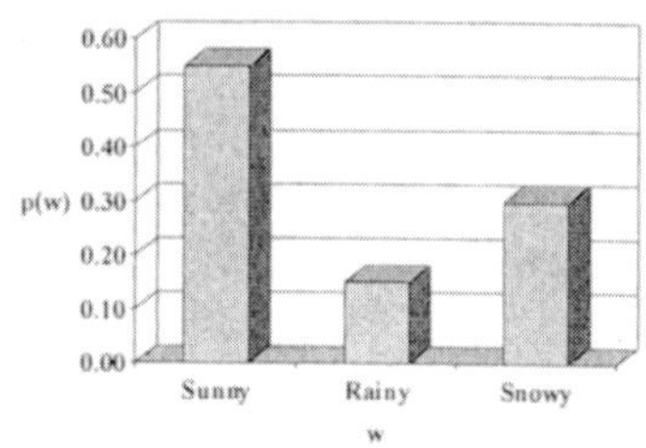

Figure 2-5: Probability density functions (PDFs) for the random variables *Dice* and *Weather*

A discrete random variable X is said to follow a *Poisson distribution* with parameter m, if it has the following probability distribution:

$$p(X=x)=\frac{m^x}{x!}e^{-m} \qquad m>0, x=0,1,2,\ldots$$

A Poisson random variable is a count of the number of events that occur in a certain time interval or spatial area. Considering the detection of a target within a given area is an event then such a count, for example, is the number of true sensor returns (not false alarms) from the area.

The *joint probability distribution* of two discrete random variables X and Y, denoted as $p(X,Y)$ or $p(XY)$, is a function whose domain is the set of ordered pairs (x,y) of events, where x and y are, respectively, possible values for X and Y. The range of the function is the set of probability values corresponding to the ordered pairs in its domain. Such a probability is denoted by $p(X=x,Y=y)$ (or simply $p(x,y)$ when X and Y are clear from the context) and is defined as

$$p(x,y) = p(X=x,Y=y) = p(X=x \text{ and } Y=y)$$

The definition of the joint probability distribution can be extended to three or more random variables. In general, the joint probability distribution of the set of discrete random variables $X_1,...,X_n$, denoted as $p(X_1,...,X_n)$ or $p(X_1...X_n)$, is given by

$$p(x_1,...,x_n) = p(X_1 = x_1,...,X_n = x_n) = p(X_1 = x_1 \text{ and...and } X_n = x_n)$$

The notion of conditional probability distribution arises when one wants to know the probability of an event, given the occurrence of another event. An example conditional probability is the probability of snowy weather later today given that the current temperature is freezing. Formally, the *conditional probability distribution* of the two random variables X and Y, denoted as $p(X|Y)$, is a function whose domain is the set of ordered pairs (x,y), where x and y, respectively, are possible values for X and Y. The range of the function is the set of probability values corresponding to the ordered pairs. The conditional probability distribution is defined as follows:

$$p(X|Y) = \frac{p(XY)}{p(Y)}, \text{ if } p(Y) > 0$$

Following are some important results for conditional probabilities that follow from this definition:

Multiplication Rule

$$p(X_0,X_1,...,X_n) = p(X_0)p(X_1|X_0)p(X_2|X_0X_1)...p(X_n|X_0X_1...X_{n-1}),$$
$$\text{if } p(X_0,X_1,...,X_n) > 0$$

Total Probability Rule

$$p(X) = \sum_{i=1}^{n} p(X|Y_i)p(Y_i), \text{ given } p(Y_i) > 0 \text{, for every } i,$$
$$\text{and given} \sum_{i=1}^{n} p(Y_i) = 1$$

Special Case: $p(X) = p(X \mid Y)p(Y) + p(X \mid \overline{Y})p(\overline{Y})$, if $0 < p(Y) < 1$

Marginalization Rule

$p(X) = \sum_{i=1}^{n} p(X, Y_i)$, given $p(Y_i) > 0$, for every i, and given $\sum_{i=1}^{n} p(Y_i) = 1$

Special Case: $p(X) = p(X, Y) + p(X, \overline{Y})$, if $0 < p(Y) < 1$

Bayes' Rule

$p(Y_j \mid X) = \dfrac{p(X \mid Y_j)p(Y_j)}{\sum_{i=1}^{n} p(X \mid Y_i)p(Y_i)}$, if $p(X) > 0$ and $p(Y_i) > 0$, for every i, and

$$\sum_{i=1}^{n} p(Y_i) = 1$$

Special Case: $p(X \mid Y) = \dfrac{p(Y \mid X)p(X)}{p(Y)}$, if $p(X) > 0$ and $p(Y) > 0$

The probability distribution of a continuous random variable is also described through its *probability density function*, or PDF. A PDF f of a continuous random variable X with domain D is defined as follows: $f(x), x \in D$, is the probability that X will assume the value d. The *expected value* or *expectation* of variable X is defined by

$$\mu = E[X] = \int_D x f(x) dx$$

The *variance* of the random variable X is defined as

$$\mathrm{var}(X) = \sigma^2 = E[X - E[X]]^2 = E[X^2] - E[X]^2$$

The expected value of a variable is known as its *population mean.* The *sample mean* is an estimator for the population mean and is commonly known as *average.* Its value depends on the data values chosen. A continuous random variable X with domain $(-\infty, +\infty)$ is said to follow a (univariate) *normal* or *Gaussian distribution* with parameters μ and σ if it has the following PDF:

$$f(X) = \frac{1}{\sigma\sqrt{2\pi}} e^{-\frac{(X-\mu)^2}{2\sigma^2}}$$

We write $X \sim N[\mu, \sigma^2]$. As shown in Figure 2-6, the PDF is a symmetrical bell-shaped curve with center at its expected value μ. The variance of a Gaussian random variable is σ^2.

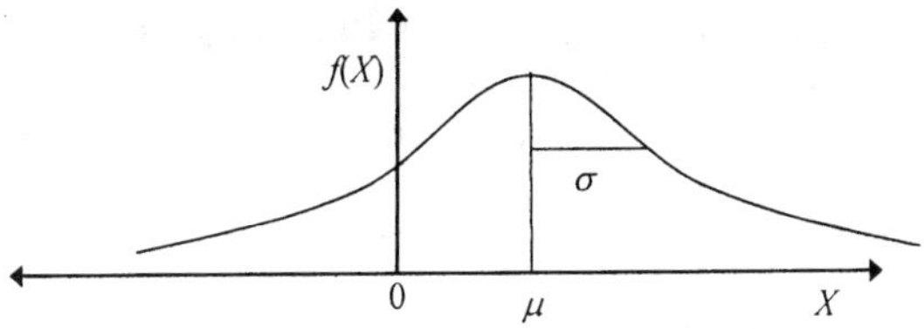

Figure 2-6: Gaussian distribution

The *Poisson distribution* is used to model the number of events occurring within a given time interval. It has the following PDF:

$$p(X,\lambda)=\frac{e^{-\lambda}\lambda^{X}}{X!},\ X=0,1,2,...$$

where λ represents the average number of events in the given time interval.

A fusion application (e.g., target tracking) deals with the attributes of a target arranged as a vector. A *vector-valued random variable* is an n-dimensional vector consisting of n random variables that are all associated with the same events. An example 3-dimensional random variable consists of the x and y coordinates of a target and its speed:

$$X=\begin{bmatrix} x \\ y \\ s \end{bmatrix}$$

The *multivariate normal distribution* model extends the univariate normal distribution model to fit vector observations. An n-dimensional random variable X is said to follow a multivariate normal distribution if it has the following PDF:

$$f(X)=\frac{1}{(2\pi)^{p/2}|\Sigma|^{1/2}}e^{-\frac{1}{2}(X-\mu)'\Sigma^{-1}(X-\mu)}$$

where μ is the vector of mean and Σ is the covariance matrix of X, that is, $\Sigma_{ij}=Cov(X_i,X_j)$. We write $X\sim N_n[\mu,\Sigma]$.

2.4 MATHEMATICAL LOGIC

This section presents a very brief introduction to the classical logics for developing approaches to logic-based (more specifically, rule-based) fusion. We start with classical propositional logic and then move to first-order logic. We left out many details for the brevity of relevance and space. There are references provided at the end of the chapter to find details, such as the model theoretic semantics of the logics, the soundness and completeness theorems, resolution theorem proving, logic programming, and logic programming language Prolog.

Mathematical logic is a study of valid arguments. Arguments are represented in logical syntax such as propositional and first-order. A *proposition* is a declarative sentence that is either *true* or *false*, but not both. Examples of propositions are:

Road mobility is a slow go

It is raining at the area of operation

Propositions will be symbolized as, for example, $P, Q, R, \ldots$, and are called *atoms* or *atomic formulae*. Compound propositions are formed by modifying with the word "not" or by connecting sentences via the words/phrases "and," "or," "if ... then," and "if and only if." These five words/phrases are called *logical connectives* and are usually symbolized as shown in Table 2-2.

Connective	Symbol
not	$\neg$
and	$\wedge$
or	$\vee$
if … then	$\rightarrow$
if and only if	$\leftrightarrow$

Table 2-2: Standard logical connectives

Example

Examples of compound propositions are as follows:

If weather is rainy then the road mobility is a slow go

Weather is not rainy

Visibility is poor or weather is rainy

The above composite propositions are symbolized respectively in the *propositional language* as:

$P \rightarrow Q$

$\neg P$

$R \vee P$

where:

P stands for "Weather is rainy"

Q stands for "Road mobility is a slow go"

R stands for "Visibility is poor"

Each of these compound propositions is called a *Well-Formed Formula* (WFF) or *formula* as part of the propositional language, and the symbols occurring in the formulae are part of the propositional alphabet. Formal definitions of these terms are given below.

A *propositional alphabet* consists of the following:

- Two parentheses "(" and ")"
- A set of propositional variables $P, Q, R, \ldots$ as atoms
- A set of logical connectives $\neg, \wedge, \vee, \rightarrow$, and $\leftrightarrow$

Well-Formed Formulae or *formulae* in propositional logic are defined as follows:

- An atomic formula is a formula, for example, $P, Q, R, \ldots$, as mentioned previously.
- If F is a formula, then $(\neg F)$ is a formula.
- If F and G are formulae, then $(F \wedge G)$, $(F \vee G)$, $(F \rightarrow G)$, and $(F \leftrightarrow G)$ are formulae.

Given a propositional alphabet, the *propositional language* comprises the set of all formulae constructed from the symbols of the alphabet. An expression is a formula only if it can be shown to be a formula by the above three conditions. A formula of the form $(\neg F)$ is called the *negation* of the formula F. Formulae of the forms $(F \wedge G)$ and $(F \vee G)$ are called the *conjunction* and *disjunction*, respectively, of the formulae F and G. A formula of the form $(F \rightarrow G)$ is called a *conditional* formula; F is called the *antecedent* and G is called the *consequent*. A formula of the form $(F \leftrightarrow G)$ is called a *biconditional* formula. We will see later that disjunctions are one way that uncertain knowledge can be represented.

The following conventions are used to avoid using parentheses in a formula. The connective $\neg$ is applied to the smallest formula following it, then $\wedge$ is to connect the smallest formulae surrounding it, and so on for the rest of the connectives $\vee$, $\rightarrow$, and $\leftrightarrow$ in that order.

Example

If parentheses are restored in the formula

$$\neg P \wedge Q \rightarrow R \leftrightarrow Q \vee R \wedge P$$

then the resulting formula would be

$$((((\neg P) \wedge Q) \rightarrow R) \leftrightarrow (Q \vee (R \wedge P)))$$

We are especially interested in assessing the "truth" of a formula as a function of the "truth" of its atoms. To do this, we first assign a *truth value* to each atom; that is, we assign "true" (denoted as $\top$) or "false" (denoted as $\perp$) to the symbolized atomic propositions that occur in a formula. Then we compute the truth value of the formula using special rules to handle the logical connectives. So for every assignment of truth values $\top$ or $\perp$ to the symbolized atomic propositions that occur in a formula, there corresponds a truth value for the formula. This can be determined using the *truth table* of the formula.

Example

Table 2-3 is the combined truth table for the fundamental formulae $\neg P$, $P \wedge Q$, $P \vee Q$, $P \rightarrow Q$, and $P \leftrightarrow Q$; it shows how to compute truth values for a formula containing these logical connectives. Using this basic truth table, the truth table for any formula can then be constructed; for example, the truth table for $(\neg P \leftrightarrow Q) \rightarrow (P \wedge (Q \vee R))$ is displayed in Table 2-4.

P	Q	$\neg P$	$P \wedge Q$	$P \vee Q$	$P \rightarrow Q$	$P \leftrightarrow Q$
$\top$	$\top$	$\bot$	$\top$	$\top$	$\top$	$\top$
$\top$	$\bot$	$\bot$	$\bot$	$\top$	$\bot$	$\bot$
$\bot$	$\top$	$\top$	$\bot$	$\top$	$\top$	$\bot$
$\bot$	$\bot$	$\top$	$\bot$	$\bot$	$\top$	$\top$

Table 2-3: Truth table in propositional logic

P	Q	R	$\neg P$	$\neg P \leftrightarrow Q$	$Q \vee R$	$P \wedge (Q \vee R)$	$(\neg P \leftrightarrow Q) \rightarrow (P \wedge (Q \vee R))$
$\top$	$\top$	$\top$	$\bot$	$\bot$	$\top$	$\top$	$\top$
$\top$	$\top$	$\bot$	$\bot$	$\bot$	$\top$	$\top$	$\top$
$\top$	$\bot$	$\top$	$\bot$	$\top$	$\top$	$\top$	$\top$
$\top$	$\bot$	$\bot$	$\bot$	$\top$	$\bot$	$\bot$	$\bot$
$\bot$	$\top$	$\top$	$\top$	$\top$	$\top$	$\bot$	$\bot$
$\bot$	$\top$	$\bot$	$\top$	$\top$	$\top$	$\bot$	$\bot$
$\bot$	$\bot$	$\top$	$\top$	$\bot$	$\top$	$\bot$	$\top$
$\bot$	$\bot$	$\bot$	$\top$	$\bot$	$\bot$	$\bot$	$\top$

Table 2-4: Truth table for $(\neg P \leftrightarrow Q) \rightarrow (P \wedge (Q \vee R))$

Given a formula F, suppose $P_1, P_2, ..., P_n$ are all atomic formulae occurring in F. Then an *interpretation* of F is an assignment of truth values to $P_1, P_2, ..., P_n$, where no P_i is assigned both $\top$ and $\bot$. Hence every row in a truth table for a formula F is an interpretation of F.

A formula F is a *tautology* or is *valid* (denoted as $\models F$) if its value is $\top$ under all possible interpretations of F.

Example

The formula $P \rightarrow (P \vee Q)$ is a tautology according to Table 2-5.

P	Q	$P \vee Q$	$P \rightarrow (P \vee Q)$
$\top$	$\top$	$\top$	$\top$
$\top$	$\bot$	$\top$	$\top$
$\bot$	$\top$	$\top$	$\top$
$\bot$	$\bot$	$\bot$	$\top$

Table 2-5: A tautology

A formula F is *false* (or *inconsistent* or a *contradiction*) if and only if its value is $\bot$ under all possible interpretations.

Example

The formula $P \vee Q \leftrightarrow \neg P \wedge \neg Q$ is false by Table 2-6.

P	Q	$\neg P$	$\neg Q$	$P \vee Q$	$\neg P \wedge \neg Q$	$P \vee Q \leftrightarrow \neg P \wedge \neg Q$
$\top$	$\top$	$\bot$	$\bot$	$\top$	$\bot$	$\bot$
$\top$	$\bot$	$\bot$	$\top$	$\top$	$\bot$	$\bot$
$\bot$	$\top$	$\top$	$\bot$	$\top$	$\bot$	$\bot$
$\bot$	$\bot$	$\top$	$\top$	$\bot$	$\top$	$\bot$

Table 2-6: A contradiction

There are various kinds of arguments that cannot be conveniently stated in the language of propositional logic. Consider the following argument as an example:

It is raining at the main road leading to the area of operation
Rain makes the mobility of every road a slow go
Therefore, the main road leading to the area of operation is a slow go

The above argument is a valid argument. However, if the three premises are symbolized as P, Q, and R, respectively, it is not possible to prove R from P and Q within the framework of propositional logic, unless the second premise "Rain makes the mobility of every road" is instantiated for the specific road leading to the area of operation. The correctness of the above argument relies upon the meaning of the expression "every," which has not been considered in the propositional logic. However, first-order logic handles this kind of argument, and also extends the propositional logic by incorporating more logical notations, such as terms, predicates, and quantifiers. The set of symbols (the *first-order alphabet*) in the case of first-order logic is defined as follows:

- Delimiter: , (comma)
- Parentheses: (,)
- Primitive connectives: $\neg$ (negation), $\rightarrow$ (implication)
- Universal quantifier: $\forall$ (for all)
- Individual variables: $x, y, z, x_1, y_1, z_1, \ldots$
- Individual constants: $a, b, c, a_1, b_1, c_1, \ldots$
- For each natural number *n*, *n*-ary predicate symbols: $P, Q, R, P_1, Q_1, R_1, \ldots$
- For each natural number *n*, *n*-ary function symbols: $f, g, h, f_1, g_1, h_1, \ldots$

Terms are expressions, which are defined recursively as follows:

- A variable or an individual constant is a term.
- If *f* is an *n*-ary function symbol and $t_1, t_2, \ldots, t_n$ are terms then $f(t_1, t_2, \ldots, t_n)$ is a term.
- An expression is a term if it can be shown to be so only on the basis of the above two conditions.

A *predicate* is a function which evaluates to either true or false, or a statement about a relation that may be true or false. If *P* is an *n*-ary predicate symbol and $t_1, t_2, \ldots, t_n$ are terms, then $P(t_1, t_2, \ldots, t_n)$ is an *atomic formula* (or *atom* or *positive literal*). A *negative literal* is a formula of the form $\neg A$, where *A* is an atom. A *literal* is either positive or negative. Based on these primitive notions, the *Well-Formed Formulae* (WFFs) or *formulae* of first-order logic are recursively defined as follows:

- Every atomic formula is a formula.
- If *F* is a formula then $\neg F$ is a formula.
- If *F* is a formula and *x* is a variable then $\forall x(F)$ is a formula.
- If *F* and G are formulae then $F \rightarrow G$ is a formula.
- An expression is a formula only if it can be generated by the above four conditions.

For convenience and improved readability of formulae, the other logical connectives, $\wedge$, $\vee$, and $\leftrightarrow$, are also introduced and defined in terms of $\neg$ and $\rightarrow$ just as in the case of propositional logic. Additionally, an *existential quantifier*, denoted as $\exists$, is introduced and defined as follows:

$$\exists x(F) \equiv \neg(\forall x(\neg F))$$

In the formulae $\exists x(F)$ and $\forall x(G)$, *F* and *G* are called the *scope* of the quantifiers $\exists x$ and $\forall x$ respectively. As in the case of propositional calculus, the same convention is made about the omission of parentheses in a formula. A formula in propositional logic can be considered as a formula in first-order logic (where the atoms are 0-ary predicates, and there are no variables, functions, or quantifiers). Hence all the results established so far in connection with propositional logic are also applicable to the set of all quantifier and variable-free formulae in first-order logic. Each ground atomic formula (no occurrence of variables) occurring in this set is considered as a propositional symbol.

Given a first-order alphabet, the *first-order language* $\mathcal{L}$ comprises the set of all formulae constructed from the symbols of the alphabet. Using the first-order language, a symbolization of the first two premises of the argument presented in the beginning of this subsection is as follows:

$$Rain(main\,road)$$

$$\forall x\left(Rain(x) \rightarrow Mobility(x, slow\,go)\right)$$

$$Mobility(main\,road, slow\,go)$$

where *Rain* and *Mobility* are unary and binary predicate symbols respectively, and *main road* and *slow go* are constants. This symbolization is within the framework of *first-order logic*. An axiomatic deduction or inferencing of the conclusion of the argument appears as follows:

Step 1: $\forall x\left(Rain(x) \rightarrow Mobility(x, slow\,go)\right)$ Given Premise

Step 2: $Rain(main\,road)$ Given Premise

Step 3: $Rain(main\,road) \rightarrow Mobility(main\,road, slow\,go)$ Axiom on Step 1

Step 4: $Mobility(main\,road, slow\,go)$ Modus Ponens on Steps 2 and 3

The premises in steps 1 and 2 are considered *proper axioms*. The above deduction is a *proof* of the conclusion $Mobility(main\,road, slow\,go)$ and therefore is a *theorem* that follows from the first-order system with the two proper axioms. Step 3 is derived from step 1 by particularization of the first premise. If the first premise is true for all roads, then it is also true for the main road. Step 4 is arrived at by the application of Modus Ponens. These steps demonstrate the basic *axiomatic* theorem-proving approach.

A procedure for determining whether or not a formula is a theorem of a particular theory $\mathcal{K}$ is called a *theorem proving procedure* or *proof procedure* for the theory $\mathcal{K}$. Theorem proving procedures deal with formulae in standard forms, for example, prenex normal form, Skolem conjunctive normal form, and clausal form. This section provides tools for obtaining these forms from given formulae.

A formula is said to be in *prenex normal form* if it is of the form

$$Q_1 x_1 Q_2 x_2 ... Q_n x_n B$$

where each Q_i is either $\forall$ or $\exists$, and the formula B is quantifier free. The formula B is called the *matrix*. A prenex normal form formula is said to be in *Skolem conjunctive normal form* if it has the form

$$\forall x_1 \forall x_2 ... \forall x_n B$$

where the matrix B is in conjunctive normal form, that is, B is a conjunction of a disjunction of literals (as defined in Section 3.1). Such a Skolem conjunctive normal form formula is said to be a *clause* if it has the form

$$\forall x_1 \forall x_2 ... \forall x_n \left(L_1 \vee L_2 \vee ... \vee L_m \right)$$

where each L_i is a literal and $x_1, x_2, ..., x_n$ are the variables of the disjunction $L_1 \vee L_2 \vee ... \vee L_m$. A formula is said to be in *clausal form* if it is a clause.

For the sake of convenience, a clause is rewritten as the disjunction $L_1 \vee L_2 \vee ... \vee L_m$ of literals without its quantifiers or as the set $\{L_1, L_2, ..., L_m\}$ of literals. Thus when a disjunction $L_1 \vee L_2 \vee ... \vee L_m$ or a set $\{L_1, L_2, ..., L_m\}$ is given as a clause C, where each L_i is a literal, then C is regarded as being of the form $\forall x_1 \forall x_2 ... \forall x_n \left(L_1 \vee L_2 \vee ... \vee L_m \right)$, where $x_1, x_2, ..., x_n$ are all the variables occurring in all the L_is. The clausal form of $\forall x \left(Rain(x) \rightarrow Mobility\left(x, SlowGo\right)\right)$ is $\neg Rain(x) \vee Mobility\left(x, SlowGo\right)$.

Every arbitrary first-order formula F can be transformed to its equivalent formula G in Skolem conjunctive normal form by applying the appropriate transformation steps. For example, the Skolem conjunctive normal form of the formula

$$\forall x \left(\neg R(x) \rightarrow P(a) \wedge \neg \exists z \neg Q(z,a) \right) \wedge \forall x \left(P(x) \rightarrow \exists y Q(y,x) \right)$$

is

$$\forall x \forall z \forall x_1 \left(\left(R(x) \vee P(a) \right) \wedge \left(R(x) \vee Q(z,a) \right) \wedge \left(\neg P(x_1) \vee Q\left(f(x_1), x_1\right) \right) \right)$$

where each conjunct in $\left\{ R(x) \vee P(a), R(x) \vee Q(z,a), \neg P(x_1) \vee Q\left(f(x_1), x_1\right) \right\}$ is a clause or in clausal form. Such a clause is then interpreted as various types of program clauses or if-then types of rules. Figure 2-7 shows the overall evolution along with examples.

Expert system rules and Prolog (Bratko, 2000; Clocksin and Mellish, 2003) rules based on logic programming (Kowalski, 1979) are just the definite version of general if-then-type rules with attached uncertainty values. By definite we mean if the condition of a rule is true then the conclusion is definite rather than a disjunction. Arguments are more general than these rules and will be considered in Chapter 9 on decision making. An argument can have a negation symbol at its consequent representing an argument against a hypothesis. Rules normally used in rule-based expert systems to represent expert knowledge are restricted to propositional syntax to avoid reasoning complexity and to enhance efficiency. Each such rule has an associated degree of uncertainty represented by elements from dictionaries (for example, probability values drawn from $[0,1]$ or certainty factor values drawn from the dictionary $[-1,+1]$ or values drawn from a qualitative dictionary such as {*high*, *medium*, *low*}). Arguments are more general than definite rules since the syntax of an argument against an option requires negative consequent.

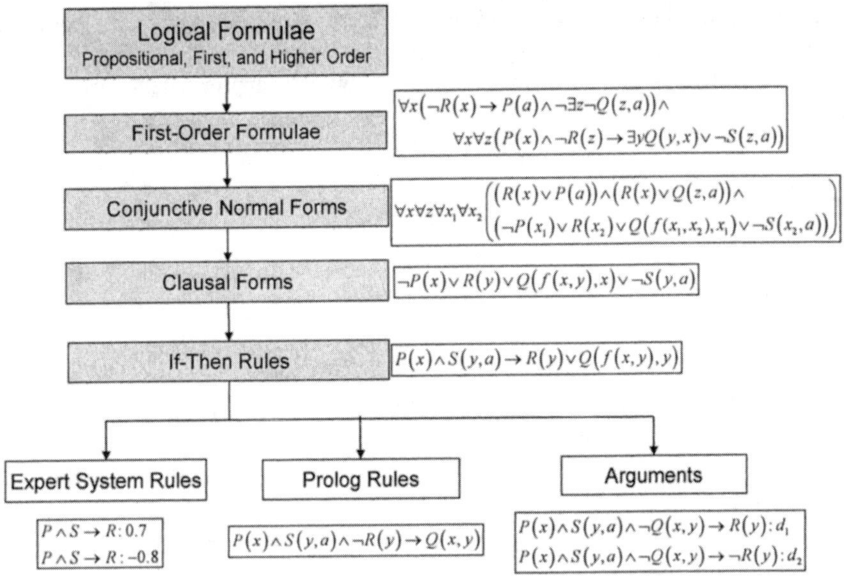

Figure 2-7: Evolution of rules with examples

The if-then-type rules in the decision-making context we will be dealing with in this chapter have the following general form:

IF *Events*
THEN *Hypothesis* (*D*)

The above rule is interpreted as "If *Events* occur, then *Hypothesis* follows with a certain degree of uncertainty *D*." In the expert system rule syntax shown in Figure 2-7, the above rule is

$$Events \rightarrow Hypothesis : D$$

In the absence of *D*, the rule becomes a definite rule, and reasoning with such rules can be performed using any of the resolution principles without the need to handle uncertainty. If *D* is a probability value, then the rule is interpreted as "If *Events* occur, then the probability that *Hypothesis* will follow is *D*." *Events* is a conjunction of propositional symbols, each representing an event, and *Hypothesis* can be a property symbol as in

IF Visibility is good
AND Mobility is go
THEN Vehicle Journey is on (0.8)

or an action symbol as in

IF Visibility is good
AND Mobility is go
THEN cancel Vehicle Journey (0.2)

But the logical reasoning presented here does not distinguish between a property symbol and an action symbol. Our objective here is to interpret rules upon receiving evidence (e.g., evidence on visibility and mobility) and then compute aggregated evidence on individual decision options (e.g., *Vehicle Journey is on* and *cancel Vehicle Journey*) by handling uncertainty appropriately. In the next chapters, we present various approaches to handling uncertainty.

2.5 ALGORITHMIC COMPLEXITY

An algorithm is a program that is guaranteed to give a correct answer to a problem within a certain time. We say an algorithm runs in *polynomial time* when there is a polynomial

$$p(x) = a_0 + a_1 x + a_2 x^2 + ... + a_k x^k$$

such that the time taken to run the algorithm is less than or equal to $p(x)$, where x is the "input length" (essentially, the amount of data needed to describe an instance of the problem), and $a_0, a_1, a_2, ..., a_k$ are nonnegative integers. Formally, if an algorithm runs in polynomial time, then we say that the algorithm complexity (or simply, the algorithm) is "*of order* x^k" or $O(x^k)$, where k is the highest power in $p(x)$ above. This theory of computational complexity started with Cook's paper (Cook, 1971), and its close relationship to combinatorial optimization can be found in (Karp, 1972).

Relationships among complexity classes are shown in Figure 2-8. A problem for which an algorithm of order x^k exists is said to be in the *polynomial class* or P *class*. Unfortunately, time-complexity functions are not always bounded this way; for example, some algorithms belong to the *exponential* class, or *EXP* class. These algorithms have "exploding" time functions which contain exponential factors, like 2^n or $n!$ (where n is again the input length of the problem). They grow extremely quickly, much faster than any polynomial function. A problem with time complexity bounded by a polynomial function is considered *tractable*; otherwise, it is *intractable*.

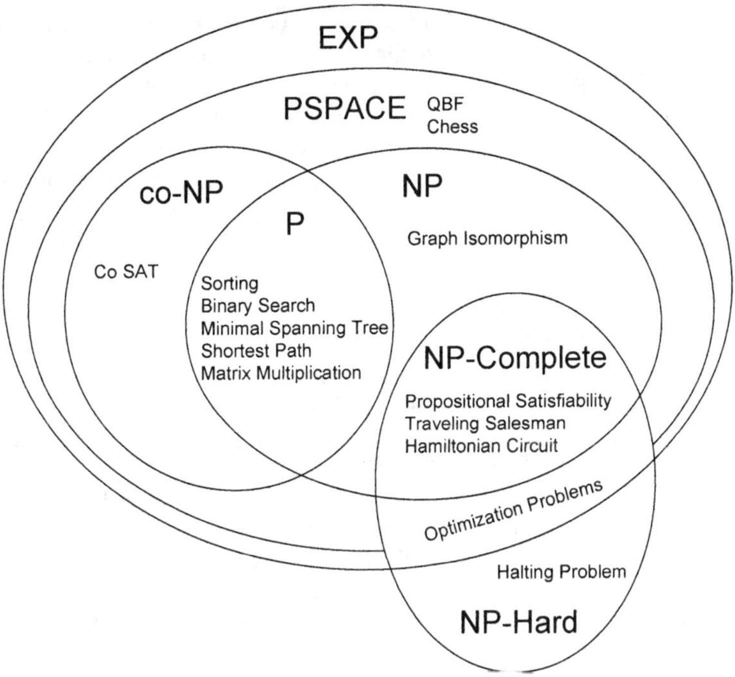

Figure 2-8: Relationships among complexity classes

A decision problem (or recognition problem) is one that takes the form of a question with a "yes" or "no" answer. Consider, for example, the Traveling Salesman Problem (TSP), which decides if there is some tour or circuit in a complete weighted graph that visits every nodc exactly once, with total path weight less than some given value. This differs from the corresponding optimization problem of finding the shortest tour of all the cities, which requires exponential running time. We say that a decision problem belongs to the NP (*Nondeterministic Polynomial*) complexity class if every "yes" instance has a certificate whose validity can be checked in polynomial time. For example, if the TSP decision problem has the answer "yes" then one certificate is a list of the orders in which the nodes should be visited. It takes only time $O(x)$ to add up the length of this tour and check it is less than the given value. By reversing the roles played by "yes" and "no" we obtain a complexity class known as *Co-NP*. In particular, for every decision problem in NP there is an associated decision problem in Co-NP obtained by framing the NP question in the negative, for example, the question "Do all traveling salesman tours have length greater than certain value?" For Co-NP problems, every "no" instance has a certificate whose validity can be checked in polynomial time.

The complexity class NP-complete is the set of decision problems that are the hardest problems in NP in the sense that they are the ones most likely not to be in P. Formally, a problem is *NP-complete* if

- It is in NP, and
- Every NP problem can be reduced to it in polynomial time.

An optimization problem for which the related decision problem is NP-complete is termed *NP-hard.* Such a problem is at least as hard or harder than any problem in NP. Thus an NP-hard problem is any problem such that every NP problem can be converted to it (reduced to it) in polynomial time. Therefore, if we are given an algorithm that solves an NP-hard problem, then the algorithm can also be used to solve any problem in NP with no more than $O\left(x^k\right)$ extra time. The class of NP-complete problems is the intersection of the classes of NP-hard and NP problems. The running time of a *P problem* on a deterministic Turing machine has a polynomial-length input, and NP problems are the polynomial-time problems on nondeterministic Turing machines.

The set of all P problems is contained in the set of all NP problems, which, in turn, is contained in the set of all EXP problems. Some examples of P problems are sorting, binary search, minimal spanning tree, shortest path, and matrix multiplication. The most well-known NP-complete problem is the propositional satisfiability (SAT) problem which determines whether a given propositional formula is satisfiable. SAT has 2^N possible solutions if there are N propositional variables, and so has exponential time complexity (assuming we must check them all to determine if there is a solution). But a possible solution is easy to check in polynomial time, and therefore it is an NP-complete problem. Another NP-complete problem is the Hamiltonian Circuit Problem which finds a circuit in a graph that passes through each vertex exactly once. The most famous outstanding question in complexity theory is whether P = NP, that is, whether the NP problems actually do have polynomial-time solutions that haven't yet been discovered. Although there is strong suspicion that this is not the case, no one has been able to prove it.

As mentioned previously, decision problems are associated with optimization problems. For example, for the TSP the associated optimization question is "What is the length of the shortest tour?" If an optimization problem asks for a certain type of structure with the minimum "cost" among such structures, we can associate with that problem a decision problem that includes a numerical bound B as an additional parameter and that asks whether there exists a structure of the required type having a cost no more than B. The problem of finding the best clique tree (which will be defined in the context of junction tree algorithms in the chapter on BNs) is such an optimization problem which is NP-hard.

The *graph isomorphism problem* (Are two graphs isomorphic?) is suspected to be neither in P nor NP-complete, though it is obviously in NP. There exists no known P algorithm for graph isomorphism testing, although the problem has also not been shown to be NP-complete. The *subgraph isomorphism problem* (Is a graph isomorphic to a subgraph of another graph?) is NP-complete. The *halting problem* (Given an algorithm and an input, will the algorithm ever stop?) is a decision problem, but is not NP-complete. It is an NP-hard problem.

The following types of algorithms are used in practice to deal with intractable problems and problems not admitting reasonably efficient algorithms:

- Approximation algorithms that settle for less than optimum solutions;
- Heuristic algorithms that are probably efficient for most cases of the problem;
- Randomized algorithms that are probably correct in typical problem instances.

For *undecidable* problems, there is no algorithm that always solves them, no matter how much time or space is allowed. The halting problem is an undecidable problem. In general, it cannot be solved in finite time. First-order logic (discussed in the last section) is also undecidable in the sense that, in general, there is no effective procedure to determine whether a formula is a theorem or not.

The *space complexity* of an algorithm is concerned with the amount of memory used but not time. The difference between space complexity and time complexity is that space can be reused. Space complexity is not affected by determinism or nondeterminism, as deterministic machines can simulate nondeterministic machines using a small amount of space (Savitch's theorem). A problem is said to be in the class PSPACE if it can be solved in space polynomial in the size of its input. Clearly, a Turing machine that uses polynomial time also uses at most polynomial space. However, a Turing machine that uses polynomial space may use an exceedingly large amount of time before halting. Therefore, NP is a subset of PSPACE. A problem is said to be PSPACE-complete if it is in PSPACE and if every other PSPACE problem is polynomial-time reducible to it. It is widely believed that PSPACE-complete problems are strictly harder than the NP-complete problems. Proving the validity of Quantified Boolean Formulae (QBF) and perfect chess playing are examples of PSPACE-complete problems.

2.6 FURTHER READING

There are plenty of well-written text books in each of the areas covered in this chapter. Here I provide only a few popular ones. The graph theory book by Deo (1974) is one of the first comprehensive texts on graph theory with applications to computer science and engineering. Wilson's book (1996) is another good introductory book on graph theory.

There are dozens of good text books on classical logics, of which three excellent ones are Copi (1979), Mendelson (1987), and Stoll (1963). Robinson's paper (1965) is highly recommended as a foundational reading on theorem proving. Chang and Lee (1973) provide a thorough coverage of automated theorem proving. For a more recent account on automated theorem proving, see

Gallier (2003). The author's own book (Das, 1992) presents a detailed overview of logic programming and its related field deductive databases.

Two good books on probability are Feller (1968) and Chung (2000). Finally, Papadimitriou (1993) is a good text book on complexity theory.

Chapter 3

Approaches to Handling Uncertainty

Handling of uncertain information is at the core of any fusion system. The questions to be addressed are how uncertainty is expressed or quantified, and how uncertain pieces of information can be aggregated or discounted from each other. This chapter surveys and categorizes different types of uncertainty and ways of expressing uncertainty, introduces computational formalisms for handling them, and then touches upon defining their interrelationships. Each of the formalisms for handling uncertainty is described in some detail without much discussion at this stage on how it can be applied to high-level fusion. Some of these formalisms will be adopted and detailed in subsequent chapters to solve various high-level data fusion problems.

3.1 IGNORANCE TO UNCERTAINTIES

Ignorance can be generally characterized as lack of information. Different types of information may be lacking in data and knowledge bases, and consequently give rise to different kinds of uncertainty. Here we present a survey of certain common forms of ignorance and of the mathematical techniques that have been suggested to quantify ignorance. We then discuss interrelationships among these techniques in the next section.

Ignorance can broadly be subdivided into the following three categories (Bonissone and Tong, 1985; Smets, 1991):

- *Incompleteness*: Incomplete information refers to cases where the value of a variable is missing. Consider a database of battlefield reports that should include information about detected unit position (latitude and longitude), category (friendly, neutral, or hostile), and type (tank, armored personnel carrier, or Humvee) for each report. If, for example, the value of the variable representing the type of a detected hostile unit at position (x, y) is missing from a report, then the information in the report becomes incomplete. But the rest of the information in the report is precise and certain (i.e., a hostile unit has been detected at (x, y)).

- *Imprecision*: Imprecise information refers to cases where the value of a variable is given, but not with enough precision. Suppose the value of the variable representing the type of a detected hostile unit at a certain location is "tank" or "Humvee." The information is complete because the values of each of the three variables representing the unit's position, category, and type are given, but the information is imprecise because there is some ambiguity as to the exact type of the detected unit.
- *Uncertainty*: Uncertain information refers to cases where the information is complete, precise, but uncertain since it might be wrong. This type of ignorance appears when the observer (human or sensor) is taken into account. It is the observer that is not certain about the available information.

Usually a database variable (i.e., a column attribute for a relational database) is constrained either by a domain, when the variable is qualitative or categorical, or by a range, when the variable is numeric. The variables representing unit type and category are qualitative variables, whereas the variable representing a unit's x or y position is a numeric one. Then it might be argued that incomplete information on qualitative variables is just a special case of imprecision.

Moreover, imprecision and incompleteness are context dependent in the sense that information that is imprecise or incomplete in one context may be precise or complete in another context. For example, the imprecise information "tank or Humvee" is sufficient for a mission to target any hostile unit, but insufficient for destroying only those targets with striking capabilities. Similarly, if the mission is to target any hostile object in the air within the 100 mile radius of the base, then the information "a hostile object is within 50 miles" is sufficient to determine that the object is a target. But if the mission is to target any hostile object within the 40 mile radius of the base then the piece of information is not sufficient to determine whether it is a target or not. Table 3-1 presents types of uncertainty and subtypes, and example instances. The rest of the section discusses how traditional classical and nonclassical logics and the theory of probability fail to handle some of these uncertainties. Specific techniques for handling these uncertainties will then be discussed in the next section.

A technique for handling uncertainty must offer a syntax for representing uncertain information. Such a technique based on classical logic will be able to express some of the uncertainties in the table. For example, existential and universal incompleteness in a knowledge base can be represented using quantifiers of first-order logic. Disjunction and negation types of imprecision can be represented using connectives of both propositional and first-order logics. First-order logic needs to be expanded with the axioms and inference rules of arithmetic to deal with intervals. There is no natural way of expressing fuzzy-valued imprecision in classical logic other than considering phrases such as "very fast" as a term of the underlying syntax of the logic. Fuzzy logic (Zadeh, 1965) has been developed to handle such imprecision. Moreover, coarse grained two-valued representation of truth in classical logic is not sufficient to express the last category of uncertainties in the table.

Type	Subtype	Example
Incompleteness	Existential	A hostile unit has been detected at a certain location, but its type is unknown
	Universal	All the units that have been detected so far at various locations are hostile, but their types are unknown
Imprecision	Disjunctive	The unit type is tank or armored personnel carrier
	Negation	The unit type is not tank
	Interval-valued	A tracked hostile unit is traveling between 25 and 40 mph
	Fuzzy-valued	A tracked unit is moving very fast
Uncertainty	Probability	The chance of the unit being tank is 70%
	Possibility	It is possible that the tracked hostile unit is traveling at 30 mph
	Credibility	0.8 is my degree of belief that the unit type is tank

Table 3-1: Classification of uncertainties and examples

The probability theory is the predominant paradigm for handling uncertainty, especially Bayesian probability where interpretation of a probability value is a subject's degree of belief. In the Bayesian approach, the probability is not based on any precise computation, but describes an individual's personal judgment (degree of belief) about how likely a particular event is to occur based on experience. But the probability theory fails to represent, for example, a disjunction naturally. One has to distribute the total probability of a disjunction among its constituent disjuncts.

Here we contrast the probabilistic and fuzzy logic approaches to handling uncertainty. Fuzzy logic handles fuzzy-valued uncertainty by allowing truth to be a matter of degree as in the membership degree in fuzzy sets. In the probabilistic approach, the information is complete, precise, but uncertain since it might be wrong. For example, if one source asserts that a target fighter aircraft of a certain type is moving at its maximum speed (known to be 1200 mph), then the information is complete and precise, but not certain since the source might be wrong. If the probability quantifying uncertainty in this case is 0.9 then it would be interpreted as "there is a 90% chance that the target is moving at its maximum speed." This is not equivalent to approximating the target speed to a value such as 1080 mph. On the other hand, in fuzzy logic there is no uncertainty about the target moving at its maximum speed, but uncertainty is about the degree to which the observed speed matches the maximum speed. The degree 0.9 is the target's degree of membership in the fuzzy set of all moving fighter aircrafts of the same type. In other words, if we order the elements of this fuzzy set then the target will be positioned 90% of the way to the maximum speed.

Two forms of possibility have been described in the literature: the epistemic and the physical (Smets, 1994). The first form, "possible that," which we have considered here, is related to our state of knowledge and is called epistemic. The second form, "possible for," deals with actual abilities independently of our knowledge about them. The distinction is similar to the one between the epistemic concept of probability and the traditional one based on chance. The Dempster-Shafer theory of belief function (Shafer, 1976) converts source credibility into probability intervals representing second-order uncertainty. Such an interval provides upper and lower bounds of the true probability.

Modal logics (Lemmon, 1977; Chellas, 1980; Hughes and Cresswell, 1996) and their epistemic interpretations (Hintikka, 1962; Meyer and van der Hoek, 1995) can be used to represent and reason with the epistemic possibility. However, this coarse grain representation of the knowledge of possibilities about assertions through this logic is quite inadequate for practical applications. That is to say, if two assertions are possible in the current world, they are indistinguishable in the modal formalism even if an agent knows that one of them is true in twice as many possible worlds (Kripke, 1963) as compared to the other one. Therefore, any epistemic interpretation of the modal operators for necessity and possibility would fail to incorporate the notion of the decision maker's "degree of belief" in something into its epistemic states (Fox and Das, 2000). We have described the Possibility Theory and Possibilistic Logic in this regard.

3.2 APPROACHES TO HANDLING UNCERTAINTIES

The general context of the uncertainty handling problem in this book can formally be stated as follows. Suppose a decision maker (e.g., an analyst or a commander) has the universe of discourse $\Omega = \{w_1, ..., w_n\}$ at hand as the set of likely hypotheses for a given decision-making problem. These hypotheses need to be associated with some values, be they probabilities, possibility measures, or any other qualitative terms suitably defined by the decision maker, representing the decision maker's quantified degrees of beliefs in their likely occurrences based on the available uncertain evidence. The last section discussed broadly various approaches to handling uncertainty. Specific technologies proposed for handling uncertainties can be categorized as shown in Table 3-2 (along the line of Pearl (1990)). This self-explanatory table provides motivation and definition of each category, along with one or more instances of seminal technologies for the category.

A technology for handling uncertainty that is model-based (see the Preface for model-based approaches) has two components: the static component describes an analyst's state of belief in a model given the information available to the analyst, and the dynamic component explains how to update beliefs given new pieces of information that become available. For the Bayesian Belief Network (BN) (Pearl, 1988; Jensen, 1996) technology, for example, the static component is represented in terms of causal graphs that provide prior distribution of variables in the graph. Propagation algorithms are available to compute posterior distributions of the variables in a graph.

Category	Motivation	Definition	Seminal Technologies
Neo-Probabilist	Traditional framework of probability is not suitable for handling large number of variables	Remains within the traditional framework, but provides computational facilities to perform fusion tasks	▪ Bayesian Belief Networks
Neo-Calculist	Probability calculus is inadequate for capturing uncertainty	New calculus for handling uncertainty	▪ Theory of Belief Functions ▪ Certainty Factors
Neo-Logicist	Monotonicity in classical logics is not suitable for handling commonsense knowledge	Deals with nonnumerical nonmonotonic inferencing for handling uncertainty in negation	▪ Default Logic ▪ Program Completion
Neo-Possibilist	Coarse grained two-valued representation of truth is not sufficient to model practical applications	Nothing is a matter of degree in two-valued Boolean logic	▪ Fuzzy Sets ▪ Fuzzy Logic ▪ Possibility Theory ▪ Possibilistic Logic

Table 3-2: Approaches to handling uncertainties

The Theory of Belief Functions (Shafer, 1976) and Certainty Factors (Shortliffe, 1976) are two prime examples of new calculus for handling uncertainty. The Theory of Belief Function or evidence theory was initiated by Arthur P. Dempster in a series of articles in the 1960s and later developed by Shafer (1976), and also known as Dempster-Shafer Theory. It uses two related measures, called belief and plausibility measures, as upper and lower probabilities. The probability measure is a special case of the belief and plausibility measures. Certainty Factors (Shortliffe and Buchanan, 1975; Heckerman and Shortliffe, 1992) is one of the early approaches to handling uncertainty in expert systems.

The monotonic nature of classical logics is not suitable for handling commonsense knowledge and related uncertainty. Neo-logicists provide nonmonotonic logics (Ginsberg, 1987) to support defeasible inferencing, the kinds of inference of everyday life in which conclusions are drawn tentatively, reserving the right to retract them in the light of further information. Various types of nonmonotonic logics have been studied over the years, of which Reiter's Default Logic (Reiter, 1980) is the most prominent one. Closed World Assumption (Reiter, 1978) is the one adopted in relational databases. Program Completion (Clark, 1978) and Circumscription (McCarthy, 1980) infer negative information from the proof-theoretic point of view. Program Completion is the theoretical foundation of the Negation as Failure rule of the logic programming language Prolog (Bratko, 2000). Closed World Assumption and Negation as Failure rules provide semantics of negative information in rule-based knowledge bases to be considered in this book. For a detailed discussion on the semantics of negative information in databases and knowledge bases, see Das (1992).

Coarse grained two-valued representation of truth is not sufficient to model practical applications. Early work in the area of many-valued logics (Łukasiewicz, 1970) deals with a multivalued representation of truth. Zadeh proposed Fuzzy Sets (1965) as a way of extending two-valued membership of ordinary sets. Fuzzy Logic allows truth to be a matter of degree as in the case of membership degree in fuzzy sets. Zadeh (1978) introduced the Possibility Theory as an extension of fuzzy sets and fuzzy logic. Possibilistic Logic (Dubois and Prade, 1988) extends the Possibility Theory with a richer representation of domain knowledge via logical sentences. The Possibility Theory is a natural tool to model the imprecision and the uncertainty when they occur together in ordinary language such as "It is likely that the vehicle is moving fast." The probability and possibility measures are special cases of the belief and plausibility measures.

The technologies listed under the four categories of Table 3-2 are described in the following four sections.

3.3 NEO-PROBABILIST APPROACH

This section presents a brief overview of the Bayesian Belief Network (BN) technology for the sake of completeness. A large chapter (Chapter 6) has been dedicated to cover details of the BN technology and its specific use in situation assessment.

3.3.1 Bayesian Belief Networks (BNs)

A BN (Pearl, 1988; Jensen, 1996) is a graphical, probabilistic knowledge representation of a collection of variables describing some domain, yielding a domain model. The nodes of the BN denote the variables and the links denote causal relationships between the variables. The topology encodes the qualitative knowledge about the domain. Conditional Probability Tables (CPTs) encode the quantitative details (strengths) of the causal relationships. The BN of Figure 3-1 encodes the relationships over the domain consisting of the binary variables,

Road, *Rain*, *Mobility*, *Transport*, *Electricity*, and *Speed*. The topology of the BN captures the commonsense knowledge that:

- *Rain* causes *Transport* disruption.
- *Rain* causes *Electricity* failure.
- *Mobility* of a road affects the *Speed* of a vehicle on the road.
- *Road* type and *Rain* determine *Mobility* of the road.

As shown in Figure 3-1, a CPT specifies the probability of each possible value of a child variable conditioned on each possible combination of its parent variable values. For example, the probability of having electricity during rain is only 0.2, whereas the probability of having electricity with no rain is 0.95.

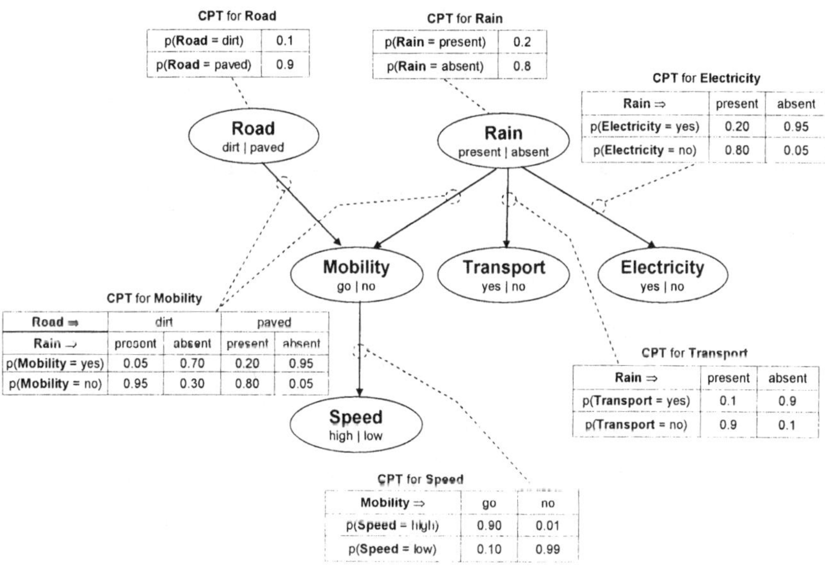

Figure 3-1: An example BN

The chance that a node of a BN is in a particular state is termed its "belief" in the state, reflecting the probability that the node is in that state given all the previous evidence received. The structure of a BN encodes other information as well. Specifically, the lack of links between certain variables represents a lack of direct causal influence, that is, they indicate conditional independence relations. This BN encodes many conditional independence relations, for example,

- *Electricity* is independent of *Transport* given *Rain*, and
- *Speed* is independent of *Rain* and *Road* given *Mobility*.

That is, once the value of *Rain* is known, the value of *Transport* adds no further information about *Electricity*. Similar conditional independence assertions hold for the other variables.

When new evidence is posted to a variable in a BN, that variable updates its own belief vector (i.e., belief represented as a vector) and then sends out messages indicating updated predictive and diagnostic support vectors to its children and parent nodes respectively. These messages are then used by other nodes to update their belief vectors and to propagate their own updated support vectors. The separation of evidence yields a propagation algorithm for polytree-type BNs (Pearl, 1988) in which update messages need only be passed in one direction between any two nodes following posting of evidence. Thus, the algorithm's complexity is proportional to the number of links in the network. This separation also prevents the possibility of double-counting evidence. Details on more complex algorithms are in Chapter 6.

3.4 NEO-CALCULIST APPROACH

This section presents two calculi, namely the Theory of Belief Function and Certainty Factors, for handling uncertainty with semantics different from probability semantics. An application of the Theory of Belief Function for target classification has been detailed in Chapter 5.

3.4.1 Theory of Belief Functions

The Theory of Belief Functions (Shafer, 1976), also known as Dempster-Shafer theory, is a generalization of the Bayesian theory of subjective probability (mainly by virtue of its explicit handling of the concept of ignorance via uncommitted beliefs) to combine accumulative evidence or to change prior opinions in the light of new evidence. Whereas the Bayesian theory requires probabilities for each question of interest, belief functions allow us to base degrees of belief for one question (for example, whether the mobility is a go) on probabilities for a related question. Arthur P. Dempster set out the basic ideas of the theory (Dempster, 1966) and then Glenn Shafer developed the theory further (Shafer, 1976). Briefly, the theory may be summarized as follows.

Suppose expert X (for example, a weather analyst) says that the mobility is not a go due to heavy rain. The decision maker's subjective probabilities for expert X being reliable or unreliable are 0.7 and 0.3. Now, expert X's statement must be true if reliable, but not necessarily false if unreliable. The expert's testimony therefore justifies 0.7 "degrees of belief" that the mobility is not a go, but only a zero (not 0.3) degree of belief that the mobility is a go. The numbers 0.7 and 0 together constitute a belief function.

Suppose subjective probabilities were based on the decision maker's knowledge of the frequency with which experts like X are reliable witnesses. Seventy percent of statements made would be true by reliable witnesses, n% would be true by unreliable ones, and $(30-n)$% would be false statements by unreliable witnesses. 0.7 and 0 are the lower bounds of true probabilities

$(70+n)/100$ and $(30-n)/100$ respectively. Thus, a single belief function is always a consistent system of probability bounds, but may represent contradictory opinions from various experts. For example, consider the belief function 0.7 and 0 from expert X's opinion of the mobility not being a go, and 0.8 and 0 from expert Y's opinion of the mobility being a go. The lower bound of the true probability for the game not being on in the first case is 0.7, but the upper bound is 0.2 in the second case, yielding contradiction.

Let Ω be a finite set of mutually exclusive and exhaustive propositions, called the *frame-of-discernment*, about some problem domain ($\Omega = \{go, no\}$ in our example decision-making problem) and $\Pi(\Omega)$ is the power set of Ω. A *basic probability assignment* (BPA) or *mass function* is the mapping

$$m : \Pi(\Omega) \to [0,1]$$

which is used to quantify the belief committed to a particular subset A of the frame of discernment given certain evidence. The probability number $m(A)$, the *mass* of A, says how much belief there is that some member of A is in fact the case, where

$$m(\Phi) = 0 \text{ and } \sum_{A \subseteq \Omega} m(A) = 1$$

The value 0 indicates no belief and the value 1 indicates total belief, and any values between these two limits indicate partial beliefs. If the probability number p for only a partial set A of hypotheses is known then the residual complementary probability number $1-p$ is assigned to the frame-of-discernment, thus allowing the representation of ignorance. A basic probability assignment m is Bayesian if $m(A) = 0$ for every nonsingleton set A. For any set $A \subseteq \Omega$ for which $m(A) \neq 0$, A is called a *focal element*.

The measure of total belief committed to $A \subseteq \Omega$ can be obtained by computing the *belief function Bel* for $A \subseteq \Omega$ which simply adds the mass of all the subsets of A:

$$Bel(A) = \sum_{B \subseteq A} m(B)$$

A *single* belief function represents the lower limit of the true probability and the following plausibility function provides the upper limit of the probability:

$$Pl(A) = \sum_{B \cap A \neq \Phi} m(B) = 1 - Bel(A^c)$$

Mass can be recovered from a belief function as follows:

$$m(B) = \sum_{A \subseteq B} (-1)^{|A-B|} Bel(A)$$

So there is a one-to-one correspondence between the two functions m and *Bel*. Two independent evidences expressed as two basic probability assignments m_1 and m_2 can be combined into a single joined basic assignment $m_{1,2}$ by Dempster's rule of combination:

$$m_{1,2}(A)=\begin{cases}\dfrac{\sum\limits_{B\cap C=A} m_1(B)m_2(C)}{1-\sum\limits_{B\cap C=\Phi} m_1(B)m_2(C)} & A\neq\Phi\\ 0 & A=\Phi\end{cases}$$

See Chapter 5 for an application of the Dempster-Shafer theory for target classification.

3.4.2 Certainty Factors

One of the early approaches to handling uncertainty in expert systems is Certainty Factors (Shortliffe and Buchanan, 1975; Heckerman and Shortliffe, 1992). The formalism provides a number to measure the decision maker's degree of confirmation or belief in hypothesis, where the maximum value for belief is +1.0 (definitely true), and the minimum value is −1.0 (definitely false). An intuitive interpretation of qualitative terms in terms of belief values is shown in Table 3-3:

Qualitative Term	Certainty Factor
Definitely	+1.0
Almost certainly	+0.8
Probably	+0.6
Maybe	+0.4
Unknown	[−0.2, +0.2]
Maybe not	−0.4
Probably not	−0.6
Almost certainly not	−0.8
Definitely not	−1.0

Table 3-3: Qualitative interpretations of certainty factors

In the certainty factor formalism, unlike in the theory of probability, a piece of evidence E could support a hypothesis H without necessarily supporting the complement of that hypothesis. In other words, although the property

$$p(H\mid E)=1-p(\neg H\mid E)$$

holds in probability theory, the certainty formalism rejects the following:

$$belief(H\mid E)=f\left(belief(\neg H\mid E)\right)$$

The approach creates the following measures of belief and disbelief:

$$MB(H,E)=\begin{cases}1 & \text{if } p(H)=1\\ \dfrac{p(H,E)-p(H)}{1-p(H)} & \text{otherwise}\end{cases}$$

$$MD(H,E)=\begin{cases} 1 & \text{if } p(H)=0 \\ \dfrac{p(H)-p(H\mid E)}{p(H)} & \text{otherwise} \end{cases}$$

The value $1-p(H)$ represents the decision maker's total doubt in the hypothesis, and thus the measure of belief *MB* based on evidence *E* computes the reduction in the decision maker's disbelief. Similarly, the measure of disbelief *MD* based on evidence *E* computes the reduction in the decision maker's belief. The measure of belief is then computed as follows:

$$cf=\frac{MB-MD}{1-\min(MB,MD)}$$

Different values of *cf* are interpreted as follows:

- *cf is positive*: evidence supports the hypothesis since $MB > MD$.
- *cf is 1.0*: evidence definitely supports the hypothesis.
- *cf is 0.0*: either there is no evidence or that the belief is canceled out by the disbelief.
- *cf is negative*: evidence favors negation of the hypothesis since $MB < MD$.

The certainty factor assigned by a rule is propagated through the reasoning chain. It involves establishing the net certainty of the rule consequent when the evidence in the rule antecedent is uncertain. Given the rule

IF *E* THEN *H* (*cf*)

and given $CF(E)$, the combined belief in the hypothesis is computed by propagating the certainty factor as follows:

$$CF(H,E)=CF(E)\times cf$$

The following are the certainty factor propagation rules for conjunctions and disjunctions:

$$CF(E_1\wedge\ldots\wedge E_n)=\max\left[CF(E_1),\ldots,CF(E_n)\right]$$

$$CF(E_1\vee\ldots\vee E_n)=\min\left[CF(E_1),\ldots,CF(E_n)\right]$$

Finally, one needs to combine certainty factors for a hypothesis implicated by more than a rule as below:

IF E_1 THEN H (cf_1)

IF E_2 THEN H (cf)

The combination of cf_1 and cf_2 for *H* given both E_1 and E_2 is

$$CF(cf_1,cf_2)=\begin{cases} cf_1+cf_2\times(1-cf_1) & \text{if } cf_1>0 \text{ and } cf_2>0 \\ \dfrac{cf_1+cf_2}{1-\min\left[\mid cf_1\mid,\mid cf_2\mid\right]} & \text{if } cf_1<0 \text{ or } cf_2<0 \\ cf_1+cf_2\times(1+cf_1) & \text{if } cf_1<0 \text{ and } cf_2<0 \end{cases}$$

A natural requirement for any formalism for handling uncertainty is that the evidence combination rule should be associative. But this rule for the certainty factor formalism is not associative. The rule should never be applied if E_1 and E_2 are deduced from a common piece of evidence E_0. To illustrate, suppose we have the following rules:

IF E_0 THEN E_1 (1.0)
IF E_0 THEN E_2 (1.0)
IF E_1 THEN H (1.0)
IF E_2 THEN H (1.0)

Let the certainty factor of E_0 be α, then both E_1 and E_2 have a certainty factor equal to α. A blind application of the combination rule leads to a certainty factor of $2\alpha - \alpha^2$ for *H*. The result is wrong as the set of four rules can be reduced into

IF E_0 THEN H (1.0)

and therefore the certainty factor of *H* given E_0 should be α. This example shows the danger of using ad hoc models blindly.

3.5 NEO-LOGICIST APPROACH

Classical logic is monotonic in the sense that inferences are deductively valid and can never be retracted in the light of new information. On the other hand, nonmonotonic logics support defeasible inferencing, the kinds of inference of everyday life in which conclusions are drawn tentatively, reserving the right to retract them in the light of further information. Thus the set of conclusions drawn on the basis of a given knowledge base does not necessarily increase with the size of the knowledge base itself. This section presents two logics, namely Default Logic and Program Completion.

3.5.1 Default Logic

Default logic (Reiter, 1980) is probably the most extensively studied nonmonotonic logic. Default logic augments classical logic by default rules, and consequently the knowledge about the world is divided into two parts, representing respectively certain knowledge **W** in closed first-order formulae and a collection **D** of special inference rules called defaults. A *default rule* has the form

$$\frac{\alpha : \beta_1, \ldots, \beta_n}{\gamma}$$

where $\alpha, \beta_1, \ldots, \beta_n, \gamma$ are well-formed formulae of classical logic. α is called the prerequisite of the default, $\beta_1, \ldots, \beta_n$ are called justifications, and γ is the consequence. $\langle \mathbf{D}, \mathbf{W} \rangle$ is a default theory. For example, the statement "sunny weather typically makes the mobility a go" can be formalized by the only default rule in the following set:

$$\mathbf{D} = \left\{ \frac{Sunny(x,y) : Mobility(x,y,go)}{Mobility(x,y,go)} \right\}$$

The rule states that if it is sunny along the road between x and y, and if it can be assumed that the mobility of the road is a go, then it can be concluded that the mobility is a go. The knowledge **W** contains some facts about

$$\mathbf{W} = \{Sunny(a,b), Sunny(c,d), \neg Mobility(c,d,go), Mobility(a,c,go)\}$$

The mobility of the road between a and b is a go because the prerequisite $Sunny(a,b)$ is true and the justification $Mobility(a,b,go)$ is consistent with what is currently known. But $Mobility(c,d,go)$ cannot be concluded from the prerequisite $Sunny(c,d)$ because it is inconsistent with what is currently known.

A default rule is *closed* if none of $\alpha, \beta_1, ..., \beta_n, \gamma$ contains free variables. A default theory is closed if all its default rules are closed. A default theory is *open* if it is not closed. The semantics of a closed default theory $\langle \mathbf{D}, \mathbf{W} \rangle$ is based on the notion of extension, which is a possible state of the world according to the knowledge base. Formally, an *extension E* can be recursively defined as follows:

$$E_0 = W$$

$$E_{i+1} = Cons(E_i) \cup \left\{ \gamma : \frac{\alpha : \beta_1, ..., \beta_n}{\gamma} \in D, \alpha \in E_i, \forall j \neg \beta_j \in E_i \right\}$$

where $Cons(X)$ denotes classical deductive closure of X. The set $E = \bigcup_{i=0}^{\infty} E_i$ is deductively closed and an extension of $\langle \mathbf{D}, \mathbf{W} \rangle$.

The definition of an extension is extended to open default theories by assuming that the default rules with free variables implicitly stand for the infinite set of closed defaults obtained by replacing the free variables with terms of the default theory. A default theory can have one, multiple, or no extensions in general. Therefore, defining entailment of a formula from a default theory is not straightforward. What can or cannot be derived from a default theory is uncertain. The standard variants are credulous entailment under which a formula F is entailed if it belongs to at least one extension of $\langle \mathbf{D}, \mathbf{W} \rangle$ and skeptical entailment under which F follows if it belongs to all extensions of $\langle \mathbf{D}, \mathbf{W} \rangle$.

3.5.2 Program Completion

Consider a knowledge base or a program of if-then rules expressed in the syntax of classical logic. The major issue of handling uncertainty in such a knowledge base is to determine whether some information can be derived from the knowledge base or not. An approach to inferring negative information (what is not true) from a program is program completion, where a program can be viewed as a knowledge consisting of if-then rules of form (in first-order logic syntax as shown in the section on mathematical logic in Chapter 2)

$$Road(x,y) \wedge \neg Rain(x,y) \rightarrow Mobility(x,y,go)$$

or facts of the form

$$Mobility(a,b,go)$$

$$Road(c,d)$$

The rule states that the mobility of a road between any two points x and y is a go if there is a road from x to y and there is no reported rain on the road. The variables in the rule are assumed to be universally quantified. The fact that $Mobility(a,b,go)$ states that the mobility of the road between the two known points a and b is known to be a go. The fact $Road(c,d)$ simply states that there is a road between c and d.

According to the program completion semantics, the program provides the if parts of the definition of the predicates. The only-if parts in a program are implicitly defined and are obtained by completing each predicate symbol. Consider a program containing only the above rule and the fact defining the predicate *Mobility*. This fact and rule jointly state that the mobility is a go from x to y if x is a and y is b or if there is a road from x to y and there is no rain on the road from x to y. Symbolically,

$$(x = a \wedge y = b \wedge z = go) \vee (Road(x,y) \wedge \neg Rain(x,y) \wedge z = go)$$
$$\rightarrow Mobility(x,y,z)$$

However, the definition leaves open the possibility that the road mobility may be a go from x to y for some other reason. What has been implicitly meant by the program is that the mobility is a go from x to y only if x is a and y is b or there is a road from x to y and there is no reported rain on the road. Symbolically,

$$Mobility(x,y,z) \rightarrow$$
$$(x = a \wedge y = b \wedge z = go) \vee (Road(x,y) \wedge \neg Rain(x,y) \wedge z = go)$$

These two together jointly give the completed definition of the predicate *Mobility* as

$$Mobility(x,y,z) \leftrightarrow$$
$$(x = a \wedge y = b \wedge z = go) \vee (Road(x,y) \wedge \neg Rain(x,y) \wedge z = go)$$

Similarly, the completed definition of the predicate *Road* is

$$Road(x,y) \leftrightarrow x = c \wedge y = d$$

One also needs to have the following completed definitions for those predicate symbols in a program, such as *Rain*, that do not appear in the head of any rule:

$$\neg Rain(x,y)$$

The completed program is the collection of completed definitions of predicate symbols in a program together with the equality theory.

Once we have a completed program, negative information such as $\neg Road(a,c)$ and $\neg Rain(c,d)$ can be derived, along with positive information $Mobility(c,d,go)$. SLDNF-resolution (Clark, 1978; Lloyd, 1987; Das, 1992) is thought of as a procedural counterpart of the declarative semantics given in terms of completed programs. This resolution scheme is, to a large extent, the basis of present-day logic programming incorporated in Prolog. SLDNF-resolution is essentially SLD-resolution augmented by the "negation as failure" inference rule.

3.6 NEO-POSSIBILIST APPROACHES

This section presents the Fuzzy Set concept, and then Fuzzy Logic based on Fuzzy Sets. The Possibility Theory is then introduced as an extension of the Fuzzy Set theory. Finally, Possibilistic Logic is presented, extending the Possibility Theory with a richer representation of domain knowledge via logical sentences.

3.6.1 Fuzzy Sets

A fuzzy set A is a subset of a referential set or universe of discourse X whose boundaries are gradual. More formally, the membership function μ_A of a fuzzy set A assigns to each element x in X its degree of membership $\mu_A(x)$ usually taking values in $[0,1]$. In a conventional set, $\mu_A(x) = 1$ (meaning that x belongs to A) or $\mu_A(x) = 0$ (meaning that x does not belong to A).

Consider the referential set as the interval $\left[-25^\circ, 120^\circ\right]$ representing temperature values, and the three associated fuzzy sets *cold*, *warm*, and *hot*. The membership functions μ_{cold}, μ_{warm}, and μ_{hot} in terms of three trapezoidal graphs have been shown in Figure 3-2.

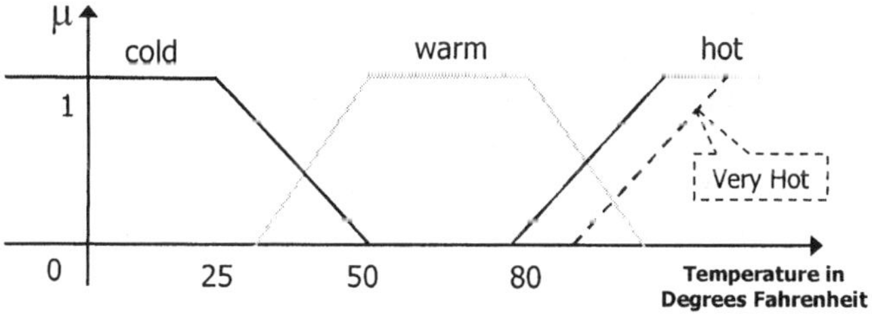

Figure 3-2: Fuzzy membership functions

Consider the membership function μ_{cold}. Any temperature value of approximately less than 25 is assumed to have the degree of membership 1.0, and this degree gradually decreases as the temperature value increases. The degree of membership is 0 for any temperature value greater than 50. The other membership functions can be interpreted in a similar manner. Note that a membership function can be a probability density function, for example, μ_{warm} could be a Gaussian with mean 75 and standard deviation 5.0.

The set theoretic operations on fuzzy sets are defined below and illustrated in Figure 3-3:

- *Union*: $\mu_{A \vee B}(X) = \max\{\mu_A(X), \mu_B(X)\}$
- *Intersection*: $\mu_{A \wedge B}(X) = \min\{\mu_A(X), \mu_B(X)\}$
- *Complement*: $\mu_{\neg A}(X) = 1 - \mu_A(X)$

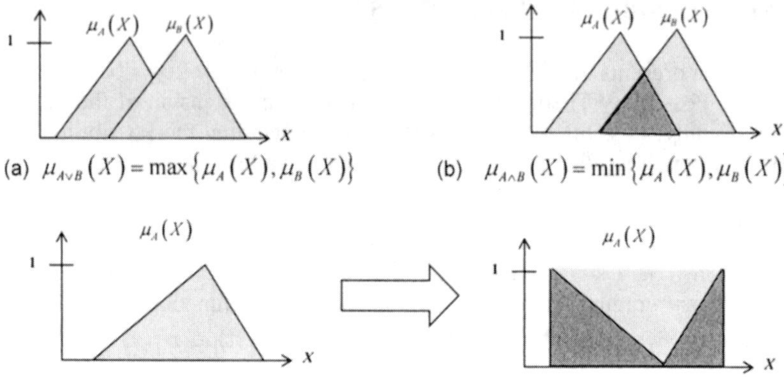

(a) $\mu_{A \vee B}(X) = \max\{\mu_A(X), \mu_B(X)\}$ (b) $\mu_{A \wedge B}(X) = \min\{\mu_A(X), \mu_B(X)\}$

(c) $\mu_{\neg A}(X) = 1 - \mu_A(X)$

Figure 3-3: Fuzzy set operations: (a) union; (b) intersection; and (c) complement

3.6.2 Fuzzy Logic

Fuzzy logic provides a foundation for approximate reasoning by extending traditional Boolean logic and by allowing truth to be a matter of degree as in the case of membership degree in fuzzy sets. It is applicable to problems where the expressive power of words is greater than that of numbers. For example, a human observer reported that an object was approaching very fast, without knowing the exact speed of the object. Words like "fast" are less precise than numbers, but imprecision can be tolerated to achieve robustness. The four facets of fuzzy logic are shown in Figure 3-4.

Here is an example traditional logic-based exact reasoning just to contrast with fuzzy reasoning:

IF Vehicle Weight is light
THEN Vehicle Speed is high

Vehicle Weight is light

Therefore, Vehicle Speed is high

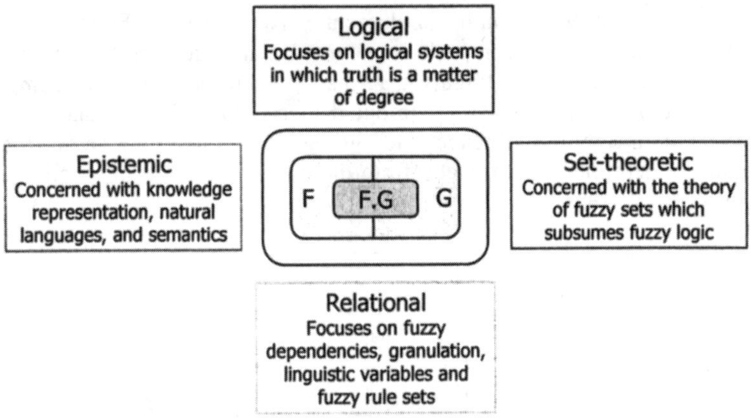

Figure 3-4: Facets of fuzzy logic (Zadeh, 2002)

In the case of fuzzy reasoning, the concept of a linguistic variable plays a central role. In the reformulation of fuzzy reasoning of the above exact reasoning, the linguistic variables are *Weight* and *Speed*. Values of linguistic variables are words or sentences in a natural or synthetic language, and are constructed by composing primary terms or predicates with modifiers. In the case of the variable *Weight*, primary terms or predicates are, for example, *light*, *medium*, *heavy*, ..., and modifiers are, for example, *not*, *very*, *quite*, *rather*, ... Thus, linguistic values of the variable *Weight* are *heavy*, *not light*, *very heavy*, and so on. A numerical or "crisp" value of the variable *Weight* is in its universe of discourse, which is the interval $[0,60]$ of values in tons. Modifiers fall into two classes:

- *Fuzzy Truth Qualifier*: quite true, very true, more or less true, etc.
- *Fuzzy Qualifier*: many, few, almost

An example fuzzy proposition is "Vehicle Weight is very heavy." Fuzzy expressions or formulae are built using fuzzy propositions, logical connectives, and quantifiers. Here is an example fuzzy formula:

IF Rainfall is medium
AND Weight is heavy
THEN Delay is long

where the linguistic variables are *Rainfall*, *Weight*, and *Delay*. A fuzzy proposition can have its truth value in the interval [0, 1] defined by a membership function.

In proposition "*X* is *P*," *P*(*X*) (or simply *P* when *X* is clear from the context) is a fuzzy set and the membership function $\mu_P(X)$ (or simply μ_P) defines the degree of membership in the set. Truth values of fuzzy propositions are defined by membership functions. Consider the three values, namely *light*, *medium*, and *heavy*, of the linguistic variable *Weight*. Then the truth value of the proposition "Weight is heavy" is defined by μ_{heavy}.

A generic fuzzy reasoning architecture is shown in Figure 3-5. The Fuzzification component computes the membership degrees for each input variable with respect to its linguistic terms. The Fuzzy Reasoning component yields the output fuzzy set using computed membership degrees and the fuzzy rules, via rule matching, inference, and aggregation. Finally, the Defuzzification component determines a crisp value from the output membership function as the final result of the solution.

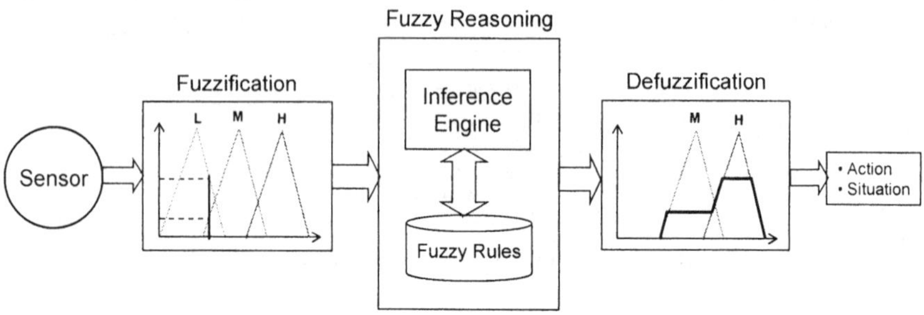

Figure 3-5: Generic fuzzy engine

We illustrate the fuzzy reasoning process in the context of an example scenario involving estimation of a vehicle's delay of arrival due to rainfall, which affects the mobility of the vehicle. There are three linguistic variables: *Weight*, *Rainfall*, and *Delay*. The underlying fuzzy rule base relating these variables is captured in the table shown in Figure 3-6. Each entry in the table is a value of the variable *Delay* given the value of the two variables *Weight* and *Rainfall* in the corresponding column and row, respectively. For example, the shaded cell corresponds to the following rule:

IF	Weight is heavy
AND	Rainfall is medium
THEN	Delay is long

Delay (Rainfall \ Weight)	Heavy	Medium	Light
Light	slight	none	none
Medium	long	slight	none
Heavy	long	long	slight

Figure 3-6: Fuzzy rule base

There will be altogether nine rules corresponding to the nine cells containing the values of the variable *Delay*. The set of rules are as follows when traversing the cells row-by-row starting from the left:

R1: IF Weight is heavy AND Rainfall is light THEN Delay is slight
R2: IF Weight is medium AND Rainfall is light THEN Delay is none
R3: IF Weight is light AND Rainfall is light THEN Delay is none
R4: IF Weight is heavy AND Rainfall is medium THEN Delay is long
R5: IF Weight is medium AND Rainfall is medium THEN Delay is slight
R6: IF Weight is light AND Rainfall is medium THEN Delay is none
R7: IF Weight is heavy AND Rainfall is heavy THEN Delay is long
R8: IF Weight is medium AND Rainfall is heavy THEN Delay is long
R9: IF Weight is light AND Rainfall is heavy THEN Delay is slight

The membership functions for each of three variables are shown in Figure 3-7. Note that the membership function for $\mu_{medium}(Weight)$ is Gaussian with mean 28.0 and variance is 6.0.

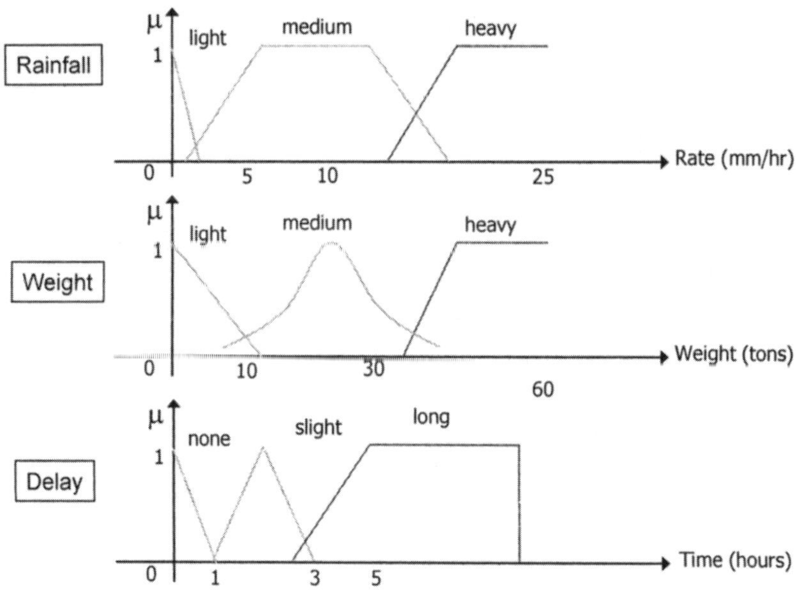

Figure 3-7: Fuzzy sets of the variables *Rainfall*, *Weight*, and *Delay*

In the fuzzification stage, one needs to determine, given the value of an input variable, the degree of membership for each term of the variable. Given the rainfall is 15.2 mm/hr and vehicle weight is 40 tons, Figure 3-8 illustrates the determination of membership degrees and the resultant values are the following:

$$\mu_{light}(15.2) = 0 \qquad \mu_{light}(40) = 0$$

$$\mu_{medium}(15.2) = 0.25 \qquad \mu_{medium}(40) = 0.1$$

$$\mu_{heavy}(15.2) = 0.70 \qquad \mu_{heavy}(40) = 0.60$$

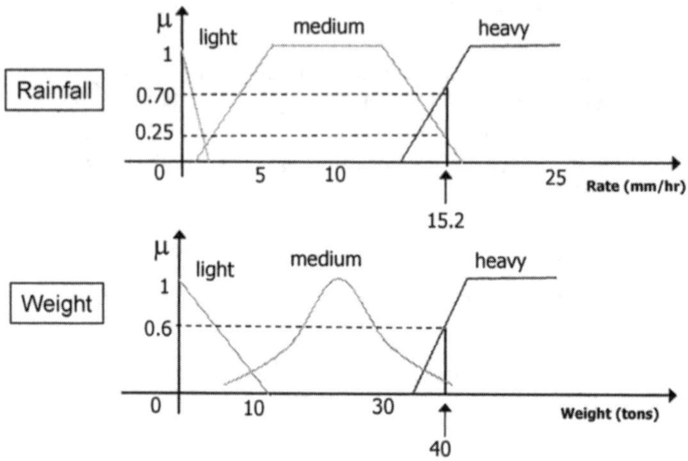

Figure 3-8: Degrees of membership for a value of the input variable

We then calculate the firing strength of every rule by combining the individual membership degree for all terms involved in the antecedent of a rule. Consider the rule

IF	Weight is heavy
AND	Rainfall is medium
THEN	Delay is long

Since we have $\mu_{heavy}(Weight) = 0.60$ and $\mu_{medium}(Rainfall) = 0.25$, and the antecedent of the rule is a conjunction, the combined membership degree is

$$\min\{\mu_{heavy}(40), \mu_{medium}(15.2)\} = 0.25$$

The next step is, for each rule, to apply the firing strength to modify its consequent fuzzy set, resulting in a new fuzzy set as the result of applying the rule. Consider again the following rule as an example:

IF	Weight is heavy
AND	Rainfall is medium
THEN	Delay is long

The modification of its consequent fuzzy set with the firing strength 0.25 is shown in Figure 3-9.

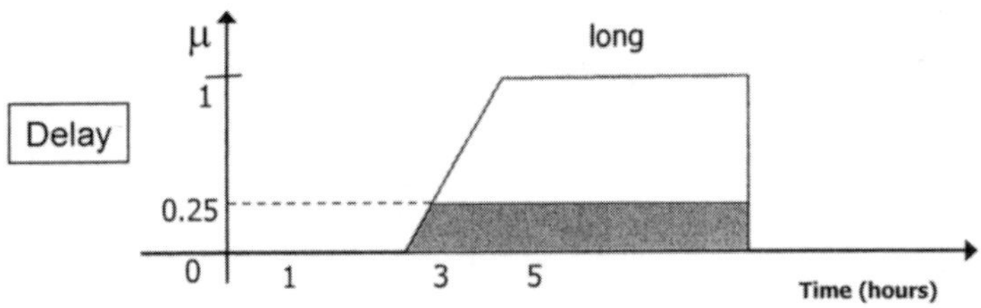

Figure 3-9: Modification of a rule's consequent fuzzy set

This modification essentially reduces the degree of membership for every value of the variable *Delay* in the conclusion. What we are interested in is the shaded area underneath. There will be nine such modifications corresponding to the consequents of the nine rules in the fuzzy rule base, and all such modifications need to be aggregated. This involves combination of responses of individual rules to yield an overall output fuzzy set using the max-operator (i.e., superimpose the shaded areas). An example aggregation involving just the three types of modifications is shown in Figure 3-10.

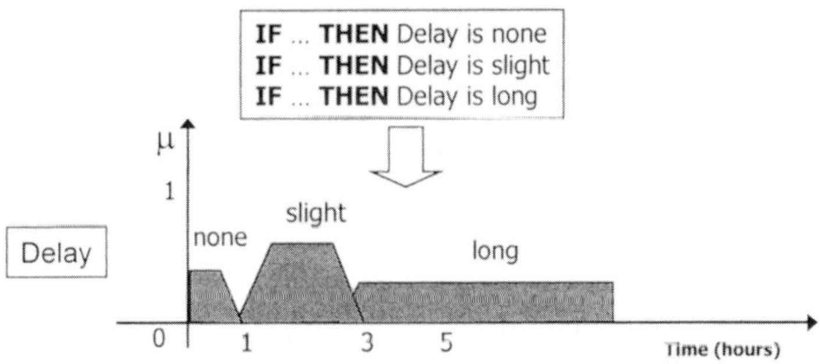

Figure 3-10: Aggregation of modified fuzzy sets

The final defuzzification process determines crisp value from the output membership function. Two of the more common defuzzification techniques are the Centroid and Maximum methods. In the Centroid method, the crisp value of the output variable is computed by finding the variable value of the center of gravity of the membership function for the fuzzy value. In the Maximum method, one of the variable values at which the fuzzy subset has its maximum truth value is chosen as the crisp value for the output variable.

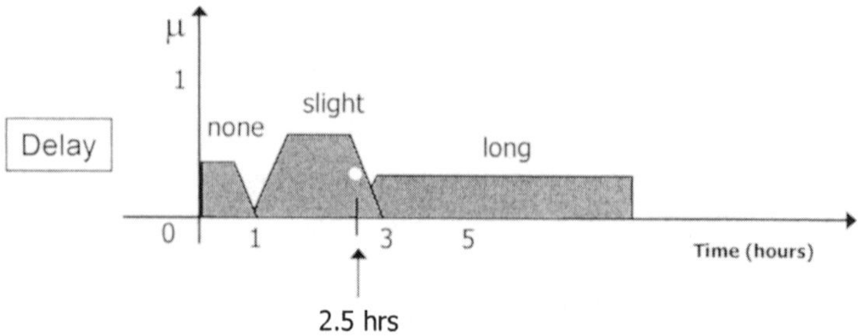

Figure 3-11: Defuzzification

Figure 3-11 shows the projection of the center of gravity of the aggregated membership function onto the time-axis, yielding the crisp value of 2.5 hours. This value is the estimated delay in arrival given the crisp values for vehicle weight and rainfall.

3.6.3 Possibility Theory

Zadeh (1978) introduced the Possibility Theory as an extension of fuzzy sets and fuzzy logic. The possibility of an event defined over a finite set of reference or universe of discourse Ω (an event is a subset of Ω) is a coefficient ranging between 0 and 1 which evaluates how possible the event is. Formally, the possibility is measured as a mapping given below:

$$\Pi : \mathcal{P}(\Omega) \to [0,1]$$

where $\mathcal{P}(\Omega)$ is the power set of Ω representing all possible events. The value of $\Pi(X)$ is 1 means the event X is completely possible and 0 means the event is impossible. The mapping must satisfy the following properties:

$$\Pi(\Phi) = 0$$

$$\Pi(\Omega) = 1$$

$$\Pi(A \cup B) = \max(\Pi(A), \Pi(B)), \text{ where } A \cap B = \Phi$$

When Ω is infinite, the last axiom can be replaced by $\Pi\left(\bigcup_i A_i\right) = \sup_i \Pi(A_i)$, where A_is are pairwise disjoint. Though possibility is compositional with respect to the union operator, it is not compositional with respect to the intersection operator. In general,

$$\Pi(A \cap B) \le \min(\Pi(A), \Pi(B))$$

The following properties of the possibility measure are deducible:

$$\max\left(\Pi(A), \Pi(\overline{A})\right) = 1$$

$$\Pi(A) + \Pi(\overline{A}) \ge 1$$

It is possible that each of $\Pi(A)$ and $\Pi(\bar{A})$ is 1 though only one of A and $\bar{A}$ can occur. The degree with the occurrence of an event A is certain is represented by a necessity measure $N : \mathcal{P}(\Omega) \to [0,1]$ with the following properties:

$$N(\Phi) = 0$$

$$N(\Omega) = 1$$

$$N(A \cap B) = \min(N(A), N(B))$$

When Ω is infinite, the last axiom can be replaced by $N\left(\bigcup_i A_i\right) = \inf_i N(A_i)$, where A_is are pairwise disjoint. The necessity measure is monotonic:

$$\text{if } B \subseteq A \text{ then } N(B) \leq N(A)$$

The following properties of the necessity measure are deducible:

$$N(A \cup B) \geq \max(N(A), N(B))$$

$$\min(N(A), N(\bar{A})) = 0$$

$$N(A) + N(\bar{A}) \leq 1$$

The necessity measure can be obtained from the possibility measure via the following:

$$N(A) = 1 - \Pi(\bar{A})$$

Thus the occurrence of an event A is certain (i.e., $N(A) = 1$) if and only if the occurrence of its complement A is impossible (i.e., $\Pi(\bar{A}) = 0$).

Like probability, a possibility measure on Ω is determined by its behavior on singletons defined as a possibility distribution attached to a variable in Ω. A possibility distribution of a variable in Ω is defined as a mapping:

$$\pi : \Omega \to [0,1]$$

where $\pi(x) = 0$ means x is impossible, $\pi(x) = 1$ means x is completely allowed, and there exists $x \in \Omega$ such that $\pi(x) = 1$. The possibility measure $\Pi(A)$ is then defined as

$$\Pi(A) = \sup_{x \in A} \pi(x)$$

3.6.4 Possibilistic Logic

Possibilistic Logic provides an approach to handling uncertainty in a logical setting. Standard Possibilistic Logic expressions are classical logic formulae associated with weights, interpreted in the framework of the Possibility Theory as lower bounds of necessity degrees. A necessity measure N on formulas is a function from the set of logical formulas of a language to a totally ordered bounded scale with 0 and 1 as bottom and top elements (e.g., the real interval $[0,1]$, which is characterized by the axioms

- $N(\top)=1$, where $\top$ stands for tautology;
- $N(\bot)=0$, where $\bot$ stands for contradiction;
- $N(P \wedge Q)=\min(N(p),N(q))$; and
- $P \equiv Q$ implies $N(P)=N(Q)$, where $\equiv$ denotes equivalence in classical logic.

A possibility measure Π is associated with *N* by duality as follows:

$$\Pi(P) = 1 - N(\neg P)$$

The above expresses that the absence of certainty in favor of $\neg P$ makes *P* possible. The following holds for all *P* and *Q*:

$$\Pi(P \vee Q) = \max(\Pi(P), \Pi(Q))$$

A first-order Possibilistic Logic formula is essentially a pair, (P,α), made of a classical first-order logic formula, *P*, and a weight, α, expressing certainty or priority. The axioms consist of all first-order axioms with weight 1. The inference rules are

- If $\vdash (P,\alpha)$ and $\vdash (P \to Q,\beta)$ then $\vdash (Q,\min(\alpha,\beta))$;
- If $\vdash (P,\alpha)$ then $\vdash (\forall x P,\alpha)$, provided the variable *x* is not bound in *P*;
- If $\vdash (P,\alpha)$ and $\vdash (P,\beta)$, where $\beta \le \alpha$.

From a semantic point of view, a possibilistic knowledge base **K** $=\{(P_1,\alpha_1),\ldots,(P_n,\alpha_n)\}$ is understood as the possibility distribution $\pi_{\mathbf{K}}$ representing the fuzzy set of models of **K**:

$$\pi_{\mathbf{K}}(w) = \min_{1..n} \max\left(\mu_{[P_i]}(w), 1-\alpha_i\right)$$

where $[P_i]$ denotes the set of models of P_i so that $\mu_{[P_i]}(w)=1$, if $w \in P_i$ (i.e., $w \vDash P_i$), and $\mu_{[P_i]}(w)$ otherwise. The degree of possibility of *w* according to the above is computed as the complement to 1 of the largest weight of a formula falsified by *w*. The model w^* maximizing $\pi_{\mathbf{K}}$ is called the best model of **K**. The best model is the most compatible with **K** among all the interpretations. The necessity measure associated with $\pi_{\mathbf{K}}$ is

$$N_{\mathbf{K}}(P) = \min_{w \in [\neg P]} \left(1 - \pi_{\mathbf{K}}(w)\right)$$

The semantic entailment is then defined as

$$\mathbf{K} \vdash (P,\alpha) \text{ iff } N_{\mathbf{K}}(P) \ge \alpha \text{ iff } \forall w\, \pi_{\mathbf{K}}(w) \le \max\left(\mu_P(w), 1-\alpha\right)$$

An important feature of Possibilistic Logic is its ability to deal with inconsistency. The level of inconsistency of a Possibilistic Logic base is defined as

$$Inc(\mathbf{K}) = \max\{\alpha : \mathbf{K} \vdash (\bot,\alpha)\}$$

More generally, $Inc(\mathbf{K})=0$ if an only if $\{P_i : (P_i,\alpha_i) \in \mathbf{K}\}$ is consistent in the classical sense. This would not be true in case α_i did represent a lower bound of the probability of P_i in a probabilistically weighted logic. The proposed system is sound and complete with respect to the inconsistency-tolerant semantics of Possibilistic Logic:

$$\mathbf{K} \vdash (F,\alpha) \text{ if and only if } \mathbf{K} \Vdash (F,\alpha), \text{ for any formula } F$$

The resolution principle for Possibilistic Logic is given as below:

$$(F_1,\alpha),(F_2,\beta) \vdash (\mathcal{R}(F_1,F_2),\min(\alpha,\beta))$$

where $\mathcal{R}(F_1,F_2)$ is the standard classical logic resolvent. The soundness of the rule can be established as follows. Let **K** be a set of possibilistic clauses, and (F,α) a possibilistic clause obtained by a finite number of successive applications of the resolution rule to **C**; then $\mathbf{K} \Vdash (F,\alpha)$.

Refutation can be easily extended to Possibilistic Logic. Let **K** be a knowledge base made of possibilistic formulas, i.e., $\mathbf{K} = \{(P_1,\alpha_1),...,(P_n,\alpha_n)\}$. Proving (F,α) from **K** amounts to adding $(\neg F,1)$, put in clausal form, to **K**, and using the above rules repeatedly until getting $\mathbf{K},(\neg F,1) \vdash (\bot,\alpha)$. We are interested here in getting the empty clause with the greatest possible weight. It holds that $\mathbf{K} \vdash (F,\alpha)$ if and only if $\mathbf{K}_\alpha \vdash F$ in the classical sense, where $\mathbf{K}_\alpha = \max\{F : (F,\beta) \in \mathbf{K} \text{ and } \beta \geq \alpha\}$.

Resolution principle for Possibilistic Logic can be applied to derive implicit relationships from Possibilistic Logic knowledge bases. A triangular norm operation can be applied for merging knowledge bases for information fusion under Possibilistic Logic. To illustrate, consider the two possibilistic knowledge bases $\mathbf{K}_1 = \{(P_1,\alpha_1),...,(P_n,\alpha_n)\}$ and $\mathbf{K}_2 = \{(Q_1,\beta_1),...,(Q_m,\beta_m)\}$. Triangular norms are associative nondecreasing symmetric operations *tn* such that $tn(1,\alpha) = \alpha$ and $tn(0,0) = 0$. The main triangular norms are *min*, *product*, and $\max(0,\alpha+\beta-1)$. Let π_{tn} be the result of the combination of $\pi_{\mathbf{K}_1}$ and $\pi_{\mathbf{K}_2}$ based on the triangular norm operation *tn*. Then, the following merged Possibilistic Logic base can be obtained:

$$\mathbf{K}_{tn} = \mathbf{K}_1 \cup \mathbf{K}_2 \cup \{(F_i \vee G_j, ct(\alpha_i,\beta_j)) : (F_i,\alpha_i) \in \mathbf{K}_1 \text{ and} (G_j,\beta_j) \in \mathbf{K}_2\}$$

where $ct(\alpha,\beta) = 1 - tn(1-\alpha,1-\beta)$. Note that if $tn = \min$ and $ct = \max$ then $\mathbf{K}_{tn}$ is semantically equivalent to $\mathbf{K}_1 \cup \mathbf{K}_2$.

3.7 TRANSFORMATION BETWEEN FORMALISMS

Table 3-4 shows some of the formalisms that we have considered in the last section, nature of their output quantifying uncertainties, and examples. The hypothesis set is $\{none, slight, long\}$ representing qualitative values for expressing the delay in arrival of a vehicle.

Formalism	Output Uncertainty	Example
Probability	Discrete or continuous probability distribution	*none*: 0.25, *slight*: 0.20, *long*: 0.55
Dempster-Shafer	Belief distribution among focal elements	{*none*}: 0.25, {*none*, *slight*}: 0.40, {*none*, *slight*, *long*}: 0.35
Certainty Factor	Certainty factor values	*none*: –0.25, *slight*: 0.20, *long*: 0.55
Classical Logic	Disjunction	*none* OR *slight*
Possibility	Possibility measure	*none*: 0.30, *slight*: 0.60, *long*: 0.40

Table 3-4: Uncertainty output types from different formalisms

There are various approaches to transformation between formalisms. Here we clarify a few transformation instances. The Dempster-Shafer theory of belief function is a generalization of Bayesian probability, and, conversely, any Bayesian approximation of belief function, such as the Transferable Belief Model (TBM) discussed below, can do the reverse transformation. A section below discusses transformation between probability and possibility approaches. The transformation between probability and certainty factors can be achieved by adjusting the scales $[0,1]$ and $[-1,1]$ appropriately. Any transformation between a user-defined symbolic dictionary for representing uncertainty (e.g., $\{low, medium, high\}$ and the rest of the formalisms has to be customized individually. Since a transformation exists between probability and Dempster-Shafer, a transformation between Dempster-Shafer and the rest of the formalism can be defined based on the transformation between probability and the rest of the formalism.

3.7.1 Transferable Belief Model

The Transferable Belief Model (TBM) (Smets, 1988; Smets and Kennes, 1994) describes a two-level mental model in order to distinguish between two aspects of beliefs: belief as weighted opinions, and belief for decision making. The two levels are: the *credal level*, where beliefs are maintained, and the *pignistic level*, where beliefs are used to make decisions (credal and pignistic derive from the Latin words "credo," I believe, and "pignus," a wage, a bet).

Beliefs can be maintained outside any decision context. TBM assumes that beliefs at the credal level are quantified by belief functions (Shafer, 1976). The credal level precedes the pignistic level in that at any time, beliefs are maintained (and revised) at the credal level. The pignistic level appears only when a decision needs to be made. When a decision must be made, beliefs held at the credal level induce a probability measure at the pignistic level. That probability measure is then used to compute the expected utilities. To map belief functions onto probability functions, TBM uses a transformation that is called the *pignistic transformation*. Note that Bayesian formalisms do not consider the credal level. They usually argue that beliefs coexist with decisions.

Smett's pignistic transformation is defined as follows. Let m be the basic belief assignment on a space Ω. Then for every element w of Ω, the pignistic probability, denoted as *BetP* to distinguish it from the subjective probability that would quantify the decision maker's beliefs according to the Bayesians, is

$$BetP(w) = \sum_{w:w \in A \subseteq \Omega} \frac{m(A)}{|A|}$$

where $|A|$ is the number of elements of Ω in A. If a person's belief at the credal level is already represented by a probability function, then the pignistic transformation just amounts to the identify operator.

3.7.2 Relating Probability and Possibility

The transformation between probability and possibility approaches to handling uncertainty has been studied by several researchers. Most of these studies examined consistency principles that must be preserved for the transformation, and devised equations. Here we introduce a few well-known ones.

We consider the universe of discourse Ω is the set $\{w_1, \ldots, w_n\}$ of singletons and A is a subset of Ω, and the probability and possibility distributions on Ω are $p = (p_1, \ldots, p_n)$ and $\pi = (\pi_1, \ldots, \pi_n)$, respectively. Table 3-5 presents a comparison of characteristics between probability and possibility approaches.

Probability	Possibility
Measure: Probability P	Measure: Possibility Π
Distribution: p	Distribution: π
$P(A) = \sum_{x \in A} p(x)$	$\Pi(A) = \sup_{x \in A} \pi(x)$
$P(A \cup B) = P(A) + P(B)$, if $A \cap B = \Phi$	$\Pi(A \cup B) = \max(\Pi(A), \Pi(B))$
$P(A \cap B) = P(A) \cdot P(B)$, if A and B are independent	$N(A \cap B) = \min(N(A), N(B))$
$P(A) + P(\bar{A}) = 1$	$\max(\Pi(A), \Pi(\bar{A})) = 1$ $\Pi(A) + \Pi(\bar{A}) \geq 1, N(A) + N(\bar{A}) \leq 1$
Ignorance: $\forall x \in \Omega, p(x) = \frac{1}{\lvert\Omega\rvert}$	Ignorance: $\forall x \in \Omega, \pi(x) = 1$

Table 3-5: Characteristics comparison between probability and possibility approaches

Zadeh (1978) stated that a high degree of possibility does not necessarily imply a high degree of probability, nor does a low degree of probability necessarily imply a low degree of possibility. He defined the degree of consistency between a probability distribution and a possibility distribution as

$$C = \sum_{i=1}^{n} p_i \cdot \pi_i$$

This degree of consistency does not define a precise relationship between probability and possibility, but rather an approximate formalization of the heuristic connection that a lessening of the possibility of an event tends to lessen its probability but not vice versa.

According to Klir (1993), the transformation from p to π must preserve some appropriate scale and the amount of information contained in each distribution expressed by the equality of their uncertainties. Assuming the elements of Ω are ordered in such a way that $p_i > 0,\ p_i \geq p_{i+1}$ and $\pi_i > 0,\ \pi_i \geq \pi_{i+1}$, where $p_{n+1} = \pi_{n+1} = 0$, Klir proposed the transformation under two scales satisfying his uncertainty preservation principle:

- Ratio scale: $\pi_i = \dfrac{p_i}{p_1},\ p_i = \dfrac{\pi_i}{n \cdot \sum_{k=1}^{n} \pi_k}$
- Log-interval scale: $\pi_i = \left(\dfrac{p_i}{p_1}\right)^{\alpha},\ p_i = \dfrac{\pi_i^{1/\alpha}}{n \cdot \sum_{k=1}^{n} \pi_k^{1/\alpha}}$ where $\alpha \in (0,1)$

The consistency principle of Dubois and Prade (1993) is guided by the principle that values of probabilities of events are bounded by the degrees of possibility and necessity:

$$\Pi(A) \leq P(A) \leq N(A)$$

The transformation $\pi \to p$ below is guided by the principle of insufficient reason, which is aimed at finding the probability distribution that contains as much uncertainty as possible but that retains the features of possibility distribution:

$$p_i = \sum_{k=i}^{n} \frac{(\pi_k - \pi_{k+1})}{k}$$

The transformation $p \to \pi$ is guided by the principle of maximum specificity, which is aimed at finding the most informative possibility distribution:

$$\pi_i = \sum_{k=i}^{n} \min(p_i, p_k)$$

The two transformations defined above are shown to be bijective.

3.8 FURTHER READING

The readers are recommended to read the seminal papers referenced in Section 3.2. The book by Pearl (1988) is still the most comprehensive account on BNs, and more generally on using probabilistic reasoning to handle uncertainty. Jensen's books (1996, 2002) are also useful guides in this field. Shafer's own book (1976) and the edited volume by Yager et al. (1994) are good sources on the Dempster-Shafer theory of belief function. There are hundreds of good books and articles available on fuzzy logic. A good book to start with is Klir and Yuan (1995). Consult Lloyd (1987) or Das (1992) on program completion, or more generally on its use in logic programming.

Chapter 4

Introduction to Target Tracking

This chapter provides a short overview of the Level 1 fusion area, mostly from the target tracking perspective. Though Level 1 fusion is not a major focus of this book, we do provide enough background materials for readers to understand interactions between Level 1 and Level 2/3 fusion processes. We start with a brief introduction of target tracking and develop a simple generic target tracking architecture. Then we describe how a tracking problem, or more generally the estimation problem in a dynamical system, is formulated via transition and observation models. Then we present the Kalman Filter (KF) technique for estimating the state of a dynamical system.

We consider various target tracking environments, namely single target single sensor, single target multiple sensors, and multiple targets and multiple sensors. Various filtering techniques for handling these cases are presented, including the Probabilistic Data Association Filter (PDAF), Joint Probability Data Association Filter (JPDAF), and Multiple-Hypothesis Tracking (MHT). We present the Interacting Multiple Model (IMM) technique for handling targets that adopt multiple motions. Finally, the Cramer-Rao Lower Bound (CRLB) performance measurement criterion is presented. References are provided throughout and at the end of the chapter for further reading.

4.1 TARGET TRACKING CONCEPT AND ARCHITECTURE

Target tracking is concerned with the state estimation of a known or an unknown number of moving objects within an environment (e.g., adversary vehicles in the battlefield environment). Available measurements via battlefield sensors may have originated from the targets, if they are detected, or due to clutter in the environment, which is considered false alarm. The main difficulty comes from the assignment of a given measurement to a target. This problem is known as the data association problem in the target tracking community (Bar-Shalom and Li, 1995). The example below illustrates the complexity of the data association problem in the presence of multiple targets within an environment.

Example

Figure 4-1 illustrates the problem of tracking (Schulz et al., 2003) two fixed adversarial vehicles in the presence of multiple sensors of two types: feature identification sensors (ID-1, ID-2, ID-3, ID-4, ID-5) and position-velocity measurement sensors (P-1, P-2, P-3, P-4). A target is detected by a sensor if it is within the circular area covered by the sensor. The solid and dotted lines are the two tracks, corresponding to the two types of vehicles A and B, respectively, observed first by the identification sensor ID-1 at time $t = 0$.

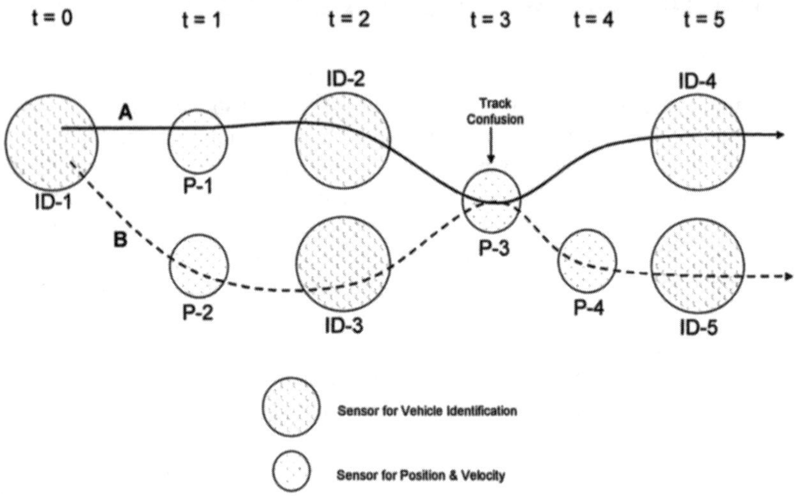

Figure 4-1: Example Multiple Target Tracking Scenario

After passing through the identification sensor ID-1, the vehicles are observed by the position and velocity sensors P-1 and P-2, respectively, at time $t = 1$. However, until they reach identification sensors ID-2 and ID-3, both tracks have the same probability of belonging to either A or B. After passing through the identification sensors ID-2 and ID-3 at time point $t = 2$, the ambiguity is resolved and the identifications are determined. After the paths cross at $t = 3$, there is confusion about the continuation of the two tracks. The position-velocity sensor P-4 observes one of the objects at time $t = 4$. The estimation algorithm may generate a higher likelihood for associating this observation with the track for A based on the direction of the vehicle detected by P-3 at time $t = 3$. However, once they reach identification sensors ID-4 and ID-5, the ambiguity is resolved and the identifications are determined. At this time point, the smoothing process will rectify any incorrect classification at the previous time point.

The example above does not consider various other possibilities that may add complexities to the association problem. For example, what happens if a position measurement is originated from a third target? Do we create a new track? What happens if an existing target stops and no further measurement is obtained? Do we delete its existing track? These are some of the hard target tracking questions.

Figure 4-2 shows the functional elements of a simple recursive target tracking system. As shown on the left of the figure, *Sensors* are placed within an *Environment* to observe targets. Various types of signals emitted and/or returned from targets, along with noise, are picked up by the sensors as observations.

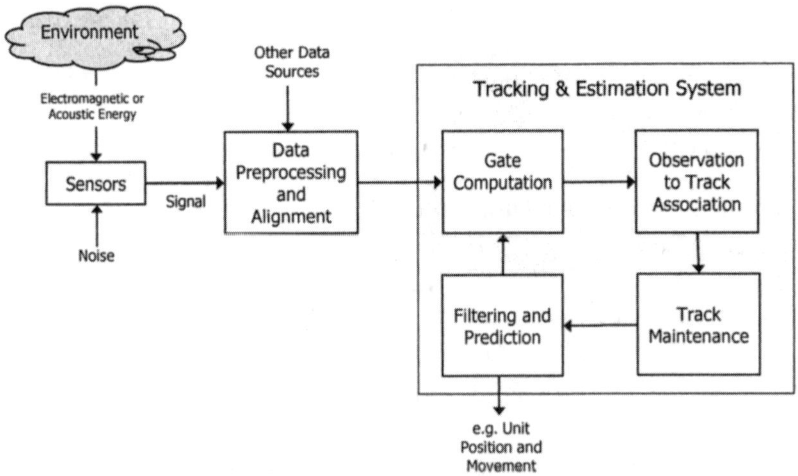

Figure 4-2: Functional architecture of a target tracking system

A *contact* by a sensor is an observation consisting of a called detection and a measurement, where a detection is *called* when the Signal-to-Noise Ratio (SNR) crosses a predefined threshold. A *measurement* associated with a detection often includes estimated position and velocity. A contact is either associated to a target or to a false alarm. A *target track* is a symbolic representation of a target moving through an area of interest. Basically, a track consists of a chronological sequence of past observations providing a likely history of movements of the target.

The observational data generated from signals are sent to the *Data Preprocessing and Alignment* component. The component permits comparisons and associations among the observations via 1) removing inconsistency, 2) addressing gaps for missing items, 3) common formatting, 4) spatiotemporal alignment, and 5) confidence normalization.

Gating involves evaluation of which possible observations to track pairings are reasonable. The *Gate Computation* component uses the uncertainty in covariances to compute the gates for associating observations to tracks. The current observations are gated with current tracks by the *Observation to Track Association* component. These two components together perform data association. The *Track Maintenance* component performs track management, that is, whenever necessary it creates new tracks, merges or stitches together existing tracks, and deletes tracks. The component initiates new tentative tracks if there are observations not assigned to existing tracks. A tentative track becomes a confirmed track upon some tests and further observations. Similarly, the Track Maintenance component deletes low-quality tracks as determined by the update history. The *Filtering and Prediction* component makes use of the observations to filter tracks and then predict maneuvers and positions.

4.2 TARGET TRACKING PROBLEM MODELING

Modeling of target tracking problems involves defining state transition and observation models to be used for estimation.

4.2.1 State Transition and Observation Models

A discrete-time state-space model of a target (or, more generally, a dynamical system) consists of the following two equations of transition (or motion) and observation (or measurement) models:

$$\left.\begin{aligned} X_{k+1} &= f(X_k, U_k) + W_k \\ Z_k &= h(X_k) + V_k \end{aligned}\right\} k = 0,1,2,\dots$$

where

- $X_k \in \mathbb{R}^{n_x}$ is the system state at the time step k (actual time is t_k);
- $Z_k \in \mathbb{R}^{n_z}$ is the measurement or observed output state at the time step k;
- $U_k \in \mathbb{R}^{n_u}$ is the known input at the time step k;
- $W_k \in \mathbb{R}^{n_w}$ is the process or state white noise;
- $V_k \in \mathbb{R}^{n_v}$ is the measurement white noise;
- $f : \mathbb{R}^{n_x} \to \mathbb{R}^{n_x}$ is the known state transition function to predict the system state one-step ahead; and
- $h : \mathbb{R}^{n_x} \to \mathbb{R}^{n_x}$ is the known observations function to relate system states to observations.

Note that the attributes of a system state vector X_k are continuous valued in this chapter, but they may also be categorical. For example, the state vector for each target in the target tracking scenario presented in the last section is likely to contain a categorical attribute defining its identification, in addition to its continuous-valued attributes for position and velocity. Techniques in Chapters 7 and 8 will be used for estimating the state of a dynamical system modeled with a hybrid state vector.

The noise variables W_k and V_k are assumed to be independent of each other, and with zero-mean normal probability distributions as

$$p(W_k) \sim N(0, Q_k)$$
$$p(V_k) \sim N(0, R_k)$$

where Q_k and R_k are, respectively, the process and measurement noise covariance matrices at time step k. Therefore,

$$E[W_k] = E[V_k] = 0 \qquad k = 0,1,2,...$$

$$\left.\begin{aligned} E\left[W_i W_j^T\right] &= \delta_{ij} Q_i \\ E\left[V_i V_j^T\right] &= \delta_{ij} R_i \end{aligned}\right\} \qquad i,j = 0,1,2,...$$

Since the process and observation noises are independent, we have

$$E\left[W_i V_j^T\right] = 0 \qquad i,j = 0,1,2,...$$

The values of the initial state and the noise variables at each time step k are uncertain and described by probability models. The transition and observation equations along with these probability models prescribe the following transition and measurement PDFs, providing a complete description of the target:

$$p(X_0)$$

$$\left.\begin{aligned} &p(X_{k+1} \mid X_k, U_k) \\ &p(Z_k \mid X_k) \end{aligned}\right\} k = 0,1,2,...$$

4.2.2 Estimation Problem

The estimate of the system state X_k at time step k is represented by the posterior conditional PDF $p(X_k \mid Z_{1:i}, U_{1:k})$, where $Z_{1:i} = \{Z_1,...,Z_i\}$ is the cumulative measurements up to the time i and $U_{1:k} = \{U_1,...,U_k\}$ is the cumulative known inputs up to the time step k. Various types of the *estimation problem* are then listed below:

- If $i = k$ then the problem of estimating $p(X_k \mid Z_{1:k}, U_{1:k})$ is called *filtering*.
- If $i < k$ then the problem is called *prediction*.
- If $i > k$ then the problem is called *smoothing*.

Often we will evaluate the filtering problem at every time step k in a Markovian fashion, that is, update sequentially $p(X_k \mid Z_{1:k}, U_{1:k})$ as $p(X_k \mid Z_k, U_k)$. In other words, the update is conditional on the observed system input and output up to the current time, but by using only the following information: 1) $p(X_k \mid Z_{k-1}, U_{k-1})$ produced at the end of the previous time step; 2) observation data from the current time step; and 3) the prescribed probabilistic noise models W_k and V_k. Unless otherwise stated, the conditioning of every PDF on the input variable U_k is left implicit.

The Bayesian approach to the solution of the prediction, filtering, and smoothing problems is specified below:

$$p(X_k \mid Z_{1:k-1}) = \int_{\mathbb{R}^{n_x}} p(X_k \mid X_{k-1})\, p(X_{k-1} \mid Z_{1:k})\, dX_{k-1}$$

$$\begin{aligned} p(X_k \mid Z_{1:k}) &= p(X_k \mid Z_k, Z_{1:k-1}) \\ &= \frac{p(Z_k \mid X_k)\, p(X_k \mid Z_{1:k-1})}{p(Z_k \mid Z_{1:k-1})} \\ &= \frac{p(Z_k \mid X_k)\, p(X_k \mid Z_{1:k-1})}{\int_{\mathbb{R}^{n_x}} p(Z_k \mid X_k)\, p(X_k \mid Z_{1:k-1})\, dX_k} \end{aligned}$$

where the current observations are independent of the past observations given the system state, that is,

$$p(Z_k \mid X_k, Z_{1:k-1}) = p(Z_k \mid X_k)$$

$$\begin{aligned} p(X_k \mid Z_{1:i}) &= p(X_k \mid Z_{1:k}, Z_{k+1:i}) \\ &= p(X_k \mid Z_{1:k}) \frac{p(Z_{k+1:i} \mid X_k, Z_{1:k})}{p(Z_{k+1:i} \mid Z_{1:k})} \\ &= p(X_k \mid Z_{1:k}) \int_{\mathbb{R}^{n_x}} \frac{p(Z_{k+1:i} \mid X_{k+1}, Z_{1:k})\, p(X_{k+1} \mid X_k, Z_{1:k})}{p(Z_{k+1:i} \mid Z_{1:k})} dX_{k+1} \\ &= p(X_k \mid Z_{1:k}) \int_{\mathbb{R}^{n_x}} \frac{p(X_{k+1} \mid Z_{1:i})}{p(X_{k+1} \mid Z_{1:k})}\, p(X_{k+1} \mid X_k)\, dX_{k+1} \end{aligned}$$

where the future states are independent of the current observations given the system's current state, that is, $p(X_{k+1} \mid X_k, Z_{1:k}) = p(X_{k+1} \mid X_k)$

The major challenge to the Bayesian approach is that the above estimation problems involving high-dimensional integrals usually do not have a closed-form solution. (If an equation, expressed in terms of generally accepted functions and mathematical operations, is a solution to a given problem then the equation is a *closed-form solution* to the problem.) A closed-form solution exists when the transition and observation models are linear and uncertainties are Gaussian. A linear system can be described by the following equations:

$$\left.\begin{aligned} X_{k+1} &= F_k X_k + G_k U_k + W_k \\ Z_k &= H_k X_k + V_k \end{aligned}\right\} k = 0, 1, 2, \ldots$$

where F_k, G_k, and H_k are, respectively, transition, input, and observation model matrices.

Example

Suppose we want to track a vehicle traveling along a straight line, and its state vector is composed of position and velocity as:

$$X_k = \begin{bmatrix} P_k \\ V_k \end{bmatrix}$$

We are able to measure the position and the vehicle in every T seconds (i.e., $T = t_{k+1} - t_k$ is the sampling or observation interval period), and also, as an input to the process, we are able to change the acceleration of the vehicle. Therefore, according to the laws of motion, the transition and observation models can be encoded as the following linear equations:

$$X_{k+1} = \begin{bmatrix} 1 & T \\ 0 & 1 \end{bmatrix} X_k + \begin{bmatrix} T^2/2 \\ T \end{bmatrix} U_k + W_k$$

$$Z_k = \begin{bmatrix} 1 & 0 \end{bmatrix} X_k + V_k$$

4.3 SINGLE SENSOR SINGLE TARGET TRACKING

This section presents two filters for single target tracking: Alpha-Beta and Kalman filters. The filters are not concerned with the data association problem to be described in the next section. In other words, one and only one measurement from a single sensor is assumed to have been obtained at every step and made associated with the only prediction to filter the track.

4.3.1 Alpha-Beta Filter

The Alpha-Beta filter (Kalata, 1992) is a widely studied technique for estimating the position of a nonaccelerating target. The filter assumes that the measurement noise is stationary (i.e., variance is constant) and the sampling interval is of a fixed time period. The state vector for the filter is composed of position and velocity as:

$$\begin{bmatrix} X_k \\ V_k \end{bmatrix}$$

The motion model of the filter is given below:

$$\begin{bmatrix} X_k \\ V_k \end{bmatrix} = \begin{bmatrix} 1 & t_k - t_{k-1} \\ 0 & 1 \end{bmatrix} \begin{bmatrix} X_{k-1} \\ V_{k-1} \end{bmatrix}$$

where $t_k - t_{k-1}$ is the time period of the sampling interval. Therefore, the motion models for position and velocity are the following:

$$X_k = X_{k-1} + (t_k - t_{k-1}) V_{k-1}$$

$$V_k = V_{k-1}$$

If Z_k is the position measurement at time step k, then estimated position and velocity are given below:

$$\hat{X}_k = X_k + \alpha (Z_k - X_k) \qquad 0 \le \alpha \le 1$$

$$\hat{V}_k = V_k + \frac{\beta}{t_k - t_{k-1}} (Z_k - X_k) \qquad 0 \le \beta \le 1$$

where

- X_k is the predicted position at time step k;
- V_k is the predicted velocity at time step k;
- $\hat{X}_k$ is the estimated position at time step k;
- $\hat{V}_k$ is the estimated velocity at time step k; and
- α and β are smoothing parameters or gain coefficients.

The performance depends upon the filter parameters α and β, which are used to weight the current measurement against the past predictions of position and velocity. The value of the parameters should be close to 1 in situations where the measurement noise is known to be small, thus placing high confidence in the measurement. On the other hand, if there is a lot of noise in the environment, the parameter values should be chosen close to 0, indicating low confidence in the measurement. Common choices of the parameters are the following:

$$\alpha = 2\sqrt{\beta} - \beta$$

$$\beta = \frac{\alpha^2}{2-\alpha}$$

An Alpha filter assumes that the target is moving with constant velocity (i.e., $\hat{V}_k = V_k$), and the Alpha-Beta-Gamma filter generalizes Alpha-Beta filters with acceleration.

4.3.2 Kalman Filter (KF)

The Kalman Filter (KF) (Kalman, 1960; Bar-Shalom and Fortmann, 1988) is based on a state-space formulation of a continuous or discrete-time system. Here we focus our attention to discrete-time KF. The system must be linear, but may be time variant. Recall that the transition and observation models for a linear system have the following form:

$$\left.\begin{aligned} X_k &= F_{k-1}X_{k-1} + G_{k-1}U_{k-1} + W_{k-1} \\ Z_k &= H_k X_k + V_k \end{aligned}\right\} k = 1, 2, \ldots$$

Given a set of observations, KF gives an estimate of the state of the target or dynamical system. Here we provide a detailed derivation of the KF for it is the most dominant and fundamental filtering technique. Readers who are not interested in this level of details may go directly to Figure 4-3 to check the overall KF steps.

For the case of Gaussian signals, and given the assumed linear model, the state estimate is optimum in the sense that it minimizes the norm of the difference between the estimate and the actual state. Symbolically,

$$\hat{X}_{i|j} = \arg\min_{\hat{X}_{i|j}} E\left[\left(X_i - \hat{X}_{i|j}\right)\left(X_i - \hat{X}_{i|j}\right)^T \mid Z_{1:j}\right]$$

where $\hat{X}_{i|j}$ is the estimate of the state at time step i given all the observations up to time step j. The solution to this minimization problem can be obtained by differentiating with respect to the expected value and equating the resultant expression to zero:

$$\frac{\partial}{\partial \hat{X}_{i|j}} E\left[\left(X_i - \hat{X}_{i|j}\right)\left(X_i - \hat{X}_{i|j}\right)^T \mid Z_{1:j}\right]$$

$$= \frac{\partial}{\partial \hat{X}_{i|j}}\left[\int_{X_i}\left(X_i - \hat{X}_{i|j}\right)\left(X_i - \hat{X}_{i|j}\right)^T p\left(X_i \mid Z_{1:j}\right)dX_i\right]$$

$$= 2\int_{X_i}\left(X_i - \hat{X}_{i|j}\right)p\left(X_i \mid Z_{1:j}\right)dX_i$$

$$= 0$$

Therefore,

$$\hat{X}_{i|j}\int_{X_i} p\left(X_i \mid Z_{1:j}\right)dX_i = \int_{X_i} X_i p\left(X_i \mid Z_{1:j}\right)dX_i$$

i.e., $\hat{X}_{i|j} = E\left[X_i \mid Z_{1:j}\right]$

Specifically,

$$\hat{X}_{k|k-1} = E\left[X_k \mid Z_{1:k-1}\right]$$

$$= E\left[F_{k-1}X_{k-1} + G_{k-1}U_{k-1} + W_{k-1} \mid Z_{1:k-1}\right]$$

$$= F_{k-1}E\left[X_{k-1} \mid Z_{1:k-1}\right] + G_{k-1}U_{k-1} + E\left[W_{k-1} \mid Z_{1:k-1}\right]$$

$$= F_{k-1}\hat{X}_{k-1|k-1} + G_{k-1}U_{k-1}$$

since the input is independent of observations, and noise is of zero-mean. For a liner filtering of the prediction $\hat{X}_{k|k-1}$ based on the current observation Z_k, the filtered estimate $\hat{X}_{k|k}$ can be written as a linear weighted sum of the prediction and observation as the following:

$$\hat{X}_{k|k} = K_k Z_k + \bar{K}_k \hat{X}_{k|k-1}$$

for some matrices K_k and $\bar{K}_k$, and their relationship is derived as follows:

$$E\left[\left(X_k - \hat{X}_{k|k}\right)Z_i^T\right] = 0,\ i = 1, \ldots, k,$$

[applying the orthogonality principle]

i.e., $E\left[\left(F_{k-1}X_{k-1} + G_{k-1}U_{k-1} + W_{k-1} - K_k Z_k - \bar{K}_k \hat{X}_{k|k-1}\right)Z_i^T\right] = 0$

[substituting the values of X_k and $\hat{X}_{k|k}$]

i.e., $F_{k-1}E\left[X_{k-1}\right]Z_i^T + G_{k-1}U_{k-1}Z_i^T - K_k E\left[Z_k\right]Z_i^T - \bar{K}_k E\left[\hat{X}_{k|k-1}\right]Z_i^T = 0$

[using $E\left[W_k\right] = 0$]

i.e., $F_{k-1}E\left[X_{k-1}\right]Z_i^T + G_{k-1}U_{k-1}Z_i^T - K_k E\left[H_k X_k + V_k\right]Z_i^T - \bar{K}_k E\left[\hat{X}_{k|k-1}\right]Z_i^T = 0$

[substituting the values of Z_k]

i.e., $F_{k-1}E\left[X_{k-1}\right]Z_i^T + G_{k-1}U_{k-1}Z_i^T - K_k H_k E\left[X_k\right]Z_i^T - \bar{K}_k E\left[\hat{X}_{k|k-1}\right]Z_i^T = 0$

$$[\text{using } E[V_k] = 0]$$

i.e., $E[X_k]Z_i^T - K_k H_k E[X_k]Z_i^T - \bar{K}_k E\left[\hat{X}_{k|k-1}\right]Z_i^T = 0$

$$[X_k = F_{k-1}X_{k-1} + G_{k-1}U_{k-1} + W_{k-1} \text{ gives}$$
$$F_{k-1}E[X_{k-1}] = E[X_k] - G_{k-1}U_{k-1}, \text{ since } E[W_k] = 0]$$

i.e., $E\left[X_k - K_k H_k X_k - \bar{K}_k \hat{X}_{k|k-1}\right]Z_i^T = 0$

i.e., $E\left[\left(X_k - K_k H_k X_k - \bar{K}_k X_k\right) + \bar{K}_k\left(X_k - \hat{X}_{k|k-1}\right)\right]Z_i^T = 0$

i.e., $E\left[\left(X_k - K_k H_k X_k - \bar{K}_k X_k\right) + \bar{K}_k\left(X_k - F_{k-1}\hat{X}_{k-1|k-1} - G_{k-1}U_{k-1}\right)\right]Z_i^T = 0$

$$[\text{since } \hat{X}_{k|k-1} = F_{k-1}\hat{X}_{k-1|k-1} + G_{k-1}U_{k-1}]$$

i.e., $E\left[\left(X_k - K_k H_k X_k - \bar{K}_k X_k\right) + \bar{K}_k\left(F_{k-1}X_{k-1} - F_{k-1}\hat{X}_{k-1|k-1} + W_{k-1}\right)\right]Z_i^T = 0$

$$[\text{since } X_k = F_{k-1}X_{k-1} + G_{k-1}U_{k-1} + W_{k-1}]$$

i.e., $\left(I - K_k H_k - \bar{K}_k\right)E[X_k]Z_i^T = 0$

$$[\text{since } E[W_{k-1}] = 0 \text{ and } E\left[\left(X_k - \hat{X}_{k|k}\right)Z_i^T\right] = 0,\ i = 1,\dots,k]$$

i.e., $I - K_k H_k - \bar{K}_k = 0$ [for any given X_k]

i.e., $\bar{K}_k = I - K_k H_k$

Therefore,

$$\hat{X}_{k|k} = K_k Z_k + \left(I - K_k H_k\right)\hat{X}_{k|k-1}$$
$$= \hat{X}_{k|k-1} + K_k\left(Z_k - H_k \hat{X}_{k|k-1}\right)$$

The above result states that the estimation is the sum of the prediction plus some fraction K_k of the difference between the actual and the predicted observation. The value of the constant K_k, known as *Kalman gain*, is yet to be derived. The estimation of the error covariance matrix

$$P_{i|j} = E\left[\left(X_i - \hat{X}_{i|j}\right)\left(X_i - \hat{X}_{i|j}\right)^T \mid Z_{1:j}\right]$$

Specifically, $P_{k|k-1} = E\left[\left(X_k - \hat{X}_{k|k-1}\right)\left(X_k - \hat{X}_{k|k-1}\right)^T \mid Z_{1:k-1}\right]$

Since $\hat{X}_{k|k-1} = F_{k-1}\hat{X}_{k-1|k-1} + G_{k-1}U_{k-1}$ and $X_k = F_{k-1}X_{k-1} + G_{k-1}U_{k-1} + W_{k-1}$, we have the following derivation:

$$P_{k|k-1} = E\left[\left(X_k - \hat{X}_{k|k-1}\right)\left(X_k - \hat{X}_{k|k-1}\right)^T \mid Z_{1:k-1}\right]$$
$$= F_{k-1}E\left[\left(X_{k-1} - \hat{X}_{k-1|k-1}\right)\left(X_{k-1} - \hat{X}_{k-1|k-1}\right)^T \mid Z_{1:k-1}\right]F_{k-1}^T + E\left[W_{k-1}W_{k-1}^T\right]$$

Therefore, $P_{k|k-1} = F_{k-1}\hat{P}_{k-1|k-1}F_{k-1}^T + Q_{k-1}$

Moreover, $P_{k|k} = E\left[\left(X_k - \hat{X}_{k|k}\right)\left(X_k - \hat{X}_{k|k}\right)^T \mid Z_{1:k}\right]$

Since $\hat{X}_{k|k} = \hat{X}_{k|k-1} + K_k\left(Z_k - H_k\hat{X}_{k|k-1}\right)$ and $Z_k = H_k X_k + V_k$, we have the following derivation:

$$\begin{aligned}
P_{k|k} &= E\left[\left(X_k - \hat{X}_{k|k}\right)\left(X_k - \hat{X}_{k|k}\right)^T \mid Z_{1:k}\right] \\
&= E\Big[\left(X_k - \hat{X}_{k|k-1} - K_k\left(Z_k - H_k\hat{X}_{k|k-1}\right)\right) \\
&\qquad \left(X_k - \hat{X}_{k|k-1} - K_k\left(Z_k - H_k\hat{X}_{k|k-1}\right)\right)^T \mid Z_{1:k}\Big] \\
&= E\Big[\left(X_k - \hat{X}_{k|k-1} - K_k\left(H_k X_k + V_k - H_k\hat{X}_{k|k-1}\right)\right) \\
&\qquad \left(X_k - \hat{X}_{k|k-1} - K_k\left(H_k X_k + V_k - H_k\hat{X}_{k|k-1}\right)\right)^T \mid Z_{1:k-1}\Big] \\
&= E\Big[\left(\left(I - K_k H_k\right)\left(X_k - \hat{X}_{k|k-1}\right) - K_k V_k\right) \\
&\qquad \left(\left(I - K_k H_k\right)\left(X_k - \hat{X}_{k|k-1}\right) - K_k V_k\right)^T \mid Z_{1:k-1}\Big] \\
&= \left(I - K_k H_k\right) E\left[\left(X_k - \hat{X}_{k|k-1}\right)\left(X_k - \hat{X}_{k|k-1}\right)^T \mid Z_{1:k-1}\right]\left(I - K_k H_k\right)^T \\
&\qquad + K_k E\left[V_k V_k^T\right] K_k^T
\end{aligned}$$

Since $P_{k|k-1} = E\left[\left(X_k - \hat{X}_{k|k-1}\right)\left(X_k - \hat{X}_{k|k-1}\right)^T \mid Z_{1:k-1}\right]$, we have the a posteriori estimation of error covariance:

$$P_{k|k} = \left(I - K_k H_k\right) P_{k|k-1} \left(I - K_k H_k\right)^T + K_k R_k K_k^T$$

We are interested in that value of the gain K_k that minimizes $P_{k|k}$. We want to partially differentiate $P_{k|k}$ with respect to K_k, and so $P_{k|k}$ needs to be rewritten to make use of the following two matrix formulae:

$$\frac{\partial}{\partial X}\left[YX^T\right] = Y, \text{ where } X \text{ and } Y \text{ are square}$$

$$\frac{\partial}{\partial X}\left[XY\right] = Y^T, \text{ where } X \text{ and } Y \text{ are square}$$

$$\frac{\partial}{\partial X}\left[XYX^T\right] = 2XY^T, \text{ where } Y \text{ is symmetric}$$

Now we rewrite $P_{k|k}$ to contain terms only of the form XY and XYX^T:

$$\begin{aligned}
P_{k|k} &= \left(I - K_k H_k\right) P_{k|k-1} \left(I - K_k H_k\right)^T + K_k R_k K_k^T \\
&= P_{k|k-1} - K_k H_k P_{k|k-1} - P_{k|k-1} H_k^T K_k^T + K_k\left(H_k P_{k|k-1} H_k^T + R_k\right) K_k^T
\end{aligned}$$

$$\frac{\partial P_{k|k}}{\partial K_k} = -\left(H_k P_{k|k-1}\right)^T - P_{k|k-1} H_k^T + 2K_k \left(H_k P_{k|k-1} H_k^T + R_k\right)^T$$
$$= -2P_{k|k-1} H_k^T + 2K_k \left(H_k P_{k|k-1} H_k^T + R_k\right)$$

Therefore, $K_k = P_{k|k-1} H_k^T \left(H_k P_{k|k-1} H_k^T + R_k\right)^{-1}$. Figure 4-3 summarizes the Kalman Filter steps as derived above.

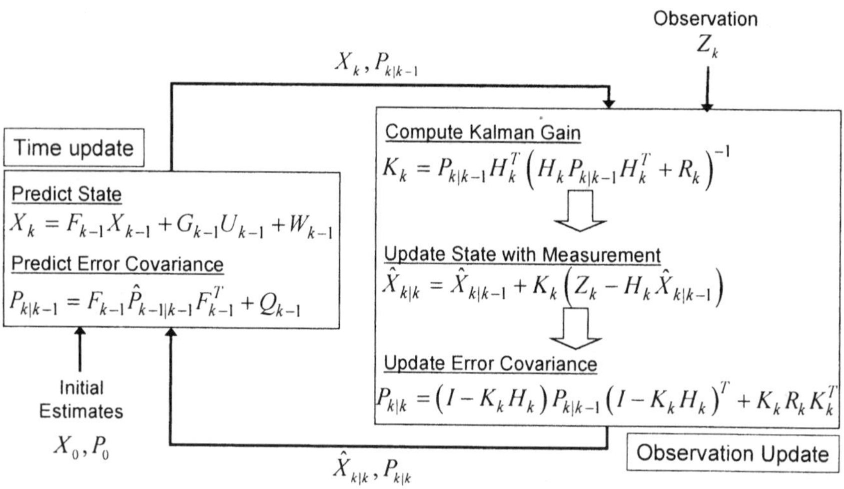

Figure 4-3: Kalman Filter steps

Example

Consider the problem of tracking a target moving on a plane at a constant velocity. The transition equation $X_{k+1} = F_k X_k + W_k$ without an input can be formulated as follows:

$$\begin{bmatrix} x_{k+1} \\ v_{x(k+1)} \\ y_{k+1} \\ v_{y(k+1)} \end{bmatrix} = \begin{bmatrix} 1 & T & 0 & 0 \\ 0 & 1 & 0 & 0 \\ 0 & 0 & 1 & T \\ 0 & 0 & 0 & 1 \end{bmatrix} \begin{bmatrix} x_k \\ v_{xk} \\ y_k \\ v_{yk} \end{bmatrix} + W_k$$

where x_k and y_k in the state vector are, respectively, x and y coordinates of the target at time step k, and v_{xk} and v_{yk} are, respectively, velocities along x and y directions. The variable T is the observation interval $t_{k+1} - t_k$. The measurement model $Z_k = H_k X_k + V_k$ is

$$\begin{bmatrix} z_{xk} \\ z_{yk} \end{bmatrix} = \begin{bmatrix} 1 & 0 & 0 & 0 \\ 0 & 0 & 1 & 0 \end{bmatrix} \begin{bmatrix} x_k \\ v_{xk} \\ y_k \\ v_{yk} \end{bmatrix} + V_k$$

The process noise W_k and the measurement noise V_k are assumed to be zero-mean white, Gaussian, and independent of each other:

$$W_k \sim N(0, Q_k)$$
$$V_k \sim N(0, R_k)$$

where the process covariance matrix Q_k and the noise covariance matrix R_k are derived as

$$Q_k = \begin{bmatrix} T^3/3 & T^2/2 & 0 & 0 \\ T^2/2 & T & 0 & 0 \\ 0 & 0 & T^3/3 & T^2/2 \\ 0 & 0 & T^2/2 & T \end{bmatrix} \begin{bmatrix} \sigma_x \\ \sigma_{vx} \\ \sigma_y \\ \sigma_{vy} \end{bmatrix}$$

$$R_k = \begin{bmatrix} \sigma_x{}^2 & 0 \\ 0 & \sigma_y{}^2 \end{bmatrix}$$

4.4 GATING AND DATA ASSOCIATION

The data association problem must be addressed when multiple measurements are obtained from multiple sensors tracking single or multiple targets within an environment. A data association technique correlates or pairs observations from multiple sensors to targets or established tracks. Usually gating involves evaluating which possible measurements to track pairings are reasonable. Gates are set up to incorporate right measurements into the state estimator of a target. Detections within a gate can be associated with the target of interest, but more than one measurement in a gate leads to an association uncertainty as shown in Figure 4-4.

Gates are drawn around a predicted target position (shown in broken line dots) in both single and multiple target tracking environments. Only those observations, some of which are noise or false alarms due to environmental clutters such as clouds and acoustic/electromagnetic interference to sensors, fall within the gate (shown in solid dots) are considered candidates to associate with the target. But some measurements can potentially fall within the common regions of two or more intersecting gates as shown in the multiple target tracking case in the figure. Measurements not associated with existing tracks (shown in grey dots) might be potential candidates for initiating new tentative tracks. A tentative track becomes confirmed when the quality of the measurements is high enough, and low-quality measurements become deleted.

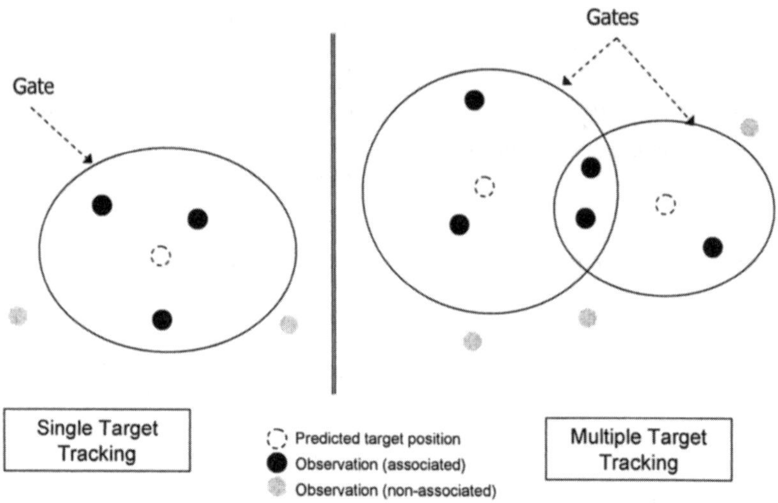

Figure 4-4: Data association in single and multiple target tracking environments

The performance of a tracking system heavily depends on the data association technique that it relies on. The technique has to appropriately handle uncertainty in sensor measurements. The Nearest-Neighbor (NN) approach (Bar-Shalom and Fortmann, 1988) to data association associates at most one observation from a sensor scan with an established track via the following normalized Mahalanobis distance of innovation:

$$\left(Z_k - \hat{Z}_{k|k-1}\right)^T S_k^{-1} \left(Z_k - \hat{Z}_{k|k-1}\right) \leq d^2$$

where

- k is time or scan index;
- Z_k is the actual measurement;
- $\hat{Z}_{k|k-1}$ is the predicted measurement;
- S_k is the covaria nce of the innovation $Z_k - \hat{Z}_{k|k-1}$; and
- d is the gate threshold.

A track-observation assignment matrix is formed where each element corresponding to a track and an observation is the normalized distance between the predicted target position of the track and the observation. If the observation does not fall within the gate then this distance is assigned a very large value to penalize. Munkres optimal assignment algorithm (Munkres, 1957; Bourgeois and Lassalle, 1971) can then be applied to a track-observation matrix and the final associations are decided based on the resultant matrix. NN approach is the simplest for data association and easy to implement. But the nearest neighbor is not always the measurement originating from the target, and hence results in the possibility of using false measurements by the underlying tracking filter in a dense environment.

A *target-oriented approach* to data association, such as the Probabilistic Data Association (PDA) (Bar-Shalom and Tse, 1975; Bar-Shalom and Fortmann, 1988) or Joint Probabilistic Data Association (JPDA) (Bar-Shalom, 1995), uses a set of established targets to form gates in the measurement space and to compute posterior probabilities. On the other hand, a *measurement-oriented approach*, such as Multiple-Hypothesis Tracking (MHT) (Reid, 1979), considers each measurement in turn and hypothesizes it to have come from some established track, a new target, or clutter. Both PDA and JPDA algorithms compute the probabilities of association only for the latest set of measurements of various targets.

Morefield (1977) views the data association problem as a discrete optimization problem for which 0-1 integer programming methods are applicable. Along this line is the *S-D* assignment approach (Deb et al., 1997), which is a discrete mathematical optimization formulation of the data association problem that systematically resembles an MHT within a window of length $S-1$. Several extensions of MHT have been proposed over the years, including fuzzy (Aja-Fernández et al., 2002) and probabilistic (Streit and Lugiubuhl, 1994) ones. The Auction approach (Bertsekas, 1988) to data association has the usual bidding and assignment phases. A belief theory approach to data association (Dang et al., 2003) constructs mass function with respect to observations and tracks, and combined data association for beliefs is produced using Dempster's combination rule.

None of the data association techniques presented here assumes that the underlying target motion or observation model is linear and Gaussian, but this assumption is often made to simplify hypothesis evaluation for target-originated measurements. For example, nonlinear models can be accommodated by their linearization using the Extended Kalman Filter (EKF) to be presented later in Chapter 8. Target state estimation in the presence of nonlinear and non-Gaussian models can be solved via approximate recursive Monte Carlo methods or particle filters (also to be presented later in Chapter 8). See Avitzour (1995) and Gordon (1997) for target tracking, and Bergman and Doucet (2000) for an application of the Markov Chain Monte Carlo (MCMC) technique for data association of multiple measurements in an over the horizon radar application.

4.5 MULTISENSOR SINGLE TARGET TRACKING (IN CLUTTER)

This section presents the Probabilistic Data Association (PDA) (Bar-Shalom and Tse, 1975; Bar-Shalom and Fortmann, 1988) and the corresponding PDA Filter (PDAF). The filter assumes that the tracking environment consist of only one target of interest and multiple sensors. But at most one validated sensor measurement can be target originated and the remaining ones are assumed to be generated from nontargets or from clutters.

4.5.1 Probabilistic Data Association Filter (PDAF)

In contrast to the NN approach, the PDA is an all-neighbors approach. The PDA tracks targets by using the weighted average of the innovations of all measurements in the gate. The PDA assumes that there is only one target of interest and at most one validated measurement can be target originated and the remaining are assumed to be false alarms. The overall approach is to form all feasible data association events with validated measurements from a validated gate and then compute the association probability of each event. The transition and observation equations in PDAF are the following:

$$\hat{X}_{k|k-1} = F_{k-1}\hat{X}_{k-1|k-1}$$

$$\hat{Z}_{k|k-1} = H_k \hat{X}_{k|k-1}$$

The predicted covariance is

$$P_{k|k-1} = F_{k-1}P_{k-1|k-1}F_{k-1}^{T} + Q_{k-1}$$

The set Z_k of all validated measurements at time step k containing m_k elements and the set $Z_{1:k}$ of all measurements up to and including time step k are expressed below:

$$Z_k = \{z_{ik}\}_{i=1}^{m_k}$$

$$Z_{1:k} = \{Z_j\}_{j=1}^{k}$$

Define the association events θ_{ik} corresponding to the observations as

$$\theta_{ik} = \begin{cases} z_{ik}\left(1 \le i \le m_k\right) \text{ is the target-originated event} \\ \text{none } \left(i = 0\right) \text{of the measurements is target originated} \end{cases}$$

The innovation v_{ik} corresponding to a measurement z_{ik} is given by

$$v_{ik} = z_{ik} - \hat{Z}_{k|k-1} = z_{ik} - H_k \hat{X}_{k|k-1}$$

The combined innovation and the innovation covariance v_k and S_k are given by

$$v_k = \sum_{i=0}^{m_k} \beta_{ik} v_{ik}$$

$$S_k = E\left[v_k v_k^{T}\right] = H_k P_{k|k-1} H_k^{T} + R_k$$

where β_{ik} is the association probability defined as

$$\beta_{ik} = p\left(\theta_{ik} \mid Z_{1:k}\right), \text{ where } \sum_{i=0}^{m_k} \beta_{ik} = 1$$

The elliptical validation region is defined as

$$v_k(\gamma) = \left\{z : \left[z - \hat{Z}_{k|k-1}\right]^{T} S_k^{-1} \left[z - \hat{Z}_{k|k-1}\right] \le \gamma\right\}$$

where γ is a threshold determined from the χ-squared distribution tables with n_z degrees of freedom, which is equal to the dimension of the measurement vector. The volume of the validated gate is the volume of the n_z-dimensional unit hypersphere:

$$V_k(\gamma) = C_{n_z} |\gamma S_k|^{1/2} = C_{n_z} \gamma^{\frac{1}{2}} |S_k|^{1/2}$$

where C_{n_z} is the volume of the n_z-dimensional unit hypersphere ($C_1 = 2$, $C_2 = \pi$, $C_3 = 4\pi/3$, ...). Therefore, an increase in γ or S_k will increase the size of the validation gate as shown in the 2-dimensional case in Figure 4-5.

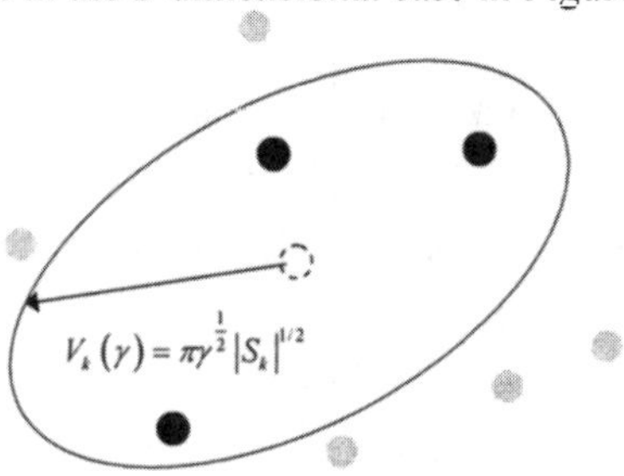

Figure 4-5: Validation gate in PDAF

The track has been initialized and the past information is summarized by

$$p(X_k \mid Z_{k-1}) = \mathcal{N}\left(X_k; \hat{X}_{k|k-1}, P_{k|k-1}\right)$$

As shown below, PDAF assumes that the best estimate of the target state is the conditional mean based upon all observations that with some nonzero probability originated from the target,

$$\begin{aligned}
\hat{X}_{k|k} &= E[X_k \mid Z_{1:k}] \\
&= \sum_{i=0}^{m_k} E[X_k \mid \theta_{ik}, Z_{1:k}] p(\theta_{ik} \mid Z_{1:k}) \\
&= \sum_{i=0}^{m_k} \hat{X}_{i(k|k)} \beta_{ik}
\end{aligned}$$

The past information $p(X_k \mid Z_{1:k-1})$ in PDAF is summarized as Gaussian and this leads to the following derivation of β_{ik}:

$$\begin{aligned}
\beta_{ik} &= p(\theta_{ik} \mid Z_{1:k}) \\
&= p(\theta_{ik} \mid Z_k, Z_{1:k-1}) \\
&= \frac{1}{c} p(Z_k \mid \theta_{ik}, Z_{1:k-1}) \, p(\theta_{ik} \mid Z_{1:k-1})
\end{aligned}$$

But, $p(Z_k \mid \theta_{ik}, Z_{1:k-1}) = \begin{cases} V_k(\gamma)^{-m_k+1} (p_G)^{-1} \mathcal{N}(v_{ik}; 0, S_k) & 1 \le i \le m_k \\ V_k(\gamma)^{-m_k} & i = 0 \end{cases}$

The state estimate is updated with the combined innovation based on probability weights:

$$\hat{X}_{k|k} = \sum_{i=0}^{m_k} \hat{X}_{i(k|k)} \beta_{ik}$$

where m_k is the number of measurements and $\hat{X}_{i(k|k)}$ is the estimate associated with the i-th measurement. The gain to be used to compute this estimate is

$$W_k = P_{k|k-1} H_k^{\ T} S_k^{\ -1}$$

The covariance associated with the updated state is

$$P_{k|k} = \beta_{0k} P_{k|k-1} + (1 - \beta_{0k}) P_{k|k}^c + \tilde{P}_k$$

where $P_{k|k}^c$ is the covariance associated with state updated with the correct measurement and $\tilde{P}_k$ is the spread of innovation

$$P_{k|k}^c = P_{k|k-1} + W_k S_k W_k^{\ T}$$

$$\tilde{P}_{k|k} = W_k \left(\sum_{i=1}^{m_k} \beta_{ik} v_{ik} v_{ik}^{\ T} - v_k v_k^{\ T} \right) W_k^{\ T}$$

The PDA algorithm steps are summarized in Figure 4-6:

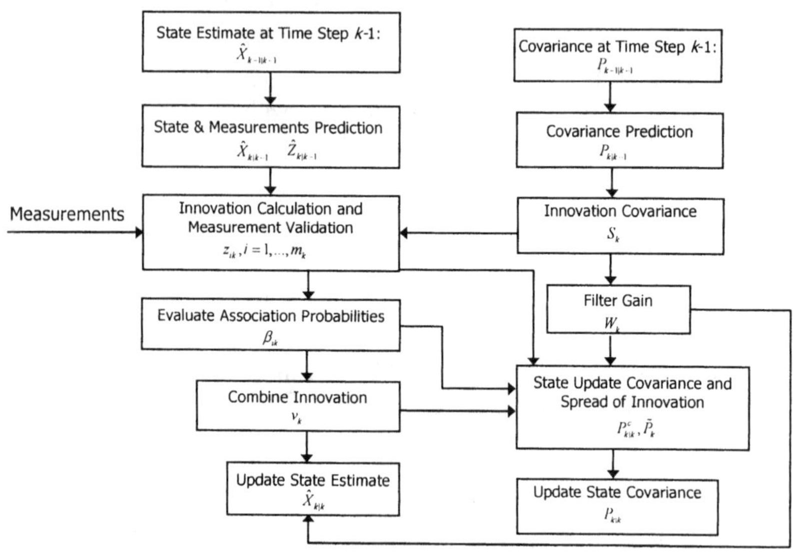

Figure 4-6: PDA algorithm flow (based on Chang (2005))

The PDA is an improvement over NN by not discarding possible target-originated observations. But in a low SNR environment, the target-originated observations may fall outside the validation gate or clutters closer to the target are likely to yield higher association probabilities.

4.6 MULTISENSOR MULTITARGET TRACKING (IN CLUTTER)

This section presents the Joint Probabilistic Data Association (JPDA) (Bar-Shalom, 1995) and Multiple-Hypothesis Tracking (MHT) (Reid, 1979) for data association in environments consisting of multiple sensors and a fixed number of targets.

4.6.1 Joint Probabilistic Data Association (JPDA)

The JPDA is an extension of the PDA to the case where there are a fixed number of targets in clutter. Each target can have different dynamics with different probabilities of detection. As shown in Figure 4-4, if there are several targets in the same region, a measurement from one target can fall in the validation gate of a neighboring target. This all-neighbors approach, like PDA, incorporates all observations within a validation region about the predicted track position into the update of that track. Also, a given observation can be used to update multiple tracks. JPDA forms a hypothesis matrix consisting of all feasible joint association events with validated measurements and then computes an association probability of each event jointly across targets. Track update is thus based on a probabilistically determined weighted sum of all observations within the validation region.

To illustrate the JPDA technique, consider the tracking scenario (Blackman and Popoli, 1999) presented in Figure 4-7, where two predicted positions P1 and P2 of tracks T1 and T2, respectively, are shown along with their gates and the observations that fall within the gates.

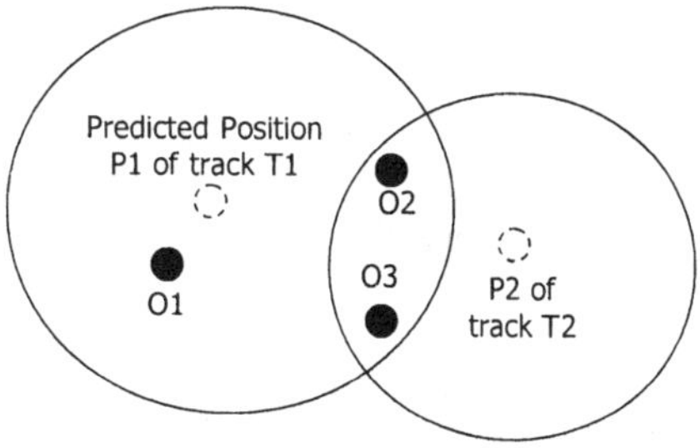

Figure 4-7: Example illustration of the JPDA

There are altogether ten hypotheses corresponding to the ten possible associations of observations to tracks. For example, one hypothesis associates O1 to T1 and O3 to T2, but there is no hypothesis that associates O1 to T2. The hypothesis matrix is shown below in Table 4-1, where

- P_D is the probability of detection (assumed 0.7);
- g_{ij} is the Gaussian likelihood function with the assignment of observation O*j* to track T*i*, which depends on how far the measurement is from the center of the gate and is formulated as

$$g_{ij} = \frac{e^{\frac{d_{ij}^2}{2}}}{(2\pi)^{\frac{M}{2}}\sqrt{|S_{ij}|}}$$

  where $d_{11}^2 = 1, d_{12}^2 = 2, d_{13}^2 = 4, d_{22}^2 = 2.5, d_{23}^2 = 3$, and the residual covariance matrix S_{ij} is ass umed to be 1; and
- β is the association probability that the validation measurement is correct, which perhaps comes from sensor confidence.

Hypothesis	Observations to Tracks 1 & 2	Hypothesis Likelihood Formulae	Likelihood and Normalized Values
H1	0 0	$(1-P_D)^2\beta^3$	$2.4\times10^{-6}(0.011)$
H2	O1 0	$g_{11}P_D(1-P_D)\beta^2$	$1.82\times10^{-5}(0.086)$
H3	O2 0	$g_{12}P_D(1-P_D)\beta^2$	$1.11\times10^{-5}(0.053)$
H4	O3 0	$g_{13}P_D(1-P_D)\beta^2$	$4.1\times10^{-6}(0.019)$
H5	0 O2	$g_{22}P_D(1-P_D)\beta^2$	$8.6\times10^{-6}(0.041)$
H6	O1 O2	$g_{11}g_{22}P_D^2\beta$	$6.47\times10^{-5}(0.306)$
H7	O3 O2	$g_{13}g_{22}P_D^2\beta$	$1.44\times10^{-5}(0.068)$
H8	0 O3	$g_{23}P_D(1-P_D)\beta^2$	$6.7\times10^{-6}(0.032)$
H9	O1 O3	$g_{11}g_{23}P_D^2\beta$	$5.04\times10^{-5}(0.239)$
H10	O2 O3	$g_{12}g_{23}P_D^2\beta$	$3.06\times10^{-5}(0.145)$

Table 4-1: JPDA hypothesis matrix

The probability p_{ij} that observation O*j* should be assigned to track T*i* is computed as below:

$$p_{10} = p(H_1) + p(H_5) + p(H_8) = 0.011 + 0.041 + 0.032 = 0.084$$
$$p_{11} = p(H_2) + p(H_6) + p(H_9) = 0.086 + 0.306 + 0.239 = 0.631$$
$$p_{12} = p(H_3) + p(H_{10}) = 0.053 + 0.145 = 0.198$$
$$p_{13} = p(H_4) + p(H_7) = 0.019 + 0.068 = 0.087$$

$$p_{20} = p(H_1) + p(H_2) + p(H_3) + p(H_4) = 0.011 + 0.086 + 0.053 + 0.019 = 0.169$$
$$p_{21} = 0$$
$$p_{22} = p(H_5) + p(H_6) + p(H_7) = 0.041 + 0.306 + 0.068 = 0.415$$
$$p_{23} = p(H_8) + p(H_9) + p(H_{10}) = 0.032 + 0.239 + 0.145 = 0.416$$

It can be observed that the most heavily weighted hypothesis is the assignment of O1 to T1. As with PDA, the performance of JPDA degrades significantly with the existence of neighboring targets or clutters that cause interference. JPDA requires a systematic way to initiate new tracks.

4.6.2 Multiple-Hypothesis Tracking (MHT)

Both PDA and JPDA algorithms compute the probabilities of association only for the latest set of measurements of various targets. The MHT technique maintains multiple different possible data association solutions as a hypotheses tree, and uses the history of sensor data to eliminate highly unlikely choices by pruning the tree, thus leaving only one best choice. To explain MHT, we denote the history $H(k)$ of a track at time step or scan k as the following chronological sequence of observations by selecting a single sensor report from each scan:

$$\left(y_{i_1}^1, y_{i_2}^2, \ldots, y_{i_k}^k\right)$$

where

- $0 \le i_j \le N_j$, where N_j is the number of observations at scan j; and
- $i_j = 0$ refers to the hypothesis that none of the observations within the gate originated from the target.

Hence a track history is just the hypothesis that the entire sequence of measurements within the history is correct with the above assumptions. A track history $H(k)$ at scan k is recursively obtained from the track history $H(k-1)$ at scan $k-1$ by selecting one of $N_k + 1$ measurements and appending to the measurement sequence specified by $H(k-1)$. Therefore, the total number of track histories at scan k is potentially the $(N_k + 1)$ times the total number of track histories at scan $k-1$. This is the main drawback of MHT because the number of hypotheses in a dense target environment can increase exponentially with each scan. Moreover, a single best estimate for the target tracking solution is not always available in MHT since the data association decision is deferred. The

MHT technique provides a systematic track initiation procedure. The technique is most likely to produce the correct association solution sooner or later as one of its hypotheses. Figure 4-8 shows an example track history.

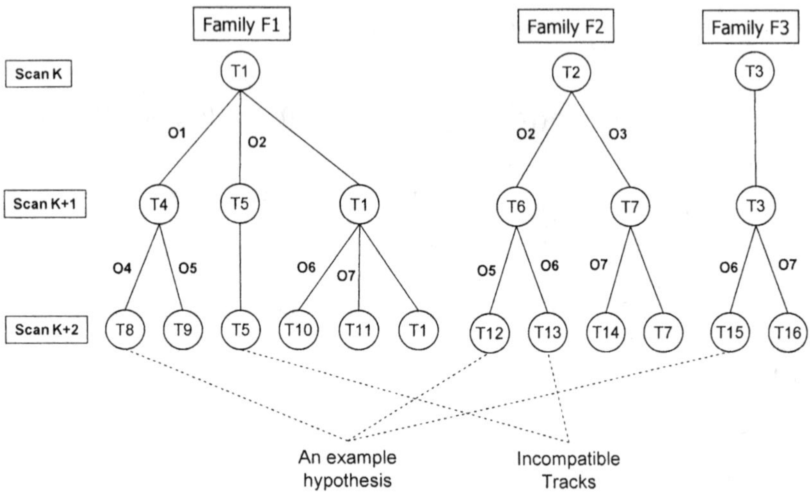

Figure 4-8: MHT tracks and hypotheses formation

As shown in Figure 4-8, there are three trees at scan k. Each tree represents a family of possible tracks corresponding to its branches, but only one branch is a true one. Tracks with new associated observations are considered as a new track and assigned a new track ID (e.g., track T1 at scan k with association observation O1 has ID T4). Tracks with the possibility of no assigned observations (i.e., no suitable observation found in their gates) would evolve according to their dynamics and maintain the old ID (e.g., track T1 at scan k or track T5 at scan $k+1$). Tracks evolved with common observations are incompatible (e.g., tracks T5 and T13 at scan $k+2$). Like JPDA, a hypothesis in MHT is defined as a collection of compatible tracks (e.g., tracks T8, T12, and T15), which can be evaluated by computing its score equal to the sum of its constituent track scores.

At each scan k, track-level pruning is performed before new tracks are created. N-scan pruning is the most widely used approach for this logic. Figure 4-9 (Blackman and Popoli, 1999) shows an instance of the N-scan pruning algorithm in which $N = 2$.

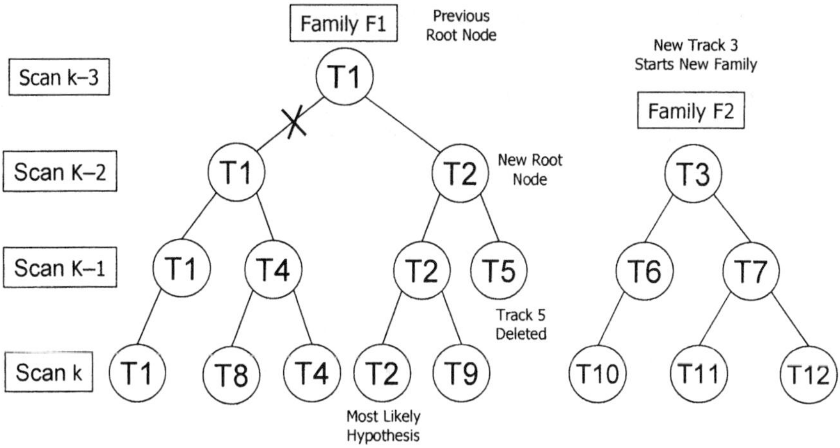

Figure 4-9: Tracks and hypotheses maintenance in MHT

The *N*-scan pruning technique first identifies the most likely hypotheses in scan *k*. Pruning is accomplished by tracing back *N* scans from each track in the most likely hypothesis and making that node the new root. Branches that do not have the same new root will be deleted. Assume, at time step *k*, track T2 is in the most likely hypotheses. Two-scan tracing back makes track 2 the new root so that the left branch in Family 1 is pruned. New families would be created as a new tree each as new targets are detected.

4.7 INTERACTING MULTIPLE MODEL (IMM)

A problem of tracking maneuvering targets with a single filter is that conventional filters, such as the KF, are designed to accommodate the largest potential maneuvering capability, resulting in large tracking errors due to measurement noise. The standard approach to detect maneuvers using increased Kalman gain suffers a delay in adapting maneuvering behavior of the target. An effective approach is to use a bank of filters, run in parallel, based on different motion models to capture different potential target maneuver modes. The Interacting Multiple Model (IMM) method (Bar-Shalom and Li, 1993) is designed to realize such an idea. Tracking target trajectory evolves in IMM according to one of a prespecified number of motion models. An overall estimate is obtained by probabilistically combining individual estimates of filters matching these models.

Suppose m_d represents the number of modes, and the mode in effect between time steps $k-1$ and k is M_k. Let μ_{jk} be the probability that the mode j is in effect between time steps $k-1$ and k, that is,

$$\mu_{jk} = p\left(M_k = j \mid Z_{1:k}\right),\ 1 \le j \le m_d$$

Further, let p_{ij} be the mode transition probability defined as

$$p_{ij} = p\left(M_k = j \mid M_{k-1} = i, Z_{1:k-1}\right)$$

Initial estimations $\hat{X}_{0j(k-1)}$ and $P_{0j(k-1)}$ at time step k for the filter matched to the jth mode are computed as follows using backward transition probabilities $\mu_{i|j(k-1)}$:

$$\hat{X}_{0j(k-1)} = \sum_{i=1}^{m_d} \hat{X}_{i(k-1)} \mu_{i|j(k-1)}$$

$$\begin{aligned}
\mu_{i|j(k-1)} &= p\left(M_{k-1} = i \mid M_k = j, Z_{1:k-1}\right) \\
&= \frac{p\left(M_k = j \mid M_{k-1} = i, Z_{1:k-1}\right) p\left(M_{k-1} = i \mid Z_{1:k-1}\right)}{\sum_{l=1}^{m_d} p\left(M_k = j \mid M_{k-1} = l, Z_{1:k-1}\right) p\left(M_{k-1} = l, Z_{1:k-1}\right)} \\
&= \frac{p_{ij} \mu_{i(k-1)}}{\sum_{l=1}^{m_d} p_{lj} \mu_{l(k-1)}}
\end{aligned}$$

$$P_{0j(k-1)} = \sum_{i=1}^{m_d} \mu_{i|j(k-1)} \left\{ P_{i(k-1)} + \left(\hat{X}_{ik} - \hat{X}_{0j(k-1)}\right)\left(\hat{X}_{ik} - \hat{X}_{0j(k-1)}\right)^T \right\}$$

The mode is updated as follows:

$$\begin{aligned}
\mu_{jk} &= p\left(M_k = j \mid Z_{1:k}\right) \\
&= p\left(M_k = j \mid Z_k Z_{1:k-1}\right) \\
&= \frac{p\left(Z_k \mid M_k = j, Z_{1:k-1}\right) p\left(M_k = j \mid Z_{1:k-1}\right)}{\sum_{l=1}^{m_d} p\left(Z_k \mid M_{k-1} = l, Z_{1:k-1}\right) p\left(M_{k-1} = l \mid Z_{1:k-1}\right)} \\
&= \frac{p\left(Z_k \mid M_k = j, Z_{1:k-1}\right) \sum_{l=1}^{m_d} p\left(M_k = j \mid M_{k-1} = l, Z_{1:k-1}\right) p\left(M_{k-1} = l, Z_{1:k-1}\right)}{\sum_{l=1}^{m_d} \left(\sum_{i=1}^{m_d} p\left(Z_k \mid M_k = i, Z_{1:k-1}\right) p\left(M_k = i \mid M_{k-1} = l, Z_{1:k-1}\right) \right) p\left(M_{k-1} = l, Z_{1:k-1}\right)} \\
&= \frac{L_{jk} \sum_{l=1}^{m_d} p_{lj} \mu_{l(k-1)}}{\sum_{l=1}^{m_d} \left(\sum_{i=1}^{m_d} L_{ik} p_{li} \right) \mu_{l(k-1)}}
\end{aligned}$$

where the likelihood function L_{jk} for mode update is evaluated using a Gaussian uniform mixture as given below:

$$\begin{aligned} L_{jk} &= p\left(Z_k \mid M_k = j, Z_{1:k-1}\right) \\ &= V_k^{-m_k}\left(1-p_D\right)V_k^{1-m_k}\frac{p_D}{m_k}\sum_{m=1}^{m_k} e_j(m) \\ &= \frac{p_D V_k^{1-m_k}}{m_k}\left(b+\sum_{m=1}^{m_k} e_j(m)\right) \end{aligned}$$

Finally, the overall estimates for the mean and covariance matrix are obtained by probabilistically combining individual estimates of filters:

$$\hat{X}_k = \sum_{j=1}^{m_d} \mu_{jk}\hat{X}_{jk}$$

$$P_k = \sum_{j=1}^{m_d} \mu_{jk}\left(P_{jk} + \left(\hat{X}_{jk} - \hat{X}_k\right)\left(\hat{X}_{jk} - \hat{X}_k\right)^T\right)$$

The IMM can be used in conjunction with PDAF or JPDAF.

4.8 CRAMER-RAO LOWER BOUND (CRLB)

A performance bound quantifies the theoretically best possible performance of an algorithm. The Cramer-Rao Lower Bound (CRLB) can tell us whether an estimator achieves the lower bound. If an unbiased estimator is found that achieves the CRLB, then we know that we have found one with minimum variance. CRLB can also be used to rule out impossible estimators.

Consider the measurement function $Z = h(X;\theta)$ in the target state estimation. We estimate the missing parameter θ from repeated n observations of the random variable Z (i.e., $1 \times n$ observation vector $\mathbf{Z} = \langle Z_1, \ldots, Z_n \rangle$) via some estimator $e(\mathbf{Z})$ of θ. Such an estimator of θ is *unbiased* if $E\left[e(\mathbf{Z})\right]$ is equal to the true value of θ. So if there is more than one estimator of θ then one with the lowest variance is the most desirable. CRLB is a lower bound on the variance of any unbiased state estimator, defined in terms of the likelihood function.

If $\hat{\theta} = e(\mathbf{Z})$ is an unbiased estimator of θ then the CRLB expression for scalar θ is

$$\operatorname{var}\left(e(\mathbf{Z})\right) = \sigma_{\hat{\theta}}^2 \geq J^{-1}(\theta)\Big|_{\substack{\theta = true \\ value}}$$

$$J(\theta) = E\left[\left(\frac{\partial \ln p(\mathbf{Z};\theta)}{\partial \theta}\right)^2\right] = -E\left[\frac{\partial^2 \ln p(\mathbf{Z};\theta)}{\partial \theta^2}\right]$$

where p is the probability density function of $\mathbf{Z}$ and J is known as the *Fisher information*. For the estimation of an $m \times 1$ vector parameter Θ and the $m \times n$ observation matrix $\mathbf{Z}$, the CRLB expression is

$$\operatorname{var}(e(\mathbf{Z})) = \sigma_{\hat{\Theta}}^{2} \geq J^{-1}(\Theta)\Big|_{\substack{\Theta = true \\ value}}$$

$$J(\Theta) = E\left[\nabla_{\mathbf{X}}(\ln p(\mathbf{Z};\Theta))\nabla_{\mathbf{X}}(\ln p(\mathbf{Z};\Theta))^{T}\right]$$

$$= -E\left[\nabla_{\mathbf{X}}\left((\nabla_{\mathbf{X}} \ln p(\mathbf{Z};\Theta))^{T}\right)\right]$$

Here is a sequence of steps for computing CRLB:

- Write log likelihood function $\ln p(\mathbf{Z};\theta)$ as a function of θ (Θ for vectors).
- Fix X and take partial derivative $\frac{\partial^2}{\partial\theta^2}\ln p(\mathbf{Z};\theta)$.
- If the result still depends on $\mathbf{Z}$ then fix θ and take the expected value with respect to $\mathbf{Z}$.
- If the result still depends on θ then evaluate at each desired specific value of θ.
- Discard negligible terms.
- Negate and form the reciprocal.

Example

Consider the following sinusoid signal model:

$$Z_k = A\cos(2\pi f_0 k + \theta) + W_k$$

where A and $f_0\ (0 < f_0 < 1/2)$ are known, and W_k is zero-mean white Gaussian noise with variance σ^2. We want to estimate the phase θ. For given data, $\mathbf{Z} = \{Z_0, ..., Z_{N-1}\}$, the log likelihood function is

$$p(\mathbf{Z};\theta) = \frac{1}{(2\pi\sigma^2)^{\frac{N}{2}}} e^{\frac{-\sum_{k=0}^{N-1}(Z_k - A\cos(2\pi f_0 k + \theta))^2}{2\sigma^2}}$$

The derivation according to the algorithm above is given below:

$$\frac{\partial}{\partial\theta}\ln p(\mathbf{Z};\theta) = -\frac{A}{\sigma^2}\sum_{k=0}^{N-1}\left(Z_k \sin(2\pi f_0 k + \theta) - \frac{A}{2}\sin(4\pi f_0 k + 2\theta)\right)$$

$$\frac{\partial^2}{\partial\theta^2}\ln p(\mathbf{Z};\theta) = -\frac{A}{\sigma^2}\sum_{k=0}^{N-1}\left(Z_k \cos(2\pi f_0 k + \theta) - \frac{A}{2}\cos(4\pi f_0 k + 2\theta)\right)$$

$$-E\left[\frac{\partial^2}{\partial\theta^2}\ln p(\mathbf{Z};\theta)\right]=\frac{A}{\sigma^2}\sum_{k=0}^{N-1}\left(E\left[Z_k\right]\cos\left(2\pi f_0 k+\theta\right)-\frac{A}{2}\cos\left(4\pi f_0 k+2\theta\right)\right)$$

$$=\frac{A}{\sigma^2}\sum_{k=0}^{N-1}\left(A\cos\left(2\pi f_0 k+\theta\right)\cos\left(2\pi f_0 k+\theta\right)-\frac{A}{2}\cos\left(4\pi f_0 k+2\theta\right)\right)$$

$$=\frac{A^2}{2\sigma^2}\left(\sum_{k=0}^{N-1}1-\sum_{k=0}^{N-1}\cos\left(4\pi f_0 k+2\theta\right)\right)\approx\frac{NA^2}{2\sigma^2}=N\times SNR$$

where *SNR* is signal-to-noise ratio. The CRLB for θ is

$$\sigma_{\hat{\theta}}^2\geq\frac{1}{N\times SNR}$$

Therefore, doubling data size or SNR halves the CRLB.

4.9 FURTHER READING

A comprehensive book on target tracking is Blackman and Popoli (1999). A good edited volume containing some seminal papers and discussions on various aspects of target tracking is Hall and Llinas (2001). Details on PDA and related works can be found in Bar-Shalom and Fortmann (1988) and Bar-Shalom and Li (1995). Stone et al. (1999) is another good book on tracking.

Chapter 5

Target Classification and Aggregation

A target classifier categorizes established tracks into known target categories and subcategories (e.g., armored vehicle and missile launcher as ground targets, fighter aircraft and commercial airline as air-based targets). Results from a target identifier help disambiguate observations to track associations, and provide input for higher-level situational assessment. Results can also be used to suppress the output of unwanted tracks. Any type of kinetic and contextual information can be used to identify a target, and the challenge is to aggregate diverse sources of uncertain information. The first half of this chapter illustrates various approaches to target classification.

The second half of this chapter introduces the spatiotemporal clustering concept to aggregate units that are moving together to create, for example, an ambush-like situation. Only the positional attribute of each tracked unit is made use of to extract such clusters. No other attribute or contextual information is necessary in the spatiotemporal algorithms presented in the chapter. Moreover, the relationship among the tracked units within an identified cluster is merely based on their correlated movements

Establishing relationships for target classification and aggregation is certainly high-level fusion, but falls short of establishing full complex relationships among targets. For these reasons, I term the target classification and aggregation approaches in this chapter as Level 1½ fusion.

5.1 TARGET CLASSIFICATION

We illustrate various approaches to target classification in the context of an example radar surveillance scenario similar to the one in Leung and Wu (2000).

5.1.1 Example Surveillance Scenario

The following are four potential features for target classification in the feature space X:

- X_1 = The target identity (target-ID or no-target-ID)
- X_2 = The estimated speed (stable or unstable)
- X_3 = The elevation angle (constant or nonconstant)
- X_4 = The estimated acceleration (maneuver or nonmaneuver)

Note that measurements required to derive the categorical values of the variables X_1 and X_3 can be obtained, respectively, from IFF (Identification Friend or Foe) system and radar sensor returns. Similarly measurements required for X_2 and X_4 are generated by the underlying tracker. The categorical values of X_4 can be derived from the fact that an aircraft must accelerate to perform a maneuver. Other categorical values, such as stability and constant angularity, can be derived by inspecting consecutive observations.

The following five are a mutually exclusive and exhaustive set Ω of target classes considered for the scenario:

- H_1 = Friendly commercial aircraft
- H_2 = Friendly military fighter aircraft
- H_3 = Hostile commercial aircraft
- H_4 = Hostile military fighter aircraft
- H_5 = False target

The target identification feature X_1 can be used to classify target tracks that belong to $\{H_1, H_2\}$ or $\{H_3, H_4, H_5\}$. In other words, if the track has a valid identification then it is for a friendly commercial or military aircraft. If the track does not have a valid identification then it is for a hostile commercial or military aircraft or a false target. Similarly, each of X_2 and X_3 classifies a track into $\{H_1, H_2, H_3, H_4\}$ or $\{H_5\}$, and X_4 classifies a track into $\{H_2, H_4, H_5\}$ or $\{H_1, H_3\}$.

The categorical values of the variables X_2, X_3, and X_4 at a time point t are derived, respectively, from observations of continuous variables Z_2, Z_3, and Z_4 over certain time intervals preceding t, rather than their instantaneous observations at t. For example, whether the elevation angle for a target at time t is constant or not is determined from measurements of the elevation angle variable Z_3 at both $t-1$ and t. Similarly, whether the speed of a target is stable or not can be determined from consecutive measurements of speed variable Z_2. This kind of conversion facilitates applications of Naïve Bayesian Classifier and Dempster-Shafer theory to target classification. Later in the application of fuzzy logic to target classification, we will make use of instantaneous observations Z_2, Z_3, and Z_4 to determine their categorical values, for example, "fast," "medium," and "slow" for the speed variable Z_2. If we had worked with these categorical values for X_2, instead of "stable" and "unstable," then the value "fast" would have classified a track into $\{H_2, H_4\}$.

5.1.2 Naïve Bayesian Classifier (NBC) for Target Classification

Naïve Bayesian Classifiers (NBCs) (Duda and Hart, 1973) are a type of BN that is used to separate a set of unlabeled data into a set of classes $C=\{C_1,\ldots,C_n\}$. NBCs are networks in which there is exactly one parent for the whole network, as shown in Figure 5-1.

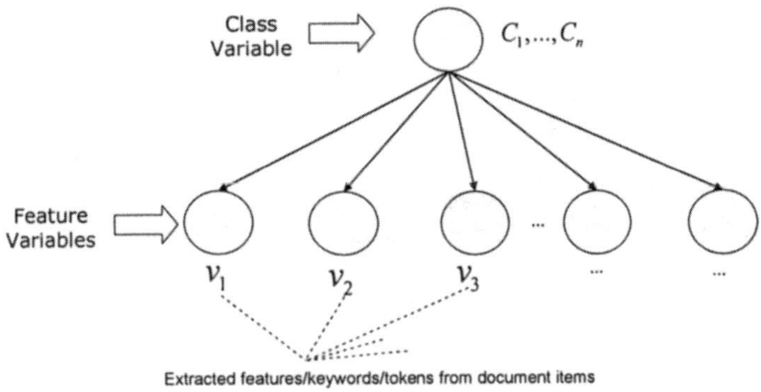

Figure 5-1: Generic structure of an NBC

An NBC, and classifiers in general, can be viewed as computing or learning a discriminant function

$$f: D \rightarrow C$$

where D is an unlabeled data or instance space (e.g., documents, articles, HUMINT) that needs to be classified and, for $d \in D$, $f(d)=C_k$ means d is in class C_k. In an NBC, each child node v_i in the network corresponds to an attribute or feature. For text based document indexing, these attributes usually indicate the presence or absence of a word in a sample of text. The parent node determines the output of the discriminant function, which is the class to which the example belongs (e.g., the five hypotheses in the example scenario).

Let d be a document that we want to classify and the given set of classes is $C=\{C_1,\ldots,C_n\}$. We want to compute $p(C_i \mid d)$, for every i.

$$p(C_i \mid d)=\frac{p(C_i)\,p(d \mid C_i)}{p(d)}=\frac{p(C_i)\prod_j p(v_j \mid C_i)}{\sum_{k=1}^{n} p(C_k)\prod_j p(v_j \mid C_k)}$$

where v_js are tokens in document d. The product of conditional probabilities comes from the assumption that tokens in a document are independent given the document class. Note that the conditional independence assumption of features does not hold in most cases. For example, word cooccurrence is a commonly used feature for text classification.

We don't need the estimated posterior $p(C_i \mid d)$ to be correct. Instead, we only need

$$\arg\max_{C_i} p(C_i \mid d) = \arg\max_{C_i} p(C_i)\prod_j p(v_j \mid C_i)$$

The score for each class can be expressed in the following tractable form for analytical purposes:

$$\log(p(C_i)) + \sum_j \log(p(v_j \mid C_i))$$

The score is not a probability value, but sufficient for the purpose of determining the most probable class. It reduces the round-off errors due to a product of small fractions caused by a large number of tokens.

An NBC for our target classification example is shown in Figure 5-2. The mutually exclusive and exhaustive set of hypotheses is the classification class and each child node of the network corresponds to a target attribute.

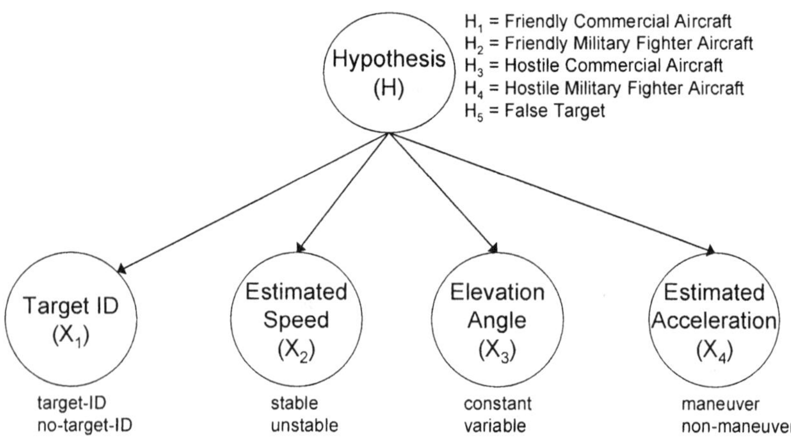

Figure 5-2: NBC for target classification

Now, observations of some subset X of the X_is (say, $X = \{X_1, X_2, X_4\}$) with certainty provide the following posteriors of the hypotheses:

$$p(H_i \mid X) = p(H_i) \prod_{j=1,2,4} p(X_j \mid H_i),\ i = 1,2,3,4$$

One would of course require the probabilities $p(X_j \mid H_i)$ to compute the above posteriors, but they can either be obtained from domain experts or learned from observations stored in a structured relational database. The decision regarding the identification of the most likely target can then use

$$\arg\max_{H_i} p(H_i \mid X) = \arg\max_{H_i} p(H_i) \prod_{j=1,2,4} p(X_j \mid H_i)$$

But if the observations generate likelihoods $p(Z_i \mid X_i)$ then the above decision criteria become

$$\arg\max_{H_i} p(H_i \mid Z) = \arg\max_{H_i} p(H_i) \prod_{j=1,2,4} \sum_{k=1,2} p(Z_j \mid X_j = x_{jk}) p(X_j = x_{jk} \mid H_i)$$

where x_{jk}s are states of the feature X_j (e.g., *stable* and *unstable* for the feature *Estimated Speed*).

Not all intelligence data for target classification are structured; in fact, much of the Human Intelligence (HUMINT) is stored in narrative, rather than structured, form. This information source is huge and complex, and a human analyst's manual interpretation can easily overlook the true semantics of a document's item resulting from misclassification. For this reason, it is necessary to apply a document indexing strategy to label such huge information sources to help analysts classify them accurately. A standard approach and reliable document indexing technique employs NBCs. Latent Semantic Indexing (LSI) (Deerwester et al., 1990; Dumais et al., 1988) is another potential approach to document classification, but its holistic approach to document clustering is a big bottleneck in applications where input term-and-document matrices are large. Chapter 10 describes how to automatically learn NBCs from a set of documents.

5.1.3 Rule-Based Expert Systems for Target Classification

This approach describes the classification options from an application as a set of possible outcomes, termed "hypotheses." In this approach, the following if-then syntax is adopted for rules in the knowledge base of an expert system:

IF *Events*
THEN *Hypothesis* (*D*)

The above rule is interpreted as "If *Events* occur, then *Hypothesis* follows with a certain degree of uncertainty *D*." The degree of uncertainty is represented as a probability value from $[0,1]$ or a certainty factor from $[-1,1]$. Bayesian inference, for example, is applied to combine incoming evidence as probability values for target classification. The inference scheme is similar to NBC except: 1) the scheme can be applied on any number of rules (as opposed to on a complete Naïve Bayesian network, 2) evidence is processed incrementally, and 3) evidence can be soft.

Bayesian inference requires an initial (prior) probability for each hypothesis in the problem space. The inference scheme then updates probabilities using evidence. Each piece of evidence updates the probability of a set of hypotheses calculated via Bayesian rule, which is defined in the following:

$$p(A \mid B) = \frac{p(B \mid A) p(A)}{p(B)}$$

where A and B are events that are not necessarily mutually exclusive, $p(A \mid B)$ is the conditional probability of event A occurring given that event B has occurred, $p(B \mid A)$ is the conditional probability of event B occurring given that event A has occurred, $p(A)$ is the probability of event A occurring, and $p(B)$ is the probability of event B occurring. In general, if there are m mutually exclusive and exhaustive hypotheses $H_1,...,H_m$ (that is, $\sum_{i=1}^{m} p(H_i) = 1$) and n possible events $X_1,...,X_n$ that can occur, then the probability of a hypothesis given some evidence is computed as follows:

$$p(H_i \mid X_j) = \frac{p(X_j \mid H_i) p(H_i)}{\sum_{k=1}^{m} p(X_j \mid H_k) p(H_k)}$$

The Bayesian inference mechanism is illustrated in the context of the following example derived from the example surveillance scenario. Consider a knowledge base consisting of the following three rules describing the chances of a friendly target based on a variety of evidence:

IF *Target Identifier* THEN *Friendly* (0.9)

IF *Stable Speed* THEN *Friendly* (0.7)

IF *Constant Elevation Angle* THEN *Friendly* (0.8)

The two mutually exclusive and exhaustive hypotheses are:

$H_1 = Friendly$

$H_2 = Hostile$

Three independent events on which evidence can be gathered are:

$X_1 = Target\ Identifier$

$X_2 = Stable\ Speed$

$X_3 = Constant\ Elevation\ Angle$

Below, we illustrate how posterior probabilities of the hypotheses are updated as evidence on speed and elevation angle is gathered.

In general, the flow of the likelihood-based computation of the Bayesian classifier is shown in Figure 5-3.

Initial State

Conditional probabilities $p(H_i \mid X_j)$ for the hypotheses are inferred as follows from the three rules of the knowledge base and $p(H_1 \mid X_j) + p(H_2 \mid X_j) = 1$:

$p(H_1 \mid X_1) = p(Friendly \mid Target\ Identifier) = 0.9$

$p(H_2 \mid X_1) = p(Hostile \mid Target\ Identifier) = 0.1$

$p(H_1 \mid X_2) = p(Friendly \mid Stable\ Speed) = 0.7$

$p(H_2 \mid X_2) = p(Hostile \mid Stable\ Speed) = 0.3$

$p(H_1 \mid X_3) = p(Friendly \mid Constant\ Elevation\ Angle) = 0.8$

$p(H_2 \mid X_3) = p(Hostile \mid Constant\ Elevation\ Angle) = 0.2$

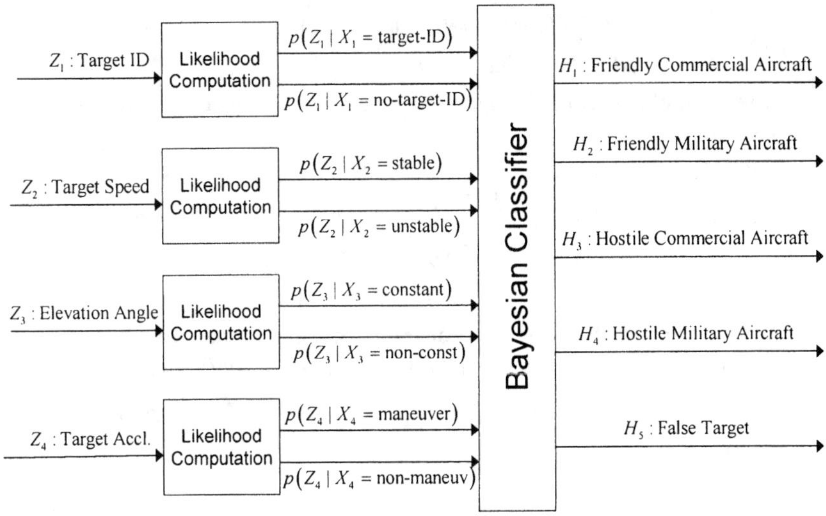

Figure 5-3: Likelihood-based computation flow in a Bayesian classifier

Given prior probabilities $p(E_j)$ and $p(H_i)$ are listed below:

$$p(X_1) = p(Target\ Identifier) = 0.01 \qquad p(H_1) = p(Friendly) = 0.2$$

$$p(X_2) = p(Stable\ Speed) = 0.2 \qquad p(H_2) = p(Hostile) = 0.8$$

$$p(X_3) = p(Constant\ Elevation\ Angle) = 0.1$$

The following conditional probabilities $p(X_j \mid H_i)$ for evidence are computed via the application of Bayesian rule:

$$p(X_1 \mid H_1) = p(Has\ TID \mid Friendly) = 0.045$$

$$p(X_1 \mid H_2) = p(Has\ TID \mid Hostile) = 0.001$$

$$p(X_2 \mid H_1) = p(Stable\ Speed \mid Friendly) = 0.7$$

$$p(X_2 \mid H_2) = p(Stable\ Speed \mid Hostile) = 0.075$$

$$p(X_3 \mid H_1) = p(Constant\ Elevation\ Angle \mid Friendly) = 0.4$$

$$p(X_3 \mid H_2) = p(Constant\ Elevation\ Angle \mid Hostile) = 0.025$$

Evidence of Speed

Assume that we first receive confirming evidence Z_2 on X_2 (i.e., Stable Speed). Then the probability $p(Friendly \mid Stable\ Speed)$ directly provides the posterior probability 0.7 for the target being friendly, which is a significant increase from the prior probability $p(Friendly) = 0.2$. If we are not completely sure about stable speed, then soft evidence e can be encoded as the following likelihood:

$$p(Z_2 \mid X_2) = \begin{bmatrix} p(Z_2 \mid Stable\,Speed) \\ p(Z_2 \mid \neg Stable\,Speed) \end{bmatrix} = \begin{bmatrix} 0.95 \\ 0.05 \end{bmatrix}$$

The posterior probability of X_2 upon receiving e is computed as shown below (α is the normalizing constant):

$$\begin{bmatrix} p(Z_2 \mid Stable\,Speed) \\ p(Z_2 \mid \neg Stable\,Speed) \end{bmatrix} = \alpha \times \begin{bmatrix} p(Z_2 \mid Stable\,Speed) \times p(Stable\,Speed) \\ p(Z_2 \mid \neg Stable\,Speed) \times p(\neg Stable\,Speed) \end{bmatrix}$$

$$= \alpha \times \begin{bmatrix} 0.95 \\ 0.05 \end{bmatrix} \times \begin{bmatrix} 0.2 \\ 0.8 \end{bmatrix} = \begin{bmatrix} 0.83 \\ 0.17 \end{bmatrix}$$

We can then compute the posterior probability of the hypotheses as follows:

$$p(Friendly \mid Z_2) = p(Friendly \mid Stable\,Speed) \times p(Stable\,Speed \mid Z_2) + p(Friendly \mid \neg Stable\,Speed) \times p(\neg Stable\,Speed \mid Z_2)$$

But we have

$$\begin{aligned} &p(Friendly \mid \neg Stable\,Speed) \\ &= \frac{p(\neg Stable\,Speed \mid Friendly) \times p(Friendly)}{p(\neg Stable\,Speed)} \\ &= \frac{(1 - p(Stable\,Speed \mid Friendly)) \times p(Friendly)}{(1 - p(Stable\,Speed))} \\ &= \frac{(1-0.7) \times 0.2}{1-0.2} = 0.07 \end{aligned}$$

Therefore,

$$p(Friendly \mid Z_2) = 0.7 \times 0.83 + 0.07 \times 0.17 = 0.59$$

$$p(Hostile \mid Z_2) = 0.41$$

Note that the probability of the target being friendly has increased significantly from the earlier prior value 0.2, but not as much as to 0.7 when evidence on X_2 was certain.

Evidence of Elevation Angle

In addition to the confirming evidence on X_2, suppose now we observe confirming evidence on X_3 (that is, constant elevation angle). The posterior probabilities are computed using the following formulae:

$$p(H_i \mid X_2 X_3) = \frac{p(X_2 X_3 \mid H_i) \times p(H_i)}{\sum_{j=1}^{2} p(X_2 X_3 \mid H_j) \times p(H_j)}, \qquad i = 1, 2$$

Since X_2 and X_3 are independent,

$$p(X_2 X_3 \mid H_i) = p(X_2 \mid H_i) \times p(X_3 \mid H_i)$$

Therefore,

$$p\left(Friendly \middle| Stable\,Speed, Constant\ Elevation\ Angle\right)$$
$$= \frac{0.7\times0.4\times0.2}{0.7\times0.4\times0.2+0.075\times0.025\times0.8} = 0.97$$
$$p\left(Hostile \middle| Stable\,Speed, Constant\ Elevation\ Angle\right)$$
$$= \frac{0.075\times0.025\times0.8}{0.7\times0.4\times0.2+0.075\times0.025\times0.8} = 0.03$$

Note that the posterior probability of the target being friendly has increased further due to evidence of both stable speed and constant elevation angle. This process of probability revision continues as evidence arrives.

When the requisite initial assumptions (for example, prior probabilities, event independence) are fairly accurate, the Bayesian approach typically provides optimum results and is difficult to beat. However, there is always some question as to how accurate our a priori assumptions are for any given situation we are modeling. Under such circumstances, where a priori assumptions are inaccurate, Bayesian methods may perform poorly. The Dempster-Shafer theory was specifically developed to mitigate these weaknesses via the modeling of ignorance as uncommitted belief.

5.1.4 Dempster-Shafer Theory for Target Classification

The Dempster-Shafer theory of belief function generates mass distribution based on source reliabilities. Table 5-1 below lists the reliabilities of the sources that provide evidence on target features based on supporting observations.

Target Feature	Measurements Source	Reliability
X_1: Target identification	IFF system	90%
X_2: Estimated speed	Radar sensor returns	70%
X_3: Elevation angle	Radar sensor returns	80%
X_4: Estimated acceleration	Target tracker	90%

Table 5-1: Reliabilities of sources providing evidence on target features

Thus, for example, if a source determines that the target has Target Identification (TID), and that the source is 90% reliable, and that a target with TID classifies target tracks that belong to $\{H_1, H_2\}$, the focal elements are $\{H_1, H_2\}$ and Ω, and $m_1(\{H_1, H_2\}) = 0.9$. We know nothing about the remaining probability so it is allocated to the whole frame of discernment as $m_1(\Omega) = 0.1$. If a target with TID classifies at a finer granularity (as opposed to $\{H_1, H_2\}$), for example, 30% of target tracks that belong to H_1 and the rest 70% of target tracks that belong to H_2, the focal elements are $\{H_1\}$ and $\{H_2\}$, and the mass distribution is at finer granularity as

$$m_1(\{H_1\}) = 0.9 \times 0.3 = 0.27,\ m_1(\{H_2\}) = 0.9 \times 0.7 = 0.63,\ m_1(\Omega) = 0.1$$

Figure 5-4 is the flow of computation in a classifier based on the Dempster-Shafer (D-S) theory of belief functions.

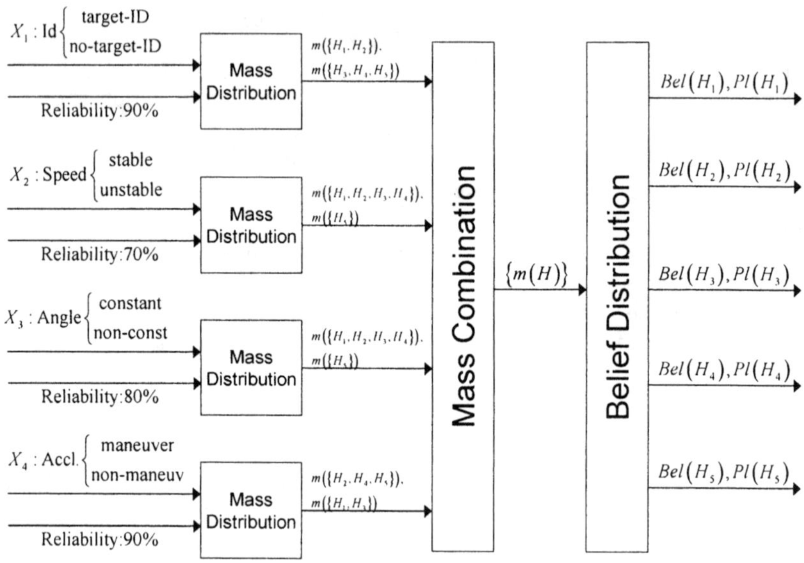

Figure 5-4: Flow of computation in a D-S theory-based target classifier

An input to the classifier generates a mass distribution, which is combined with the previous mass distribution via Dempster's combination rule. Beliefs and plausibilities on focal elements are then derived from the mass distribution.

Suppose sources provide evidence that a target has TID and its speed is stable. The new evidence provides the focal elements $\{H_1, H_2, H_3, H_4\}$ and Ω, with $m_2(\{H_1, H_2, H_3, H_4\}) = 0.7$. The remaining probability, as before, is allocated to the whole frame of discernment as $m_2(\Omega) = 0.3$. Because the evidence is independent of each other, Dempster's rule can then be used to combine the masses as follows:

	$m_2(\{H_1, H_2, H_3, H_4\}) = 0.7$	$m_2(\Omega) = 0.3$
$m_1(\{H_1, H_2\}) = 0.9$	$m_{1,2}(\{H_1, H_2\}) = 0.63$	$m_{1,2}(\{H_1, H_2\}) = 0.27$
$m_1(\Omega) = 0.1$	$m_{1,2}(\{H_1, H_2, H_3, H_4\}) = 0.07$	$m_{1,2}(\Omega) = 0.03$

The basic probability assignments m_1 and m_2 are different but consistent, and therefore, for example, the degrees of belief in both $\{H_1, H_2\}$ and $\{H_1, H_2, H_3, H_4\}$ being true is the product of $m_1(\{H_1, H_2\})$ and $m_2(\{H_1, H_2, H_3, H_4\})$, that is, 0.63. The revised focal elements and their beliefs and plausibilities are shown in the following table:

Focal Element (A)	$Bel(A)$	$Pl(A)$
$\{H_1, H_2\}$	0.90	1.0
$\{H_1, H_2, H_3, H_4\}$	0.97	1.0
Ω	1.0	1.0

Finally, the nonconstant elevation angle suggests the focal elements are $\{H_5\}$ and Ω, so that $m_3(\{H_5\}) = 0.8$, $m_3(\Omega) = 0.2$. Dempster's rule of combination applies as shown in the table below, but with one modification. When the evidence is inconsistent, their products of masses are assigned to a single measure of inconsistency, say k.

	$m_3(\{H_5\}) = 0.8$	$m_3(\Omega) = 0.2$
$m_{1,2}(\{H_1, H_2\}) = 0.90$	$k = 0.72$	$m_{1,2,3}(\{H_1, H_2\}) = 0.18$
$m_{1,2}(\{H_1, H_2, H_3, H_4\}) = 0.07$	$k = 0.056$	$m_{1,2,3}(\{H_1, H_2, H_3, H_4\}) = 0.014$
$m_{1,2}(\Omega) = 0.03$	$m_{1,2,3}(\{H_5\}) = 0.024$	$m_{1,2,3}(\Omega) = 0.006$

The total mass of evidence assigned to inconsistency k is 0.72 + 0.056 = 0.776. The normalizing factor is $1 - k = 0.224$. The resulting masses of evidence are as follows:

$$m_{1,2,3}(\{H_1, H_2\}) = 0.18 / 0.224 = 0.8$$
$$m_{1,2,3}(\{H_1, H_2, H_3, H_4\}) = 0.014 / 0.224 = 0.06$$
$$m_{1,2,3}(\{H_5\}) = 0.024 / 0.224 = 0.11$$
$$m_{1,2,3}(\Omega) = 0.006 / 0.224 = 0.03$$

The revised focal elements and their beliefs and plausibilities are shown in the following table:

Focal Element (A)	$Bel(A)$	$Pl(A)$
$\{H_1, H_2\}$	0.8	0.89
$\{H_1, H_2, H_3, H_4\}$	0.86	0.89
$\{H_5\}$	0.11	0.14
Ω	1.0	1.0

We consider two examples to illustrate two special cases for evidence aggregation. Consider the case when the set of focal elements of the basic probability distribution m_2 is exactly the same as m_1 (e.g., evidence of X_2 and X_3). The evidence combination table is shown below.

$H = \{H_1, H_2, H_3, H_4\}$	$m_2(H) = 0.8$	$m_2(\Omega) = 0.2$
$m_1(H) = 0.7$	$m_{1,2}(H) = 0.56$	$m_{1,2}(H) = 0.14$
$m_1(\Omega) = 0.3$	$m_{1,2}(H) = 0.24$	$m_{1,2}(\Omega) = 0.06$

Now,

$$Bel(\{H_1, H_2, H_3, H_4\}) = 0.56 + 0.14 + 0.24 = 0.94$$
$$= 0.7 + 0.8 - 0.7 \times 0.8$$

In general, when two mass distributions m_1 and m_2 agree on focal elements, the combined degree of belief on a common focal element is $p_1 + p_2 - p_1 \times p_2$, where p_1 and p_2 are evidences on the focal element by the two distributions. This formula coincides with the noisy-or technique in BNs (discussed in Chapter 6) for combining probabilities of variables that have certain properties.

As opposed to agreeing on focal elements, if m_2 is contradictory to m_1 then an example evidence combination is shown in the table below.

$H=\{H_1,H_2,H_3,H_4\}$	$m_2(H_5)=0.6$	$m_2(\Omega)=0.4$
$m_1(H)=0.7$	$k=0.42$	$m_{1,2}(H)=0.28$
$m_1(\Omega)=0.3$	$m_2(H_5)=0.18$	$m_{1,2}(\Omega)=0.12$

Now,

$$Bel(\{H_1,H_2,H_3,H_4\}) = 0.28/(1-0.42)$$
$$= 0.7(1-0.6)/(1-0.42)$$

In general, when two mass distributions m_1 and m_2 are contradictory, then the combined degree of belief on the focal element for m_1 is $p_1(1-p_2)/(1-p_1 \times p_2)$ and the combined degree of belief on the focal element for m_2 is $p_2(1-p_1)/(1-p_1 \times p_2)$, where p_1 and p_2 are evidences on the focal element by the two distributions.

5.1.5 Fuzzy Logic for Target Classification

The fuzzy rules for the classification of radar track data are shown in Table 5-2 below:

	Altitude	Speed	Range	Target
Rule 1	medium	fast	near	Fighter Aircraft
Rule 2		medium	far	Passenger Aircraft
Rule 3	very high	very fast		Missile
Rule 4	low	slow		Passenger Aircraft
Rule 5	low	very fast		Missile

Table 5-2: Fuzzy rules

The first row in the table, for example, states the rule

IF *Altitude* is *medium*
AND *Speed* is *fast*
AND *Range* is *near*
THEN *Target* is *Fighter Aircraft*

Figure 5-5 shows the various stages of a classification system for the example problem. Possible inputs to a fuzzy logic-based classification system are the observed altitude, speed, and range of the target we intend to classify. An input is fuzzified in the first step. This means that for each linguistic variable the degree is determined to which the input fulfills it. This is done by evaluating the values of the corresponding membership functions. For example, for the input speed 600 m/s the degree of "medium" is $\mu_{fast}(600)=0.3$. The degree of the rule antecedents is then computed by taking the minimum of all such degrees.

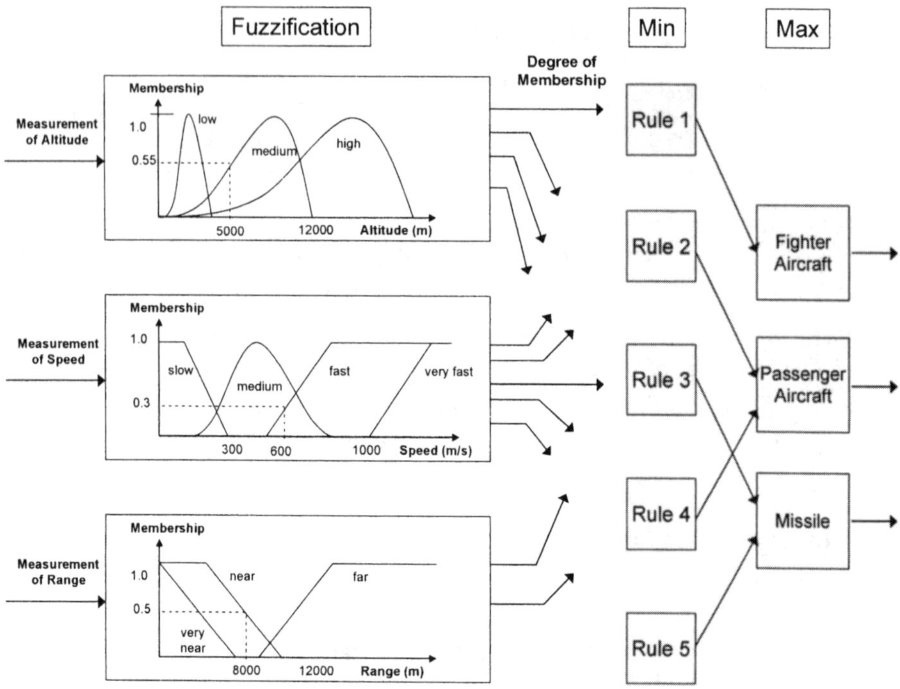

Figure 5-5: Flow of computation in a fuzzy logic-based target classifier

Consider, for example, the first rule. For the input values 5000 m, 600 m/s, and 6000 m, respectively, of the variables *Altitude*, *Speed*, and *Range*, the following degrees are obtained:

$$\mu_{medium}(5000) = 0.55$$

$$\mu_{fast}(600) = 0.3$$

$$\mu_{near}(8000) = 0.5$$

So the degree of rule antecedent is $\min\{0.55, 0.3, 0.5\} = 0.3$, which is the degree with which the rule is fulfilled, i.e., the degree with which the target belongs to the class *Fighter Aircraft*. The degree with which the target belongs to each of the three classes is then computed by taking the maximum of all the rules describing the class as follows:

degree(*Fighter Aircraft*) = max{degree(Rule 1)}
degree(*Passenger Aircraft*) = max{degree(Rule 2), degree(Rule 4)}
degree(*Fighter Aircraft*) = max{degree(Rule 3), degree(Rule 5)}

The output class is then the class corresponding to the maximum of these three degrees.

5.2 TARGETS AGGREGATION

This section introduces the spatiotemporal clustering (STC) concept to determine clusters of units that are moving together to create, for example, an ambush-like situation. We develop various algorithms for extracting spatiotemporal clusters from target tracks.

5.2.1 Spatiotemporal Clustering (STC) Concept

The purpose of STC (Das et al., 2006; Kanjilal et al., 2007) is to identify and track groups of objects in time that may indicate some threat, such as a convoy or a coordinated set of vehicles as part of an ambush operation. This approach is based on the assumption that: 1) in dense urban environments a group of multiple entities following the same path indicate the possibility of a military formation on maneuvers, and 2) multiple formations that appear to be converging to some future location of the blue unit(s) together indicate a strong potential for enemy ambush. The advantage of this technique is that no semantic interpretation of data is required, as only space and time information is used.

In STC, more generally, at least one attribute of the entities must vary with time. In general, STC are temporal extensions of spatial clustering, thus recognizing units that are moving together, where this "togetherness" property is observable only when unit movement is considered over some period of time. We are also interested in recognizing units that follow or parallel each other's course. For example, Figure 5-6 illustrates (in an abstract, but essentially nonurban environment) some cluster patterns of interest in such a military context and also their combination patterns as defined below:

- *Follows*: One cluster of objects traces the same (or similar) spatial route as another cluster at a later time (spatial trajectories are similar but shifted in time). For example, Cluster B follows Cluster A in Figure 5-6.
- *Convergent*: One cluster of objects traces a spatial path convergent with the path of another cluster with respect to a known target (occur at the same time but are shifted in space). For example, Cluster C movement converges with that of Cluster E in Figure 5-6 in that both are moving towards the high-value target.
- *Parallels*: One cluster of objects traces the same (or a similar) spatial pattern but is offset in space (spatial trajectories are similar and occur at the same time but are shifted in space). For example, Cluster D parallels Cluster A in Figure 5-6.

Any implementation of spatiotemporal clustering has to work very closely with the underlying Level 1 fusion engine performing target tracking. The fusion engine provides the necessary tracking information of individual units, which usually is a chronological set of observations. Each observation includes at a minimum the unit's location and the time observed.

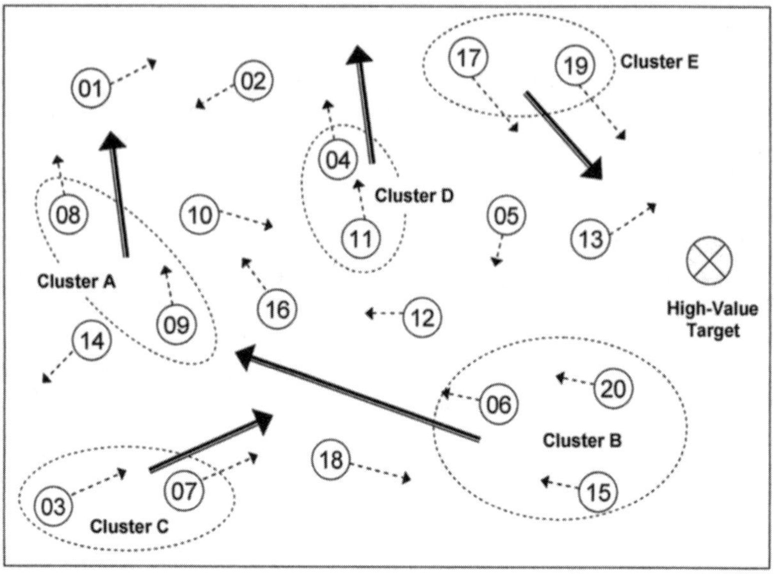

Figure 5-6: Spatiotemporal cluster patterns

A simple hierarchical clustering algorithm will be able to identify clusters of units as they appear and move across the environment. The clustering algorithm can use both location and heading to determine clusters, and the user may fine-tune the algorithm by specifying maximum distance and minimum membership to determine the size of identified clusters. One output of the clustering process is the units that were identified as belonging to a cluster, and the other is the cluster itself. While not required, the cluster algorithm can benefit from user interaction at run time. This is because multiple clusters may be found and the user may be interested in processing only one on the basis of position and heading.

The above clustering technique requires formation of an initial set of clusters; it then tracks those clusters over time. For our purposes, we have developed the following four algorithms that could robustly cluster entities, specifically vehicles, in both urban and nonurban environments:

- Manhattan distance-based grid-constrained clustering
- Directivity- and displacement-based clustering
- Orthogonality-based clustering
- Singular Value Decomposition (SVD)-based clustering

The four subsections that follow describe these algorithms in detail.

5.2.2 Manhattan Distance-Based Grid-Constrained Clustering

Consider an urban scenario where the target is known, and the vehicles are moving on predefined street grids. The overall steps for clustering are as follows.

Algorithm

- Step 1: Order entities in terms of increasing Manhattan distance from known target along street grid.
- Step 2: Generate initial convergent clusters.
- Step 3: Refine clusters based on positional coordinates.

Consideration of Manhattan distance alone leads to unilateral clustering (e.g., the vehicles moving in the same direction towards the target) or convergent clustering (e.g., the vehicles converging to the target from different directions). These two types of clusters can be further segmented into subclusters using positional coordinates.

One characteristic feature of this algorithm is that the application expects constrained movements of the entities and the variations of at least one attribute (e.g., position) over specified grids.

5.2.3 Directivity- and Displacement-Based Unconstrained Clustering

Consider the land-based movements of objects as shown in Figure 5-7.

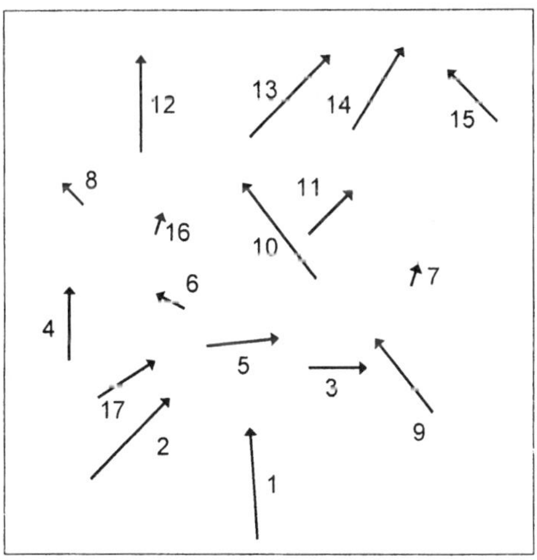

Figure 5-7: Movements of objects

In directivity- and displacement-based clustering, the land-based (i.e., 2-dimensional) movements of n entities over a period of m time instants are observed in terms of x and y coordinates, giving rise to a $2 \times n \times m$ data set. The objective is to identify the r clusters the entities belong to. In Figure 5-7, entity 1 will be the first vector selected, as it has the largest magnitude. Entities 8, 16, 6, and 7 are rejected because of insignificant movements.

The four clusters, as shown in Figure 5-8 (marked with the same arrow pattern and thickness), successively selected are Cluster 1: {1, 12, 4}, Cluster 2: {13, 2, 14, 17, 11}, Cluster 3: {10, 9, 15}, and Cluster 4: {5, 3}.

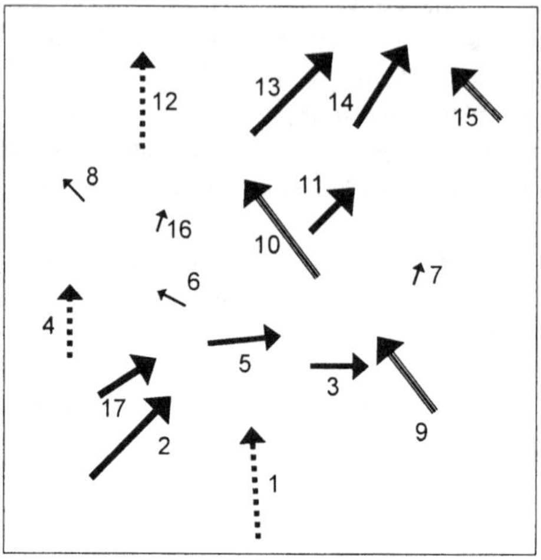

Figure 5-8: Clusters based on length of movement and direction

Algorithm

- *Step 1*: For all the n entities, compute the Δx and Δy over the specified minimum time interval, forming matrix **A**. Identify the column X_1 of maximum norm, which becomes the first seed vector.
- *Step 2*: Detect all column vectors of **A**, which are aligned in the direction of X_1, irrespective of the column norm. These form a cluster $\mathbf{C}_1$.
- *Step 3*: Identify the next seed vector X_i among the rest of the available vectors, based on maximum norm.
- *Step 4*: Repeat Steps 1 and 2 to identify the additional clusters $\mathbf{C}_i$ until all the vectors are exhausted.
- *Step 5*: Move the time-window and repeat Steps 1 to 4.

If in addition to the displacement, the actual locations of the objects are incorporated in the data vector, the algorithm can extract additional knowledge about the entities. For example, within the Cluster 2 in Figure 5-8, positional attributes will identify the two subclusters constituting the objects {17, 2} and the objects {13, 14, 11} (as shown in Figure 5-9), where the former follow the latter.

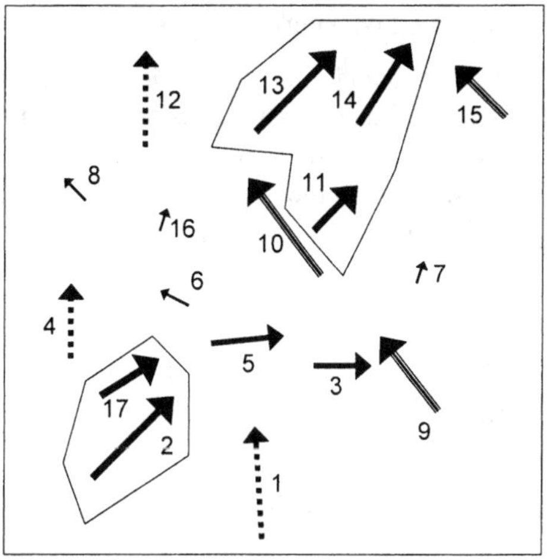

Figure 5-9: Clustered based on length of movement, direction, and position

In a scenario involving a large number of objects, there is sense in substituting a sub-cluster with an equivalent vector. One possible way to generate such a vector is to consider the typical positional attributes (e.g., $x, y, \Delta x, \Delta y$) of each entity arranged in the rows of matrix **A**. Figure 5-10 shows the equivalent vectors of the two subclusters in Figure 5-9. The resultant vector in each subcluster is the average energy pattern. One attractive feature of this approach is that it can lead to hierarchical clustering, which can lead to significant reduction in the dimension of the problem.

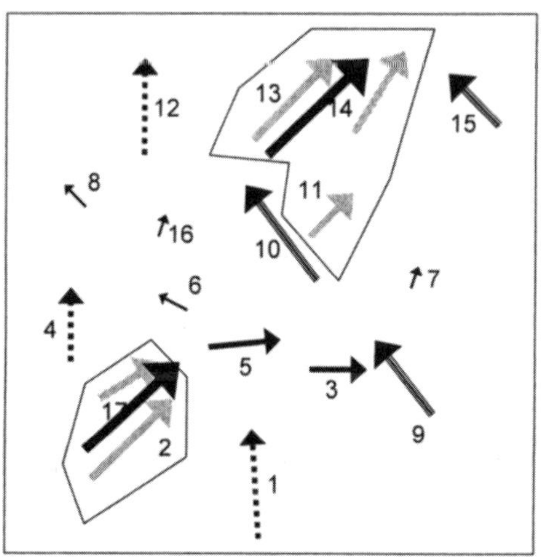

Figure 5-10: Equivalent vectors for subclusters

5.2.4 Orthogonality-Based Clustering

Consider an $m \times n$ matrix **A** containing the spatiotemporal information vectors, where each column is a candidate X_i for clustering and the column elements contain the temporal information associated with each vector. The objective is to form $k(< n)$ clusters based on candidates' affiliations to successive orthogonal directions. We assume that all the vectors X_i are normalized to unit norm.

Algorithm

- *Step 1*: Let the choice for a starting point for a cluster X_p be known. (Swap X_1 with X_p choice).
- *Step 2*:
  - For each of the rest of the vectors X_i in **A**, compute $s_i = \left|X_1^T X_i\right|$.
  - Select all the (say, $n_1 - 1(\geq 0)$) vectors of **A**, for which the magnitude of s_i is larger than a predefined threshold δ, and rearrange **A** to place these $n_1 - 1$ vectors in the 2 to n_1 columns of **A**. So, the first n_1 rows of **A** will constitute the first cluster. The rest of $n - n_1$ vectors of **A** are the unresolved vectors, which are the candidates for further clustering.
  - Let the vector X_j correspond to the minimum value of s_i. This will be the seed vector for the cluster farthest from the first cluster.
  - Let Q_1 be the unit norm resultant vector for Cluster 1.
- *Step 3*:
  - Rotate all the $n - n_1$ vectors of **A** in columns $n_1 + 1$ to n and call it **B**, where columns are $Y_i = X_i - Q_i^T X_i Q_1$. Y_is are scaled to unit norm. These will be components of the original candidate vectors in a direction orthogonal to the direction of Cluster 1.
  - As in Step 2, compute $s_i = \left|X_j^T Y_i\right|$.
  - Select the n_2 vectors of **B** for which s_i are larger than the threshold δ.
  - Let Q_2 be the unit norm resultant vector for the new cluster, and let X_j be redefined as the vector which corresponds to the minimum value of s_i. This is the seed vector for the subsequent cluster.

- *Step 4*:
  - Thus in the selection of the k-th cluster, form **B**, where columns $Y_i = X_i - \left(Q_1{}^T X_i Q_1 + Q_2{}^T X_i Q_2 + \ldots + Q_{k-1}{}^T X_i Q_{k-1}\right)$, and compute $s_i = \left|X_j{}^T Y_i\right|$.
  - Select the n_k vectors in **B** for which s_i are larger than the threshold δ. Redefine X_j as the vector which corresponds to the minimum value of s_i.
- *Step 5*:
  - The clustering is terminated when $\|Y_i\|$ is insignificantly low. The corresponding candidate vector X_i in the original data set may either be rejected or may be assigned to a cluster for which $\left|Q_i{}^T X_i\right|$ is maximum.

While the orthogonally distinct vectors are distinct with respect to directivity, the converse may not be true. For the entities, if it is desired to apply differential importance to the characteristic attributes, the data matrix may be premultiplied by a weighting matrix $\left\{w_{ij}\right\}_{ij}$, where typically $w_{ij} = 0$ for $i \neq j$, and w_{ij} may be different or even time-varying.

This algorithm is particularly applicable where orthogonality of the attributes is of particular significance. This algorithm is also expected to be relatively immune to additive noise in the data.

5.2.5 Singular Value Decomposition-Based Clustering

The Singular Value Decomposition (SVD) (Golub and Van Loan, 1996) is featured in one of the two STC algorithms and in one of the two preprocessing stages. The SVD of an $m \times n$ matrix **A** is given by $\mathbf{A} = \mathbf{USV}^T$, where **U** and **V** are orthogonal matrices, and **S** is diagonal. The number of nonzero singular values $\left(s_1, \ldots, s_p\right)$ is arranged in decreasing order in the diagonal matrix **S**, and gives the number of independent rows (or columns) of **A**. The maximum rank p of **A**, denoted as $rank(\mathbf{A})$, is $\min(m, n)$. The left singular vectors U_i of **U** and the right singular vectors V_i of **V** span the range and domain spaces of **A**, respectively. Some of the features of SVD relevant to the present work are as follows:

- If all the m objects make identical displacements over n intervals, the rows of the matrix **A** will be identical, and the number of nonzero singular values or p will be 1.
- If the rows are nearly identical (that is, $s_1 \gg s_i$), the least squares estimation of the most dominant orthogonal mode will be given by $U_1 S_1 V_1^T$, where U_1 and V_1 are the first left and the first right singular vectors, respectively.

- If instead the objects make random movements, the number of nonzero singular values will be p.

For any case in between these two extreme cases, the number of nonzero singular values will be between 1 and p. This feature has been used for detection of significant movements in a cluttered space.

In this algorithm, the clustering is explored in the singular vector space. The spatiotemporal movement is considered in terms of the x, y, or $\|(x,y)\|$ coordinates arranged row-wise for each entity, forming the data matrix ($\mathbf{A}_x$, $\mathbf{A}_y$, $\mathbf{A}_{xy}$, respectively), which is singular value decomposed. As the column vectors in $\mathbf{U}$ span the entity space, the elements of the first few singular vectors are considered to be the attributes of the entities. The number of singular vectors (U_i) to be considered may be based on the relative magnitude of the singular values. The clustering is performed as illustrated in the following example.

We considered a scenario described earlier with added complexity (see Figure 5-11). There are altogether 100 entities, moving in unconstrained space (no street grid). In the scenario, 20 entities move from left to right diagonally, with 10 entities following the other 10; 20 entities move diagonally in a nearly orthogonal direction with 10 entities following the other 10; 10 entities move in a spirally shaped pathway; the remaining 50 entities make local spurious movements. It is assumed that data for 100 time steps are available.

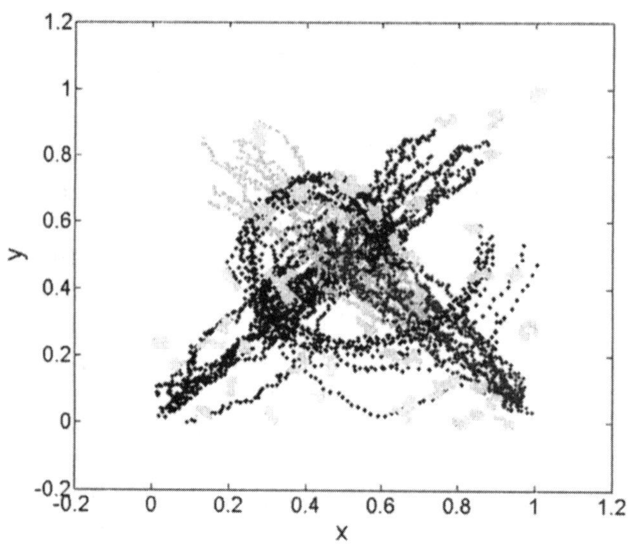

Figure 5-11: The x and y coordinates of the 100 entities over 100 time intervals

As shown in Table 5-3, in all three cases of data configuration, the first three orthogonal modes contain the major part of the information in the data matrix. Hence we use the three corresponding left singular vectors as the basis to map the process, as shown in Figure 5-12.

Data Configuration	1st SV	2nd SV	3rd SV	4th SV	5th SV	$\|\mathbf{A}\|_F / \|(s_1, s_2, s_3)\|$
$\mathbf{A}_x$	57.09	9.50	5.26	0.58	0.38	99.99
$\mathbf{A}_y$	47.91	9.62	2.72	0.76	0.34	99.98
$\mathbf{A}_{xy}$	75.27	9.36	4.55	0.99	0.43	99.99

Table 5-3: Singular values for configurations with x, y, and $\|(x, y)\|$

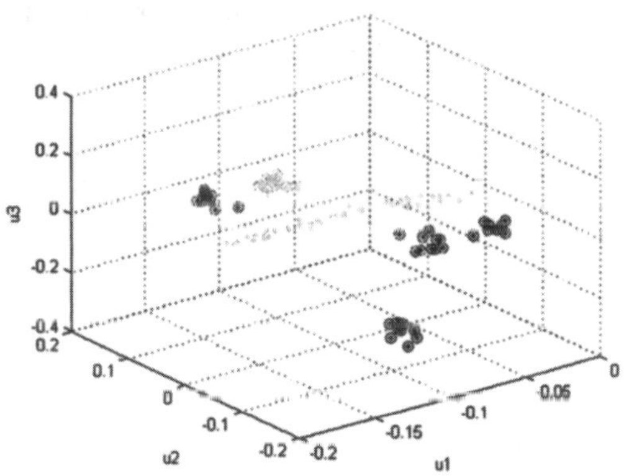

Figure 5-12: Segregation of clusters of 100 entities in a singular vector space with x data sets

When only x coordinates (i.e., data matrix $\mathbf{A}_x$) have been used for analysis, it is observed that the clusters are comparatively less segregated when y coordinates (i.e., data matrix $\mathbf{A}_y$) or $\|(x, y)\|$ attributes (i.e., data matrix $\mathbf{A}_{xy}$) are used, as shown in Figure 5-13 and Figure 5-14 respectively.

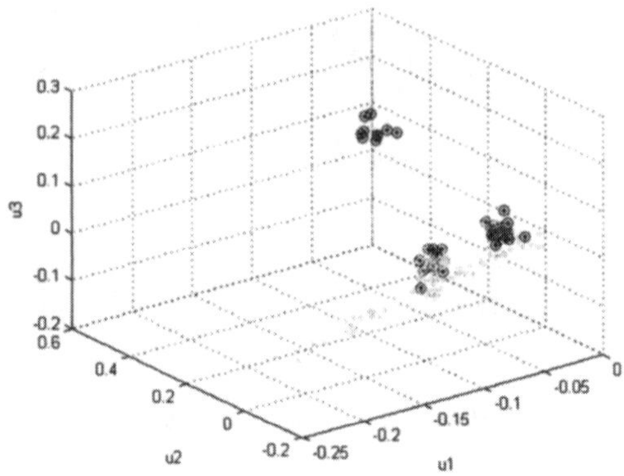

Figure 5-13: Singular vector space mapping of 100 entities when *y* coordinates are used

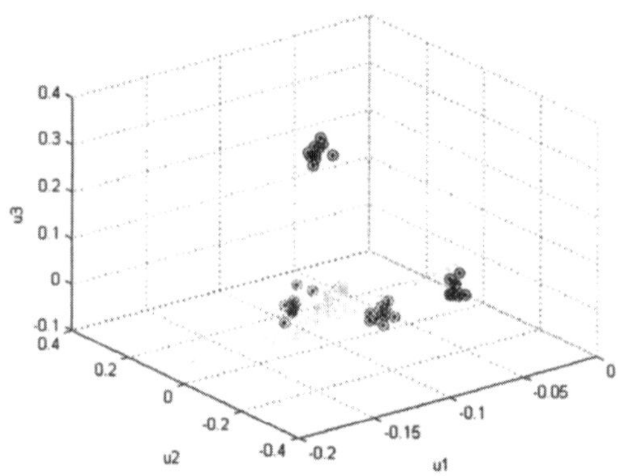

Figure 5-14: Singular vector space mapping of 100 entities when $\|(x, y)\|$ attributes are used

Generally speaking, the SVD approach absorbs the temporal variations, and thereby simplifies the problem of clustering. However, supervisory probing may be necessary to decide the best choice of vectors. It is worth noting that the spirally moving entities cluster distinctly in all three modes of SVD-based clustering. Thus the SVD approach is able to distinctly segregate entities making uniquely characteristic movements.

5.2.6 Preprocessing through Entropy Measure

A preprocessing stage is an essential prelude to STC for the following reasons:

- In a high-dimensional problem, it is necessary to eliminate data with little meaningful information before clustering in order to limit the clutter as well as the computational load.
- It is necessary to unearth best sets of explicit or implicit attributes of the entities.
- Proper transformation of the data or conversion of the problem to a mathematically well-posed problem can substantially improve the numerical stability and the clustering performance.

5.3 FURTHER READING

Leung and Wu (2000) present a detailed approach to target classification starting from low-level radar observations. Further details on spatiotemporal clustering and preprocessing can be found in Das et al. (2006) and Kanjilal et al. (2007).

Chapter 6

Model-Based Situation Assessment

This chapter presents the Bayesian Belief Network (BN) technology, which is perhaps the most widely used technology for model-based Situation Assessment (SA). We start by defining BNs and illustrating the concept of conditional independence within BNs. Then we present the evidence, belief, and likelihood concepts. Next, we present two propagation algorithms to compute variables' posterior probabilities based on evidence: one for the class of networks without cycles, and another for a class of networks with restrictive cycles, and analyze the complexities of these two algorithms. We provide some guidance for acquiring network probabilities, including the noisy-or technique. We summarize the overall advantages and disadvantages of BN technology and show its application to SA in the Theater Missile Defense (TMD) domain. Finally, we conclude the chapter by providing a selected list of commercially available BN tools and packages.

6.1 BAYESIAN BELIEF NETWORKS

A Bayesian Belief Network (BN) is a network (that is, a graph consisting of nodes and links) with the following interpretation:

- Each node in the network represents a random variable that can take on multiple discrete values; these values are mutually exclusive events constituting the sample space over which the random variable is defined. The terms *node* and *variable* are synonymous in the context of a BN and are often used interchangeably.
- Each link in the network represents a relation or *conditional dependence* between the two variables it connects, and an associated *conditional probability table* (CPT) quantifies the relation associated with the link.
- There is a notion of causality between two linked variables, so links with explicit direction (represented by arrows) are drawn from "cause" nodes to "effect" nodes.

- The network is based on some *marginal* and *conditional independence* assumptions among nodes. These assumptions are described in the next section.
- The state of a node is called a *belief*, and reflects the posterior probability distribution of the values associated with that node, given all the a priori evidence.

The probabilities in a CPT are typically acquired from subject matter experts in the domain, but can also be learned automatically given a large enough number of training instances (details on BN learning are in Chapter 10). The causality restriction can sometimes be relaxed by allowing links between a pair of nodes that are simply correlated, and the direction of the arrow between the two nodes is decided based on the ease of acquisition of the probabilities in the CPT. A typical causal relationship between two variables inevitably brings the temporal dimension into the modeling problem. We avoid building any formal dynamic or temporal BNs within which the state of a variable is allowed to change over time by modeling a simple snapshot of the problem domain at a particular time (dynamic BNs are discussed in Chapter 7).

Example

An example BN is shown in Figure 6-1. It illustrates causal influences on whether or not a vehicle is going to start a journey, and how much delay is going to occur if it starts, given the current weather, road condition, visibility, and other related sources of evidence. Note that there are many ways to approach modeling an SA or decision-making problem with BNs. The selection of random variables and their granularities and interdependences is largely subjective, but should be driven by the problem-solving requirements. (Please note that we are not emphasizing any specific modeling methodology; the specific BN structure in Figure 6-1 is constructed to illustrate the technology via a single unified example. The conditional probabilities that will follow are made up and may not accurately reflect the reality.)

Each node in the BN in Figure 6-1 and its associated mutually exclusive and exhaustive states (possible values of the random variable representing the node) are described below (from top to bottom):

- *Weather*: Overall weather condition during the journey – {*normal*, *rainy*, *foggy*}. A weather state is normal if it is neither rainy not foggy.
- *Eyesight*: Eyesight of the driver, especially during the night-time – {*good*, *bad*}.
- *Roads*: Condition of roads that the vehicle is going to take for the journey – {*dry*, *slippery*, *muddy*}. Here the dry condition is synonymous to the normal road condition.
- *Visibility*: Overall visibility of the day during the journey – {*clear*, *poor*}.

- *Journey*: Whether the vehicle's journey is to start or cancel – {*start*, *cancel*}. The variable provides a recommendation to the human analyst to make an appropriate decision about the journey.
- *Weight*: Weight of the vehicle – {*heavy*, *light*}.
- *Mobility*: Mobility along the route of the vehicle – {*go*, *slow go*, *no go*}.
- *Delay*: Delay in arrival – {*none*, *short*, *long*}.

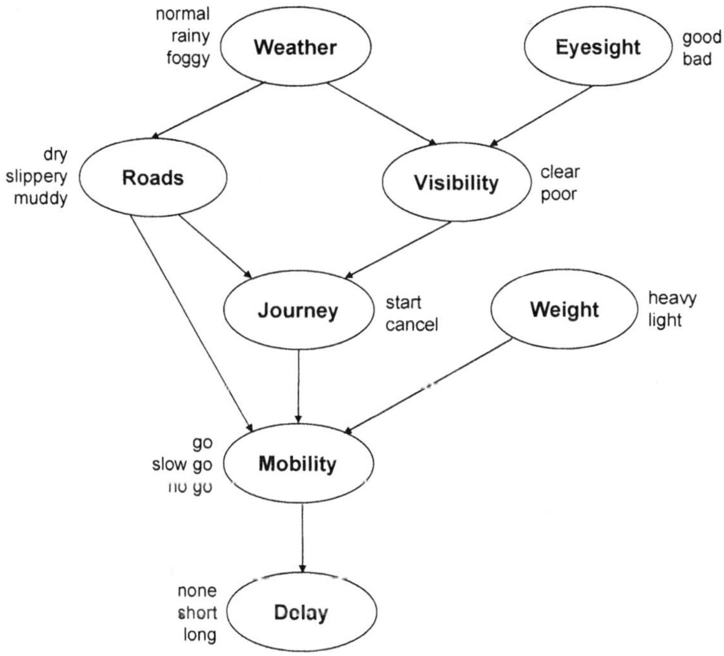

Figure 6-1: An example BN

All states in the random variable *Weather*, for example, are mutually exclusive, that is, the overall weather condition cannot be both normal and rain at the same time. But a probability density function f of the variable *Weather* with the domain {*normal*, *rainy*, *foggy*} can be defined and interpreted as "$f(x)$ is the probability that *Weather* will assume the value x." One such function can be defined as follows:

$$f(normal) = 0.80, f(rainy) = 0.05, f(foggy) = 0.15$$

The causal influences by numbers (as shown in Figure 6-2) along the directions of the arrows are defined as follows:

1) Weather condition determines the condition of the roads. For example, rainy weather is likely to cause roads to be muddy or slippery (or could even be both, but I have not considered that state for the sake of simplicity).

2) Weather condition and the driver's eyesight together determine the overall visibility along the route. For example, a foggy day is likely to cause very poor visibility and therefore may be unsuitable for driving. A driver with bad eyesight is likely to have the same effect on a rainy night.
3) Road condition and visibility together determine whether the vehicle is allowed to start its journey or not. For example, a very slippery road condition is likely to cause the cancellation of the journey because vehicles lose traction on a slippery surface. Poor visibility may cause cancellation of a journey due to the safety of the driver.
4) Condition of the roads, journey status, and the vehicle's weight together determine the mobility. For example, when the road is slippery and the vehicle is light, the mobility is likely to be a slow go, since these conditions might slow down the vehicle. A muddy road is likely to have a similar effect irrespective of the weight of the vehicle.
5) Mobility determines the amount of delay in arrival, since the slow or no go mobility is likely to cause a delay in arrival.

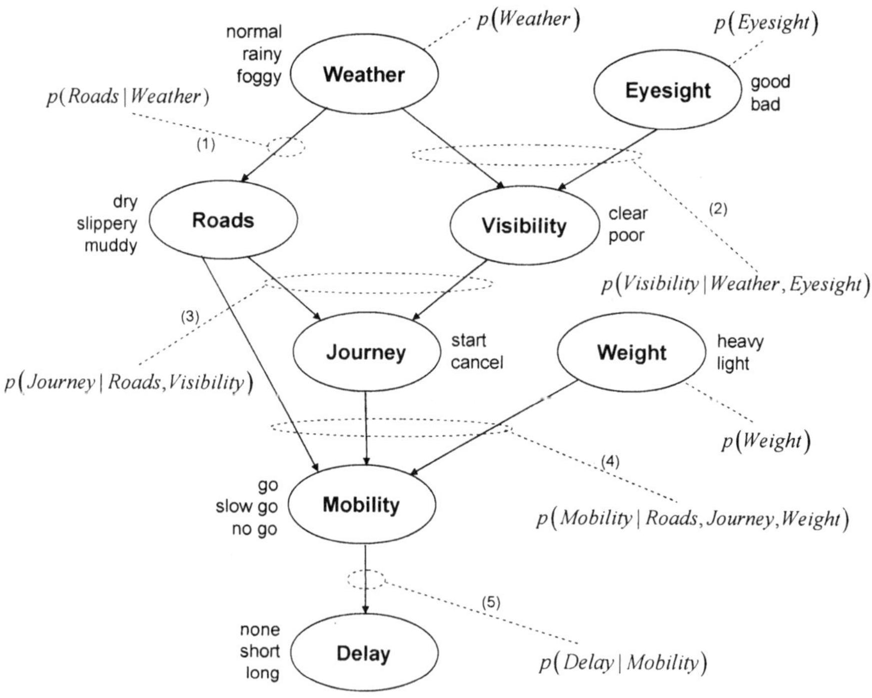

Figure 6-2: BN with prior and conditional probabilities

The three prior probabilities (corresponding to the three nodes without parents) and five conditional probabilities of the BN in Figure 6-2 are provided in Table 6-1 through Table 6-8.

Weather		
	normal	0.55
	rainy	0.15
	foggy	0.3

Table 6-1: $p(Weather)$

Eyesight		
	good	0.99
	bad	0.01

Table 6-2: $p(Eyesight)$

Weight		
	heavy	0.6
	light	0.4

Table 6-3: $p(Weight)$

Weather		normal	rainy	foggy
Roads	dry	0.9	0.7	0.8
	slippery	0.09	0.2	0.15
	muddy	0.01	0.1	0.05

Table 6-4: $p(Roads \mid Weather)$

Weather		normal		rainy		foggy	
Eyesight		good	bad	good	bad	good	bad
Visibility	clear	0.95	0.6	0.3	0.1	0.4	0.2
	poor	0.05	0.4	0.7	0.9	0.6	0.8

Table 6-5: $p(Visibility \mid Weather, Eyesight)$

Roads		dry		slippery		muddy	
Visibility		clear	poor	clear	poor	clear	poor
Journey	start	0.99	0.1	0.8	0.1	0.01	0
	cancel	0.01	0.9	0.2	0.9	0.99	1

Table 6-6: $p(Journey \mid Roads, Visibility)$

Roads		dry				slippery				muddy			
Journey		start		cancel		start		cancel		start		cancel	
Weight h – heavy, l – light		h	l	h	l	h	l	h	l	h	l	h	l
Mobility	go	0.9	0.8	0	0	0.2	0.1	0	0	0.09	0	0	0
	slow go	0.1	0.2	0	0	0.8	0.9	0	0	0.9	0.9	0	0
	no go	0	0	1	1	0	0	1	1	0.01	0.1	1	1

Table 6-7: $p(Mobility \mid Roads, Journey, Weight)$

Mobility		go	slow go	no go
Delay	none	1	0	0
	short	0	1	0
	long	0	0	1

Table 6-8: $p(Delay \mid Mobility)$

This BN and its associated CPTs shown above are used throughout the rest of this chapter to illustrate algorithms and other related concepts. Note that the CPT in Table 6-8 is in the form of an identity matrix, and guarantees the perfect causal relationship between the states of *Delay* and *Mobility* variables.

6.2 CONDITIONAL INDEPENDENCE IN BELIEF NETWORKS

Two random variables Y and Z are said to be (marginally) *independent*, denoted as $Y \perp Z$, if

$$p(Y,Z) = p(Y)p(Z)$$

for any combination of values for the variables Y and Z. The variable Y is *conditionally independent* of Z given another variable X, denoted as $Y \perp Z \mid X$, if

$$p(Y,Z \mid X) = p(Y \mid X)p(Z \mid X)$$

Therefore,

$$\begin{aligned} p(Y \mid Z,X) &= \frac{p(Y,Z \mid X)}{p(Z \mid X)} \\ &= \frac{p(Y \mid X)p(Z \mid X)}{p(Z \mid X)} \\ &= p(Y \mid X) \end{aligned}$$

Similarly, $p(Z \mid Y,X) = p(Z \mid X)$. Note that marginal independence (no conditioning) does not imply conditional independence; nor does conditional independence imply marginal independence.

Figure 6-3 represents conditional independence in a chain fragment of a BN where a node X is between two other nodes Y and Z. We factorize the joint probability distribution of the variables X, Y, and Z as follows:

$$\begin{aligned} p(X,Y,Z) &= p(Z \mid X,Y)p(X,Y) \\ &= p(Z \mid X)p(X \mid Y)p(Y) \end{aligned}$$

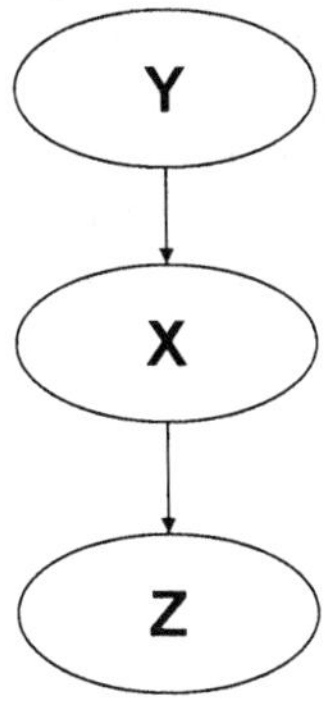

Figure 6-3: Conditional independence in a chain fragment: *Z* is conditionally independent of *Y* given *X*

Example

Figure 6-4 shows an example instantiation of conditional independence in a chain network fragment as shown in Figure 6-3.

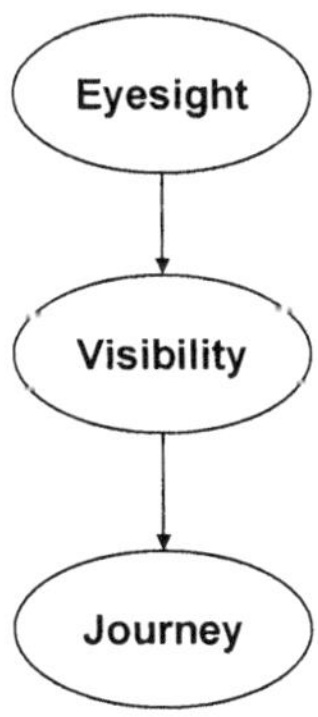

Figure 6-4: Example conditional independence in a chain network fragment: *Eyesight* and *Journey* are conditionally independent given *Visibility*

The variables *X*, *Y*, and *Z* represent, respectively, the driver's eyesight, visibility condition, and the status of the journey, respectively. If we observe, with complete certainty, that the visibility condition is poor, then the probability of the journey being canceled is determined. Therefore, confirmation that the driver's eyesight is good or bad will not change the probability of the journey status, and vice versa.

Figure 6-5 represents conditional independence in a tree network fragment of a BN where the node X is the parent of two other nodes Y and Z. In this case, we factorize the joint probability distribution of the variables X, Y, and Z as follows:

$$p(X,Y,Z) = p(Z \mid X,Y)\, p(X,Y) = p(Z \mid X)\, p(Y \mid X)\, p(X)$$

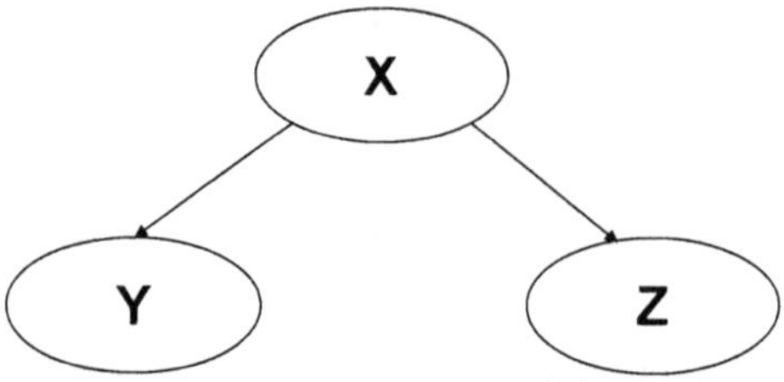

Figure 6-5: Conditional independence in a tree network fragment: *Z* is conditionally independent of *Y* given *X*

Example

Figure 6-6 shows an example instantiation of conditional independence in a tree network fragment. The variables X, Y, and Z represent the weather, roads, and visibility conditions, respectively. If we observe rainy weather then the probabilities of the road condition being slippery and the visibility condition poor are determined, and the confirmation of the road condition being slippery will not change the probability of the visibility being poor, and vice versa.

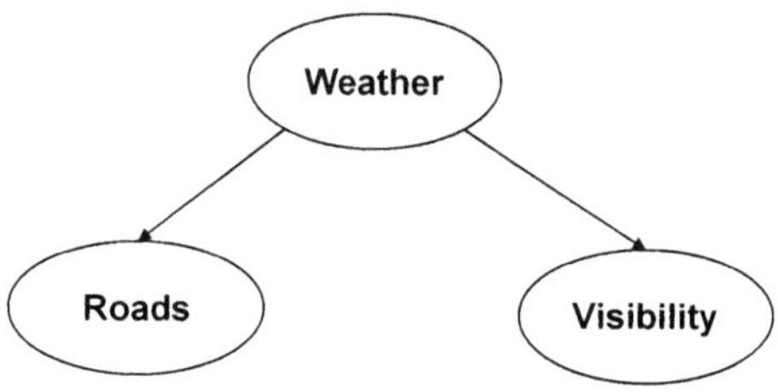

Figure 6-6: Example conditional independence in a tree network fragment: *Roads* and *Visibility* are conditionally independent given *Weather*

Figure 6-7 shows conditional dependence in a polytree network fragment between the nodes Y and Z, given that we know about X; the two variables are marginally independent if we know nothing about X. For a polytree fragment as shown in Figure 6-7, the probability distribution of the variables Y, Z, and X can be factorized as follows:

$$p(X,Y,Z) = p(X \mid Y,Z)\, p(Y,Z) = p(X \mid Y,Z)\, p(Y)\, p(Z)$$

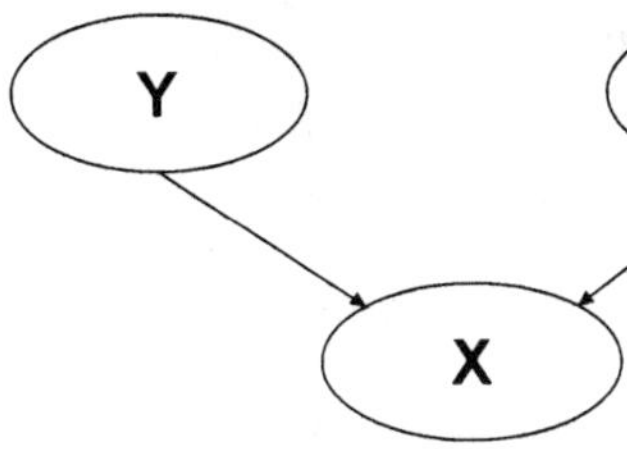

Figure 6-7: Conditional dependence in a polytree fragment: *Y* is conditionally dependent on *Z* given *X*

Example

Figure 6-8 shows an example instantiation of the conditional dependence in a polytree network fragment. Both the weather condition and the driver's eyesight can affect the overall visibility. Before any evidence is obtained on the visibility, the probability of the weather condition to be in a particular state is independent of the probability of the driver's eyesight. However, once some evidence on the overall visibility, say the poor visibility condition, is obtained, the weather condition may influence the probability of the driver's eyesight. For example, observation of foggy weather (thus explaining why the visibility is poor) may decrease the probability of bad eyesight. This phenomenon is termed as *explaining away*. In other words, observation of the status of one parent "explains away" the other, given a value of the child node.

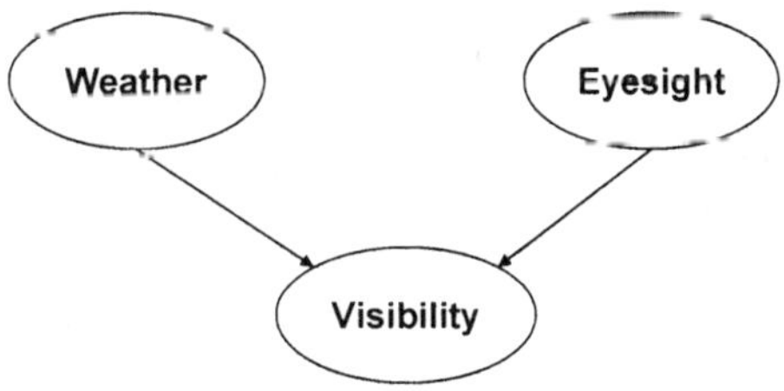

Figure 6-8: Example conditional independence in a polytree network fragment: *Weather* is conditionally dependent on *Eyesight* given *Visibility*

In view of the joint distribution formulae for these three types of network fragments, the joint probability distribution in a directed acyclic graph (DAG) ("directed" means the links have an explicit direction represented by arrows, and "acyclic" means that the arrows may not form a directional cycle in the network) type of network can be factored into conditional probabilities, where each factor involves only a node and its parents. This is stated through the following result:

Proposition 6-1: Consider a network consisting of variables $X_1, X_2, ..., X_n$. The joint probability distribution $p(X_1, X_2, ..., X_n)$ is the product of all conditional probabilities specified in the network:

$$p(X_1, X_2, ..., X_n) = \prod_{i=1}^{n} p(X_i \mid pa(X_i))$$

where $pa(X_i)$ denotes the parent variables of X_i.

Example

Consider the network shown in Figure 6-1. For any combination of values *w*, *e*, *r*, *v*, *j*, *wt*, *m*, *d* of the variables *Weather*, *Eyesight*, *Roads*, *Visibility*, *Journey*, *Weight*, *Mobility*, and *Delay*, respectively, the joint probability is

$$p(w, e, r, v, j, wt, m, d) =$$

$$p(w) p(e) p(r \mid w) p(v \mid w, e) p(j \mid r, v) p(wt) p(m \mid r, j, wt) p(d \mid m)$$

Influence in a BN is only allowed to flow along the links given in the network. Therefore, independence between two nodes is represented by the absence or blocking of links between the two nodes. Whether a link between a pair of nodes exists or not is determined by a property called *d-separation*. Before we formally define *d*-separation, we need to introduce three kinds of connections between a node *X* and two of its neighbors *Y* and *Z*. The three possibilities are shown in Figure 6-3, Figure 6-5, and Figure 6-7. Their example instantiations are shown in Figure 6-4, Figure 6-6, and Figure 6-8, respectively.

In Figure 6-3, there are links from *Y* to *X* and from *X* to *Z*. In other words, *Y* has an influence on *X*, which in turn influences *Z*. The connection or path between *Y* and *Z* is called *linear*. In this case, causal or deductive evidence from *Y* will change the belief of *X*, which then changes the belief of *Z*. Similarly, diagnostic or abductive evidence from *Z* will change the belief of *X*, which then changes the belief of *Y*. But if the state of *X* is known, then the connection between *Y* and *Z* is blocked and they cannot influence each other any more. Thus, *Y* and *Z* become independent given *X*. We say that *Y* and *Z* are *d*-separated given *X*.

In Figure 6-5, there are links from *X* to *Y* as well as from *X* to *Z*. In other words, *X* has influence on both *Y* and *Z*, and *Y* and *Z* will influence each other via *X*. The connection or path between *Y* and *Z* is called *diverging* and the node *X* is said to have diverging arrows. As in the linear case, if the state of *X* is known then *Y* and *Z* cannot influence each other any more, and we say that *Y* and *Z* are *d*-separated given *X*.

The third case, shown in Figure 6-7, is the opposite of the previous case. In this case, there are links from *Y* to *X* as well as from *Z* to *X*. In other words, both *Y* and *Z* have influence on *X*. The connection or path in this case is called *converging* and the node *X* is said to have converging arrows. In this case, if nothing is known about X then Y and Z are independent, and therefore cannot influence each other. But if the state of X is known then Y and Z can influence each other. In other words, Y and Z are already *d*-separated, but not when X is given.

In general, two nodes Y and Z in a DAG type of network are *d*-separated if for all paths between Y and Z, there is an intermediate node X such that either

- The path between Y and Z is serial or diverging at node X and the state of X is known, or
- The path between Y and Z is converging at node X and neither X nor any of its descendants have received evidence.

Two nodes Y and Z in a DAG are *d-connected* if they are not *d*-separated. The following proposition establishes a connection between conditional independence and *d*-separateness.

Proposition 6-2: If any two nodes Y and Z in a DAG are *d*-separated with evidence e entered, then Y and Z are conditionally independent given e (i.e., $Y \perp Z \mid e$ or $p(Y \mid Z, e) = p(Y \mid e)$).

Example

Consider the network shown in Figure 6-1. Let Y = *Roads* and Z = *Eyesight*. The two paths between the two nodes Y and Z are:

$$Roads \leftarrow Weather \rightarrow Visibility \leftarrow Eyesight$$

$$Roads \rightarrow Journey \leftarrow Visibility \leftarrow Eyesight$$

The first path contains a diverging node (*Weather*) and the second path contains a converging node (*Journey*). If the state of the variable *Weather* is known and the variable *Journey* and its descendants *Mobility* and *Delay* have not received evidence, then the nodes *Roads* and *Eyesight* are *d*-separated. Alternatively, if the variable *Visibility* and its descendants *Journey*, *Mobility*, and *Delay* have not received evidence, then the nodes *Roads* and *Eyesight* are *d*-separated.

The above definition of *d*-separation between two nodes takes into account the evidence entered into the network. Here we present a more generalized definition of *d*-separation that identifies a set of nodes, instead of a single node, that could potentially separate two nodes in a network. Moreover, the definition provided here is between two sets of nodes rather than between two nodes.

For any three disjoint subsets S_X, S_Y and S_Z of nodes, S_X is said to *d*-separate S_Y and S_Z if in every path between a node in S_Y to a node in S_Z there is a node X satisfying one of the following two conditions:

- X has converging arrows and none of X or its descendants is in S_X, or
- X does not have converging arrows and X is in S_X.

Example

Consider the network shown in Figure 6-1. Let $S_Y = \{Roads\}$ and $S_Z = \{Eyesight\}$. The set of all paths from a node in S_Y to a node in S_Z is:

$$Roads \leftarrow Weather \rightarrow Visibility \leftarrow Eyesight$$

$$Roads \rightarrow Journey \leftarrow Visibility \leftarrow Eyesight$$

Suppose $S_X = \Phi$. The first path contains node *Visibility* with converging arrows and none of its three descendants *Journey*, *Mobility*, and *Delay* is in S_X. The second path contains the node *Journey* with converging arrows and none of its two descendants *Mobility* and *Delay* is in S_X. Therefore, S_Y and S_Z are *d*-separated by the empty set. But, if we consider $S_X = \{Mobility\}$, the first path contains the node *Visibility* with converging arrows and its descendant *Mobility* is in S_X. Also, the second path contains the node *Journey* with converging arrows and its descendant *Mobility* is in S_X. Although, the first path contains the node *Weather* without converging arrows, the node does not belong to S_X. Therefore, $\{Mobility\}$ does not *d*-separate S_Y and S_Z Note that Φ does not *d*-separate $\{Roads\}$ and $\{Visibility\}$, but $\{Weather\}$ does, and so does $\{Weather, Mobility\}$.

The generalized set-theoretic definition of *d*-separation above yields the following proposition.

Proposition 6-3: For any three disjoint subsets S_X, S_Y and S_Z of variables in a DAG, S_X *d*-separates S_Y and S_Z if and only if S_Y and S_Z are conditionally independent given S_X (that is, $S_Y \perp S_Z \mid S_X$).

Once we have built BNs, either with the help of domain experts or via automated learning from past observations, we need to reason with them, that is, examine how the variables in a network change their beliefs when observations are propagated into the network as evidence.

6.3 EVIDENCE, BELIEF, AND LIKELIHOOD

Evidence on a variable is a statement of certainties of its states based on certain observations. Since the states of a BN variable are mutually exclusive, such a statement of certainty of a state variable is usually made with a percentage that represents the chance of being in that state. If the statement constituting evidence for a variable gives the exact state of the variable (that is, 100%), then it is *hard* evidence (which is also called *instantiation*); otherwise, the evidence is called *soft*. As an example, consider the variable *Weather* whose states are *normal*, *rainy*, and *foggy*. If the evidence *e* is based on someone's direct observation of the weather and states that the weather is normal, then it is hard evidence and is denoted by *Weather* = *normal*. In general, if $E = \{X_1, ..., X_n\}$ is the set of all variables whose values are known as $X_1 = a_1, ..., X_n = a_n$ then

$$e = \{X_1 = a_1, ..., X_n = a_n\}$$

where each a_i is hard evidence of the state of X_i. For example, if E = {*Weather*, *Roads*, *Journey*} and the evidence states that the weather is normal, the roads are dry, and the journey is started, then

$$e = \{Weather = normal, Roads = dry, Journey = start\}$$

On the other hand, consider the situation when the source of evidence on the variable *Weather* is based on the observation of output from a sensor (the sensor is not explicitly modeled as a BN variable). The statement constituting evidence states that there is an 80% chance that the weather is normal, 15% chance that the weather is rainy, and 5% chance that the weather is foggy. The evidence in this case is inexact and therefore *soft*. Evidence on a variable X yields a *likelihood vector*, denoted as $\lambda(X)$, expressed in terms of probability measures. For example, the above soft evidence on the variable *Weather* yields the likelihood vector

$$\lambda(Weather) = \begin{bmatrix} 0.80 \\ 0.15 \\ 0.05 \end{bmatrix}$$

The hard evidence $e = \{Weather = sunny\}$ yields the likelihood vector $\begin{bmatrix} 1 \\ 0 \\ 0 \end{bmatrix}$.

Usually, total evidence accumulated on the states of a variable is expected to be more or less than 100%. An example of such type of evidence obtained by observing output from a sensor (for example, the sensor displays green for normal weather, blue for rainy, and red for foggy; ambiguities occur because a faulty sensor or a noisy environment can cause display of different lights for a given weather state) states that there is a 70% chance that the weather is normal, a 50% chance that the weather is rainy, and a 15% chance that the weather is foggy. The likelihood vector for this evidence is the following:

$$\lambda(Weather) = \begin{bmatrix} 0.70 \\ 0.50 \\ 0.15 \end{bmatrix}$$

The above evidence states that if all 100 weather circumstances similar to the current one are normal, the sensor prediction is likely to be correct (that is, displays green) 70 times; if all are rainy, it is likely to be correct (blue) 50 times; and if all are foggy, it is likely to be correct (red) 15 times.

How do we then relate evidence to probability? Observe that $70:50:15$ is the ratio of the number of times the sensor is likely to produce $e_{Weather}$ if all 100 weather circumstances are normal, to the number of times it is likely to produce $e_{Weather}$ if all 100 are rainy, to the number of times it is likely to produce $e_{Weather}$ if all 100 are foggy. This relation yields the following likelihood ratio:

$$p\left(e_{Weather} \mid Weather = sunny\right) : p\left(e_{Weather} \mid Weather = rainy\right) :$$
$$p\left(e_{Weather} \mid Weather = foggy\right) = 70:50:15$$

This ratio gives the following likelihood:

$$\begin{aligned} \lambda\left(Weather = sunny\right) &= p\left(e_{Weather} \mid Weather = sunny\right) = 0.70 \\ \lambda\left(Weather = rainy\right) &= p\left(e_{Weather} \mid Weather = rainy\right) = 0.50 \\ \lambda\left(Weather = foggy\right) &= p\left(e_{Weather} \mid Weather = foggy\right) = 0.15 \end{aligned}$$

The likelihood vector $\lambda\left(Weather\right)$ is $p\left(e_{Weather} \mid Weather\right)$, and we therefore have the following:

$$\lambda\left(Weather\right) = p\left(e_{Weather} \mid Weather\right) = \begin{bmatrix} 0.70 \\ 0.50 \\ 0.15 \end{bmatrix}$$

If the sensor always displays green when the weather is normal, then the above likelihood vector changes to

$$\lambda\left(Weather\right) = \begin{bmatrix} 1.0 \\ 0.5 \\ 0.15 \end{bmatrix}$$

But, conversely, the green display does not necessarily mean the weather is normal though it certainly indicates a high probability of normal weather.

The CPTs of a BN remain unchanged upon the arrival of evidence. When evidence is *posted* to the designated node to compute posterior probabilities of the nodes in the networks, the node state certainties, or probability distribution, change. After receiving evidence e, the posterior probability of node X is $p\left(X \mid e\right)$. The *belief* of the node X of a BN, denoted as $Bel\left(X\right)$, is the overall belief of the node X contributed by all evidence so far received. Therefore, if e is the evidence received so far then $Bel\left(X\right) = p\left(X \mid e\right)$.

Consider the network fragment as shown in Figure 6-9. Suppose e_X^+ and e_X^- are the total evidence connected to X through its parents and children, respectively. In other words, e_X^+ and e_X^- are the evidence contained in the upper and lower sub-networks with respect to the node X. We then define the following two π and λ vectors:

$$\pi(X) = p(X \mid e_X^+)$$
$$\lambda(X) = p(e_X^- \mid X)$$

The vectors $\pi(X)$ and $\lambda(X)$ represent the distributions of the total supports among the states of X through its parents and children, respectively. If the network that contains the fragment is a tree, then the vectors $\pi(X)$ and $\lambda(X)$ represent the distributions of the total causal and diagnostic supports among the states of X by all its ancestors and descendants, respectively.

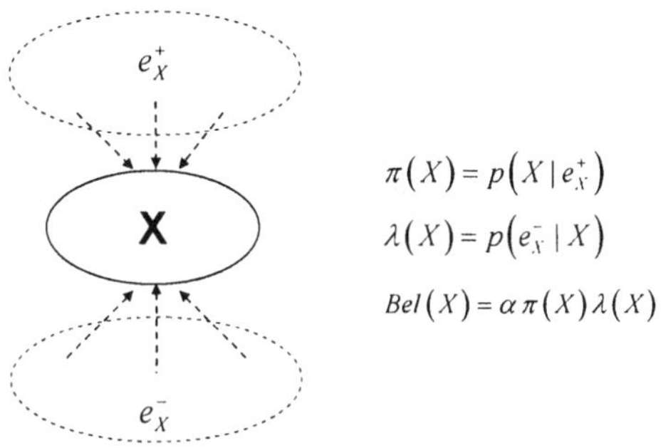

Figure 6-9: Network fragment containing node X

6.4 PRIOR PROBABILITIES IN NETWORKS WITHOUT EVIDENCE

Evidence propagation and belief updating start with "fresh" networks without any observed evidence. Then the π and λ vectors of the variables, and hence the belief vectors, are updated incrementally as evidence is accumulated. In this section, we detail how these initial vectors of the variables in a fresh network are computed.

If no evidence has yet been propagated in a network, then, for every variable X in the network, $\pi(X)$ is $p(X)$, since e_X^+ is empty. Therefore, $\pi(X)$ is simply the prior probability of the variable X. Since e_X^- is the empty set, $\lambda(X)$ is $p(\Phi \mid X)$. Since Φ is a constant, each $p(\Phi \mid x)$ is equal to $\frac{1}{n}$, where n is the number of states of X. For the purpose of simplicity, we will write an initial λ vector simply as a nonnormalized n-vector $(1,1,\ldots,1)$.

Recall that the relationships among the variables in a network are quantified via CPTs of the form $p(X \mid U_1, U_2, \ldots, U_n)$ for a variable X with parents $U_1, U_2, \ldots, U_n$. Therefore, if X has no parent (that is, is a *root node*) then its CPT is just $p(X)$, which is its prior probability. With this in mind, we present a simple recursive routine to compute the beliefs and the π and λ vectors in a fresh network.

First, mark all the root nodes. Then recursively compute $p(X)$ of a node X, each of whose parents is already marked, and mark the node X itself. If X has parents $U_1, U_2, ..., U_n$ then

$$\begin{aligned} p(X) &= \sum_{U_1,...,U_n} p(X \mid U_1,...,U_n) p(U_1,...,U_n) \\ &= \sum_{U_1,...,U_n} p(X \mid U_1,...,U_n) p(U_1 \mid U_2,...,U_n) ... p(U_{n-1} \mid U_n) p(U_n) \end{aligned}$$

Since $U_1, U_2, ..., U_n$ are marginally independent

$$p(X) = \sum_{U_1,...,U_n} p(X \mid U_1,...,U_n) \prod_{i=1}^{n} p(U_i)$$

Thus, $p(X)$ can be computed using its CPT and the prior probabilities of its parents.

Example

Consider the network in Figure 6-10, along with the prior probabilities $p(Roads)$ and $p(Visibility)$ of the root nodes *Roads* and *Visibility*, respectively, and the two CPTs $p(Journey \mid Roads, Visibility)$ and $p(Mobility \mid Journey)$ of the two other nodes of the network. The network also shows the initial π and λ vectors, and hence belief vectors, of each of the two root nodes.

The prior probability of the node *Journey* and then of the node *Mobility* is computed as follows:

$$\begin{aligned} p(Journey) &= \sum_{Roads, Field} p(Journey \mid Roads, Visibility) p(Roads) p(Visibility) \\ &= \begin{bmatrix} 0.67 \\ 0.33 \end{bmatrix} \end{aligned}$$

$$\begin{aligned} p(Mobility) &= \sum_{Game} p(Mobility \mid Journey) p(Journey) \\ &= \begin{bmatrix} 0.50 \\ 0.17 \\ 0.33 \end{bmatrix} \end{aligned}$$

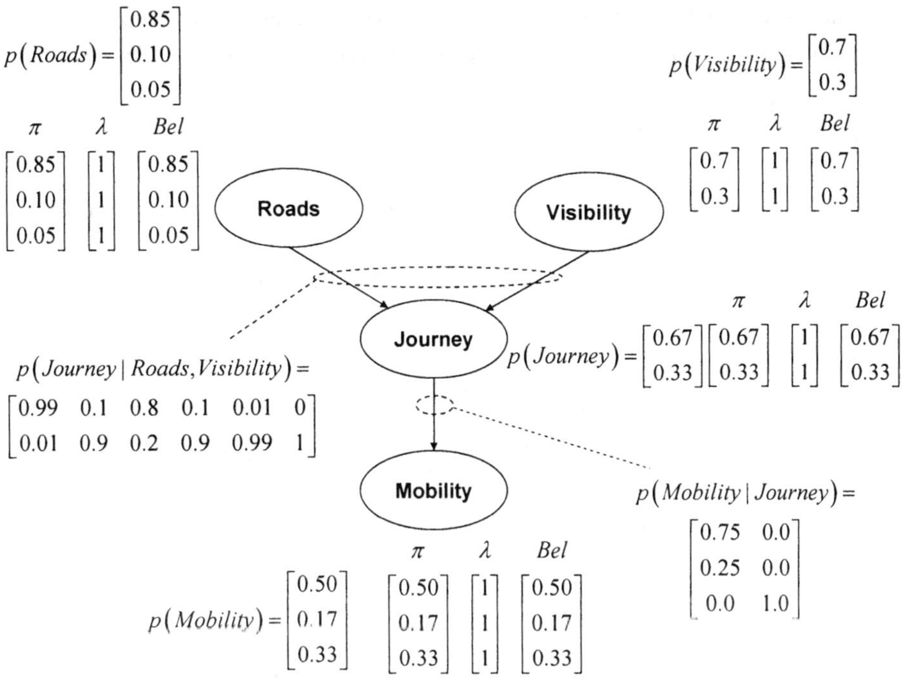

Figure 6-10: Initial probabilities, beliefs, and π and λ vectors

6.5 BELIEF REVISION

In this section, we describe how a node revises its own belief upon receiving evidence on itself. Suppose a node X receives evidence e_X and the probability vector $p(X)$ is its current state certainties. Then its posterior probability is defined as:

$$p(X \mid e_X) = \frac{p(X)\, p(e_X \mid X)}{p(e_X)} = \alpha\, p(X)\lambda(X)$$

where the normalizing constant α is computed by summing over mutually exclusive and exhaustive states of the variable X (that is, $\sum_{X=x} p(X) = 1$):

$$\alpha = \frac{1}{p(e_X)} = \frac{1}{\sum_X p(X, e_X)} = \frac{1}{\sum_X p(e_X \mid X)\, p(X)}$$

Therefore, the belief of the node X after receiving evidence e_X becomes the normalized product of its prior probability vector $p(X)$ with the likelihood vector $\lambda(X)$

Example

Consider the node *Weather* whose prior probability $p(Weather)$ and posted evidence $e_{Weather}$ are shown in Figure 6-11. A particular evidence e_X on a variable X in a network will be hypothetically considered as a binary child node of the node X, where the CPT $p(e_X \mid X) = \lambda(X)$.

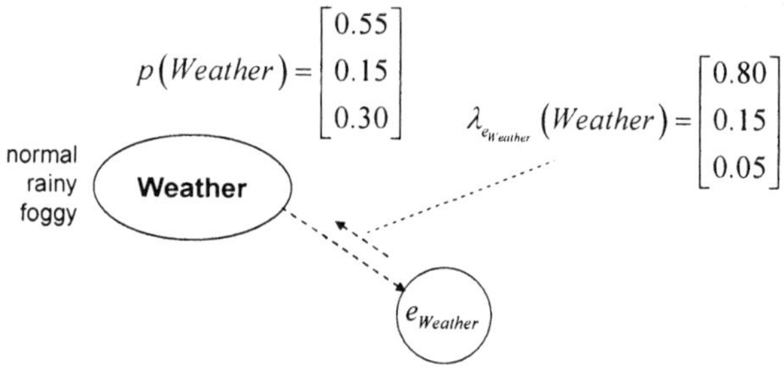

Figure 6-11: Posting evidence on a node

The posterior probability of the variable *Weather*, for example, is computed as follows:

$$p(Weather \mid e_{Weather}) = \begin{bmatrix} \alpha \times 0.55 \times 0.80 \\ \alpha \times 0.15 \times 0.15 \\ \alpha \times 0.30 \times 0.05 \end{bmatrix} = \begin{bmatrix} 0.92 \\ 0.05 \\ 0.03 \end{bmatrix}$$

where $\alpha = \dfrac{1}{0.55 \times 0.80 + 0.15 \times 0.15 + 0.30 \times 0.05} = 2.09$.

Now consider belief revision in the case illustrated in Figure 6-9. The posterior probability of the node X upon receiving all the evidence is computed as follows:

$$
\begin{aligned}
Bel(X) &= p\left(X \mid e_X^+, e_X^-\right) \\
&= \frac{p\left(e_X^+, e_X^-, X\right)}{p\left(e_X^+, e_X^-\right)} \\
&= \frac{p\left(e_X^- \mid e_X^+, X\right) p\left(X \mid e_X^+\right) p\left(e_X^+\right)}{p\left(e_X^+, e_X^-\right)} \\
&= \alpha\, p\left(X \mid e_X^+\right) p\left(e_X^- \mid X\right) \\
&= \alpha\, \pi(X) \lambda(X)
\end{aligned}
$$

where

$$
\begin{aligned}
\alpha &= \frac{p\left(e_X^+\right)}{p\left(e_X^-, e_X^+\right)} = \frac{1}{p\left(e_X^- \mid e_X^+\right)} \\
&= \frac{1}{\sum_{X=x} p\left(e_X^- \mid e_X^+, X\right) p(X)} \\
&= \frac{1}{\sum_{X=x} p\left(e_X^- \mid e_X^+, X\right) p\left(X \mid e_X^+\right)} \\
&= \frac{1}{\sum_{X=x} \pi(X) \lambda(X)}
\end{aligned}
$$

Note that $p\left(e_X^- \mid e_X^+, X\right) = p\left(e_X^- \mid X\right)$ because X separates e_X^+ and e_X^-. The node belief is therefore the normalized product of its λ and π vectors, which can be factorized as follows.

Consider a concrete instantiation of the above case as shown in Figure 6-12. The node X has parents U, V, and W, through which it received evidence e_U^+, e_V^+, and e_W^+ respectively. Node X has children Y, Z, and Q, through which it received evidence e_Y^-, e_Z^-, and e_Q^-, respectively. Thus,

$$
\begin{aligned}
e_X^+ &= \left\{e_U^+, e_V^+, e_W^+\right\} \\
e_X^- &= \left\{e_Y^-, e_Z^-, e_Q^-\right\}
\end{aligned}
$$

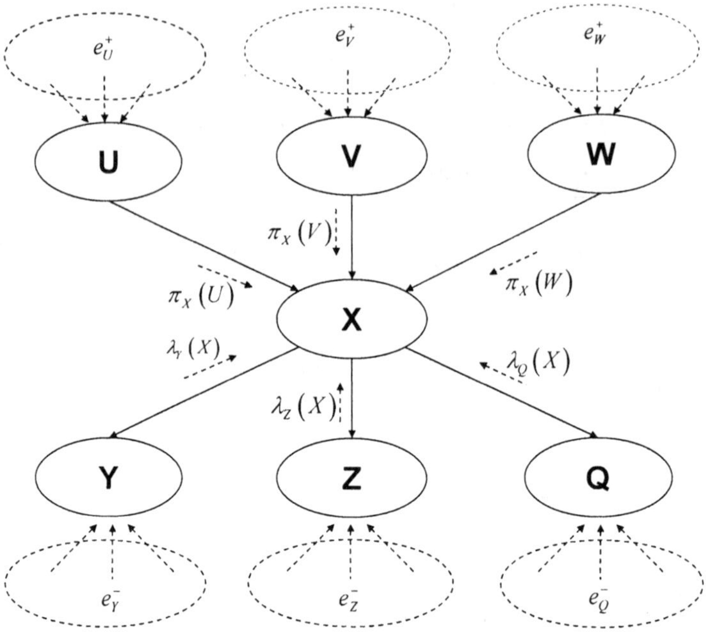

Figure 6-12: Node *X* has multiple parents and children

$$
\begin{aligned}
\pi(X) &= p\left(X \mid e_X^+\right) \\
&= p\left(X \mid e_U^+, e_V^+, e_W^+\right) \\
&= \sum_{U,V,W} p\left(X \mid U,V,W,e_U^+,e_V^+,e_W^+\right) p\left(U,V,W \mid e_U^+,e_V^+,e_W^+\right) \\
&= \sum_{U,V,W} p\left(X \mid U,V,W\right) p\left(U \mid V,W,e_U^+,e_V^+,e_W^+\right) \\
&\qquad p\left(V \mid W,e_U^+,e_V^+,e_W^+\right) p\left(W \mid e_U^+,e_V^+,e_W^+\right) \\
&= \sum_{U,V,W} p\left(X \mid U,V,W\right) p\left(U \mid e_U^+\right) p\left(V \mid e_V^+\right) p\left(W \mid e_W^+\right) \\
&= \sum_{U,V,W} p\left(X \mid U,V,W\right) \pi_X(U) \pi_X(V) \pi_X(W)
\end{aligned}
$$

The above derivation uses the conditional assumption since *U*, *V*, and *W* separate *X* from e_U^+, e_V^+, and e_W^+, respectively. The derivation also uses independence relationships, such as *U* is independent of *V*, *W*, e_V^+, and e_W^+; *V* is independent of *U*, *W*, e_U^+, and e_W^+; and *W* is independent of *U*, *V*, e_U^+, and e_V^+. Similarly,

$$
\begin{aligned}
\lambda(X) &= p(e_X^- \mid X) \\
&= p(e_Y^-, e_Z^-, e_Q^- \mid X) \\
&= p(e_Y^- \mid e_Z^-, e_Q^-, X)\, p(e_Z^- \mid e_Q^-, X)\, p(e_Q^- \mid X) \\
&= p(e_Y^- \mid X)\, p(e_Z^- \mid X)\, p(e_Q^- \mid X) \\
&= \lambda_Y(X)\, \lambda_Z(X)\, \lambda_Q(X)
\end{aligned}
$$

This derivation uses the conditional independence assumptions that given *X*, e_Y^- is independent of e_Z^- and e_Q^-, and e_Z^- is independent of e_Q^-.

Now consider the case shown in Figure 6-13, when a given node *X* receives evidence e_X. In addition, suppose e_X^+ and e_X^- are the total evidence connected to *X* through its parents and children, respectively. Then the revised $\lambda(X)$ can be computed by using the conditional independence assumption derived from the fact that *X* separates e_X and e_X^-:

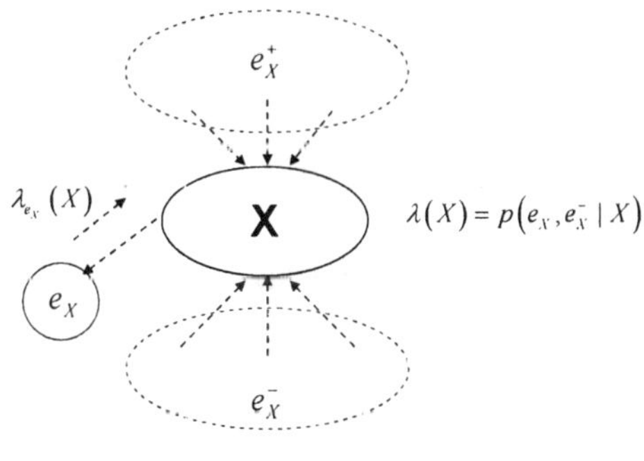

Figure 6-13: Node *X* receives evidence e_X

$$
\begin{aligned}
\lambda_{new}(X) &= p(e_X\, e_X^- \mid X) \\
&= p(e_X^- \mid e_X, X)\, p(e_X \mid X) \\
&= p(e_X^- \mid X)\, p(e_X \mid X) \\
&= \lambda(X)\, \lambda_{e_X}(X)
\end{aligned}
$$

Thus, a node revises its λ vector by multiplying its λ vector with the likelihood vector for the evidence. Note that $\pi(X)$ remains unchanged as e_X^+ is unchanged. The revised belief of *X* is computed as follows, using the necessary independence assumption derived from the fact that *X* separates e_X^+ from e_X^-, and e_X:

$$
\begin{aligned}
Bel_{new}(X) &= p\left(X \mid e_X^+, e_X^-, e_X\right) \\
&= \frac{p\left(e_X^+, e_X^-, e_X, X\right)}{p\left(e_X^+, e_X^-, e_X\right)} \\
&= \frac{p\left(e_X^-, e_X \mid e_X^+, X\right) p\left(X \mid e_X^+\right) p\left(e_X^+\right)}{p\left(e_X^+, e_X^-, e_X\right)} \\
&= \alpha\, p\left(X \mid e_X^+\right) p\left(e_X^-, e_X \mid X\right) \qquad \left[\text{where } \alpha \text{ is } \frac{1}{p\left(e_X^-, e_X \mid e_X^+\right)}\right] \\
&= \alpha\, \pi(X) \lambda_{new}(X)
\end{aligned}
$$

Therefore, the revised belief is simply the product of the revised $\lambda(X)$ with unchanged $\pi(X)$.

Example

Consider the network fragment at the top half of Figure 6-14 along with the π, λ, and belief vectors.

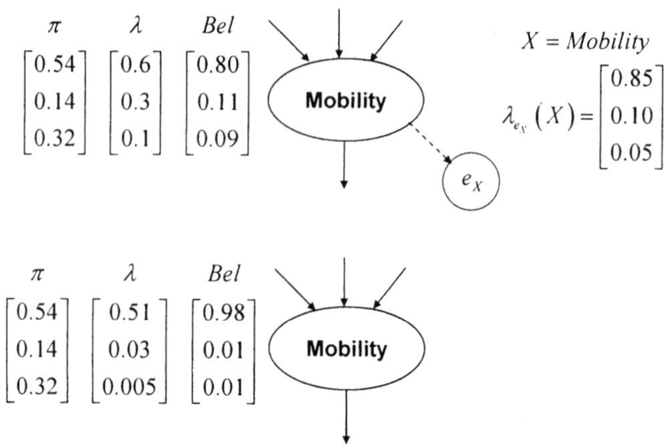

Figure 6-14: Example of a belief revision

As shown in the figure, evidence e_X has been posted into the node *Mobility*. Then the revised λ and belief vectors are computed below (also shown at the bottom half of the figure).

$$\lambda_{new}(X) = \lambda(X)\lambda_{e_X}(X) = \begin{bmatrix} 0.6 \\ 0.3 \\ 0.1 \end{bmatrix} \begin{bmatrix} 0.85 \\ 0.10 \\ 0.05 \end{bmatrix} = \begin{bmatrix} 0.51 \\ 0.03 \\ 0.005 \end{bmatrix}$$

$$\pi(X) = \begin{bmatrix} 0.54 \\ 0.14 \\ 0.32 \end{bmatrix}$$

$$Bel(X) = \alpha\pi(X)\lambda_{new}(X) = \alpha \begin{bmatrix} 0.54 \\ 0.14 \\ 0.32 \end{bmatrix} \begin{bmatrix} 0.51 \\ 0.03 \\ 0.005 \end{bmatrix} = \begin{bmatrix} 0.98 \\ 0.015 \\ 0.005 \end{bmatrix}$$

6.6 EVIDENCE PROPAGATION IN POLYTREES

In the previous section, we detailed how a node updates its own belief upon receiving evidence on itself. This section discusses how a node *X* in a polytree updates its own beliefs when evidence is observed on one of its neighboring nodes and how it propagates the effects of that evidence to the neighboring nodes to help update their beliefs. We consider six different cases corresponding to possible polytree fragments around the node *X* that contain its immediate neighbors:

- *Upward* propagation in a *linear* fragment: *X* has only one parent *U*, and one child *Y*, and the child *Y* receives evidence.
- *Downward* propagation in a *linear* fragment: *X* has only one parent *U*, and one child *Y*, and the parent *U* receives evidence.
- *Upward* propagation in a *tree* fragment: *X* has only one parent *U*, and three children *Y*, *Z*, and *Q*, and one of the children, say *Y*, receives evidence.
- *Downward* propagation in a *tree* fragment: *X* has only one parent *U*, and three children *Y*, *Z*, and *Q*, and the parent *U* receives evidence.
- *Upward* propagation in a *polytree* fragment: *X* has parents *U*, *V*, and *W*, and three children *Y*, *Z*, and *Q*, and one of the children, say *Y*, receives evidence.
- *Downward* propagation in a *polytree* fragment: *X* has parents *U*, *V*, and *W*, and three children *Y*, *Z*, and *Q*, and one of the parents, say *U*, receives evidence.

6.6.1 Upward Propagation in a Linear Fragment

This case is illustrated in Figure 6-15, where the node *X* has only one parent *U*, and one child *Y*, and the child *Y* receives evidence e_Y. The node *Y* updates its belief and sends the message $\lambda_Y(X)$ to *X*. The node *X* updates its belief upon receiving the message from *Y* and, in turn, sends the message $\lambda_X(U)$ to *U* to help update its belief. All the π vectors remain unchanged, as there is no new causal evidence. Next, we compute the values of $\lambda_Y(X)$, $\lambda_X(U)$, and their relations to the new beliefs of X and U.

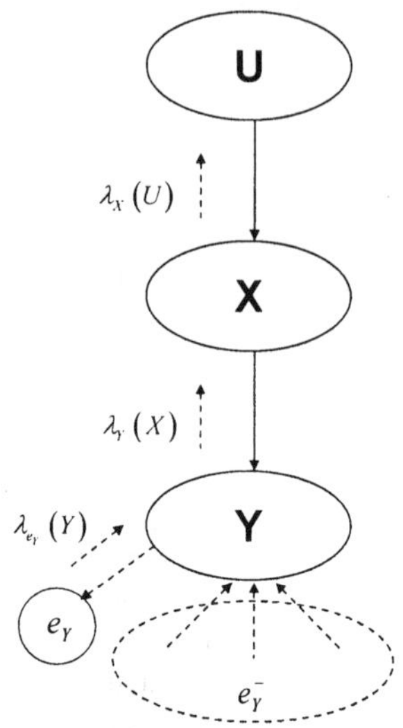

Figure 6-15: Upward propagation of evidence e_Y in a linear fragment

Let e_Y^- be the total evidence connected to Y, but not through its parent X. If Y now receives evidence e_Y, then

$$\lambda_{new}(Y) = p(e_Y, e_Y^- \mid Y)$$

Since Y separates X from e_Y and e_Y^-, the revised $\lambda(X)$ can now be computed as follows:

$$\begin{aligned}\lambda_{new}(X) &= p(e_Y\, e_Y^- \mid X)\\ &= \sum_Y p(e_Y\, e_Y^- \mid Y, X)\, p(Y \mid X)\\ &= \sum_Y p(e_Y\, e_Y^- \mid Y) p(Y \mid X)\\ &= \sum_Y \lambda_{new}(Y)\, p(Y \mid X)\end{aligned}$$

Therefore, the revised $\lambda(X)$ can be computed at the node Y by taking the product of the revised $\lambda(Y)$ and the CPT $p(Y \mid X)$. The revised value $\lambda_{new}(X)$ is then sent to the node X from Y as the message $\lambda_Y(X) = \lambda_{new}(X)$. Note that $\pi(X)$ remains unchanged as e_X^+ is unchanged. Since X separates e_X^+ from e_Y and e_Y^-, the node X revises its belief as follows:

$$
\begin{aligned}
Bel_{new}(X) &= p\left(X \mid e_Y, e_Y^-, e_X^+\right) \\
&= \alpha\, p\left(X \mid e_X^+\right) p\left(e_Y, e_Y^- \mid X\right) \\
&= \alpha\, \pi(X) \lambda_{new}(X)
\end{aligned}
$$

Therefore, X revises its belief by multiplying the revised $\lambda(X)$, sent as a message by Y, with its unchanged $\pi(X)$. The revised $\lambda(U)$ can now be computed as follows:

$$\lambda_{new}(U) = \sum_X \lambda_{new}(X)\, p(X \mid U)$$

X sends $\lambda_{new}(U)$ as a message $\lambda_X(U)$ to U.

Example

Consider the linear fragment shown in Figure 6-16 along with the π, λ, and belief vectors.

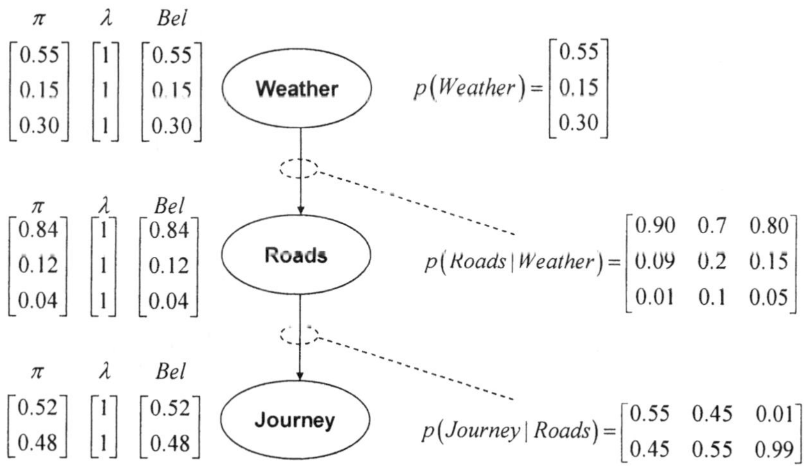

Figure 6-16: Example linear fragment

The upward propagation of evidence due to the posting of evidence $e_{Journey}$ at the node *Journey* is shown in Figure 6-17.

The two λ-values $\lambda_{Game}(Roads)$ and $\lambda_{Roads}(Weather)$ in the figure are computed as follows:

$$\lambda_{Journey}(Roads) = \lambda_{new}(Roads)$$
$$= \sum_{Journey} \lambda_{new}(Journey)\, p(Journey \mid Roads)$$
$$= \begin{bmatrix} 0.9\times0.55+0.1\times0.45 \\ 0.9\times0.45+0.1\times0.55 \\ 0.9\times0.01+0.1\times0.99 \end{bmatrix} = \begin{bmatrix} 0.54 \\ 0.46 \\ 0.11 \end{bmatrix}$$

$$\lambda_{Roads}(Weather) = \lambda_{new}(Weather)$$
$$= \sum_{Roads} \lambda_{new}(Roads)\, p(Roads \mid Weather)$$
$$= \begin{bmatrix} 0.54\times0.90+0.46\times0.09+0.11\times0.01 \\ 0.54\times0.7+0.46\times0.2+0.11\times0.1 \\ 0.54\times0.80+0.46\times0.50+0.11\times0.05 \end{bmatrix} = \begin{bmatrix} 0.53 \\ 0.48 \\ 0.51 \end{bmatrix}$$

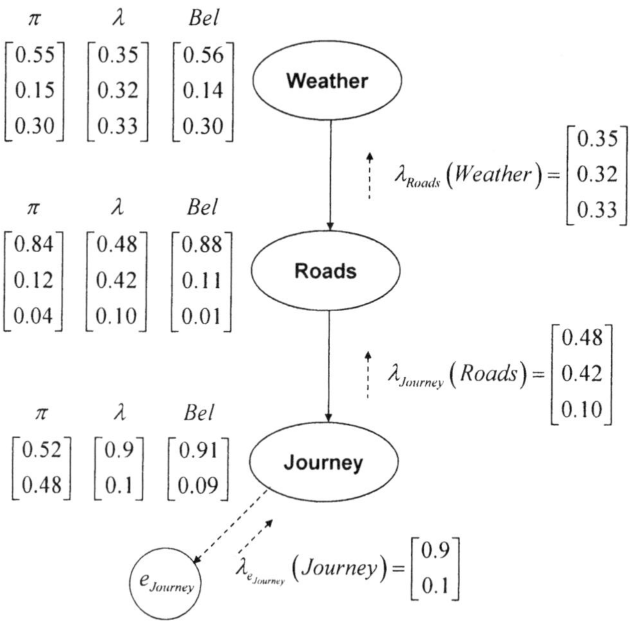

Figure 6-17: Example upward propagation in a linear fragment

6.6.2 Downward Propagation in a Linear Fragment

This case is illustrated in Figure 6-18, where the node X has only one parent U, and one child Y, and the parent node U receives evidence e_U. The node U updates its belief and sends the message $\pi_X(U)$ to X. The node X updates its belief upon receiving the message from U and, in turn, sends the message $\pi_Y(X)$ to X to help update its belief. Next, we compute the values of $\pi_X(U)$, $\pi_Y(X)$, and their relations to the new beliefs of X and Y, respectively.

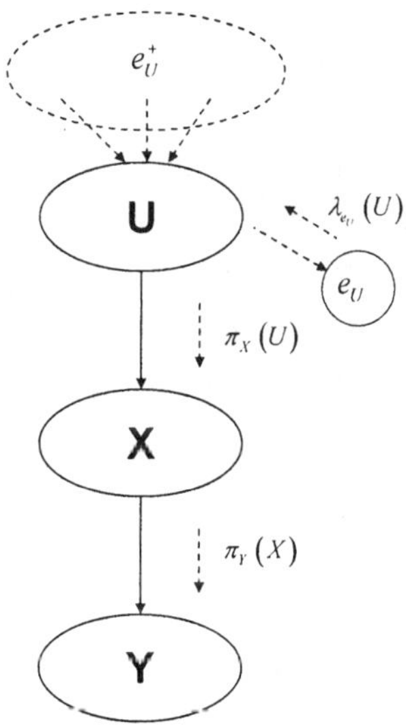

Figure 6-18: Downward propagation of evidence e_U in a linear fragment

Let e_U^+ be the total evidence connected to U but not through its child X. If U now receives evidence e_U, then $\lambda_{new}(U) = \lambda_{e_U}(U)\lambda(U)$. Note that $\pi(U)$ remains unchanged since there is no change in evidence connected through the parents of U. The revised $\pi(X)$ can now be computed as follows:

$$\begin{aligned}\pi_{new}(X) &= p(X \mid e_U e_U^+) \\ &= \sum_U p(X \mid U, e_U, e_U^+) p(U \mid e_U, e_U^+) \\ &= \sum_U p(X \mid U) p(U \mid e_U e_U^+) \quad [U \text{ separates } X \text{ from } e_U \text{ and } e_U^+] \\ &= \sum_U \pi_X(U) p(X \mid U)\end{aligned}$$

where $\pi_X(U) = p(U \mid e_U e_U^+)$ is simplified as follows:

$$\pi_X(U) = p(U \mid e_U\, e_U^+)$$
$$= \frac{p(e_U, e_U^+, U)}{p(e_U, e_U^+)}$$
$$= \frac{p(e_U \mid e_U^+, U)\, p(U \mid e_U^+)\, p(e_U^+)}{p(e_U, e_U^+)}$$
$$= \alpha\, p(U \mid e_U^+)\, p(e_U \mid U) \quad \left[\text{since } U \text{ separates } e_U \text{ and } e_U^+\right]$$
$$= \alpha\, \pi(U)\, \lambda_{e^U}(U) \quad \left[\alpha \text{ is } \frac{1}{p(e_U \mid e_U^+)}\right]$$

The node U can compute $\pi_X(U)$ by multiplying its likelihood vector for the evidence with its π vector. Therefore, the revised $\pi(X)$, $\pi_{new}(X)$, can be computed at the node X by taking the product of $\pi_X(U)$ and the CPT $p(X \mid U)$. The revised value $\pi_{new}(X)$ is then sent to the node Y from X as the message $\pi_Y(X)$. Note that $\lambda(X)$ remains unchanged since e_X^- is unchanged. The node X revises its belief as follows:

$$Bel_{new}(X) = p(X \mid e_U, e_U^+, e_U^-)$$
$$= \alpha\, p(e_U, e_U^+ \mid X)\, p(X \mid e_U^-) \left[\text{since } X \text{ separates } e_U^- \text{ from } e_U \text{ and } e_U^+\right]$$
$$= \alpha\, \pi_{new}(X)\, \lambda(X)$$
$$= \alpha\, \pi_X(U)\, p(X \mid U)\, \lambda(X)$$

Therefore, X revises its belief by multiplying message $\pi_X(U)$ sent by U, with its unchanged λ vector $\lambda(X)$ and the CPT $p(X \mid U)$. Similarly, X sends a message $\pi_Y(X)$ to Y to help revise its belief.

Example

Consider the linear fragment shown in Figure 6-19 along with the π, λ, and belief vectors.

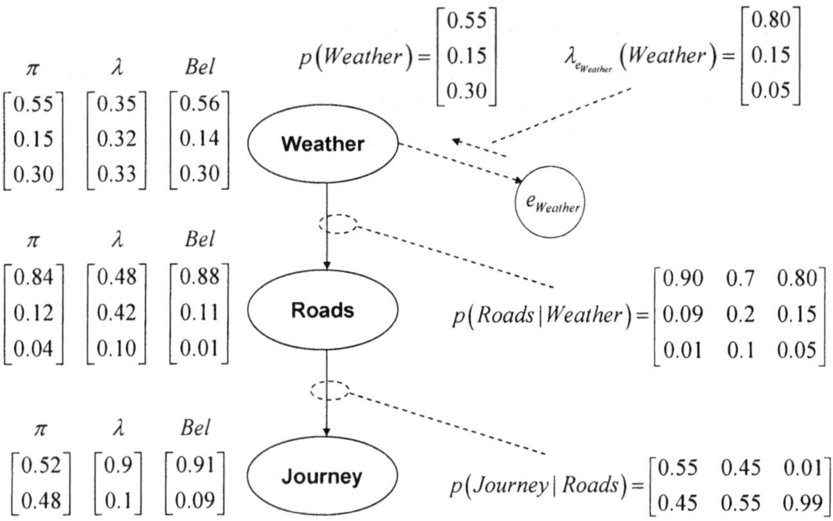

Figure 6-19: Example linear fragment

The downward propagation of evidence due to the posting of evidence $e_{Weather}$ at the node *Weather* is shown in Figure 6-20.

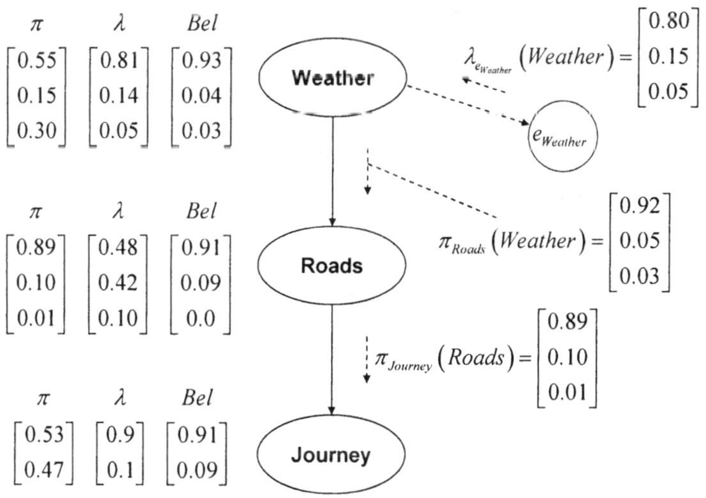

Figure 6-20: Example downward propagation in a linear fragment

6.6.3 Upward Propagation in a Tree Fragment

This case is illustrated in Figure 6-21, where the node X has only one parent U, and three children Y, Z, and Q, and the child Y receives evidence e_Y. The node Y updates its belief and sends the message $\lambda_Y(X)$ to X. The node X updates its belief upon receiving the message from Y and, in turn, sends the diagnostic message $\lambda_X(U)$ to U and the causal messages $\pi_Z(X)$ and $\pi_Q(X)$ to Z and Q, respectively, to help update their beliefs. The messages $\lambda_Y(X)$ and $\lambda_X(U)$ are computed as above in the case of upward propagation in a linear fragment. Next, we compute $\pi_Z(X)$, $\pi_Q(X)$, and their relations to the new beliefs of Z and Q.

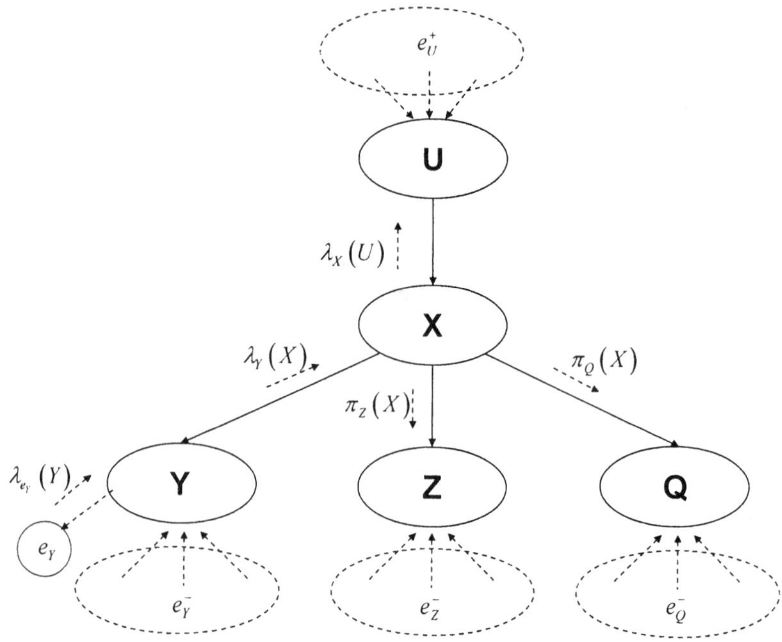

Figure 6-21: Upward propagation of evidence e_Y in a tree fragment

Let e_Y^-, e_Z^-, and e_Q^- be the total diagnostic evidence associated with Y, Z, and Q, respectively, that is not obtained through their parent X, and e_U^+ be the total causal evidence associated with U that is not obtained through its child X.

$$
\begin{aligned}
\pi_{new}(Z) &= p\left(Z \mid e_Y, e_Y^-, e_Q^-, e_U^+\right) \\
&= \sum_X p\left(Z \mid X, e_Y, e_Y^-, e_Q^-, e_U^+\right) p\left(X \mid e_Y, e_Y^-, e_Q^-, e_U^+\right) \\
&= \sum_X p\left(Z \mid X, e_Y, e_Y^-, e_Q^-\right) p\left(X \mid e_Y, e_Y^-, e_Q^-, e_U^+\right) \\
&\qquad\qquad \left[\text{given } X,\ Z \text{ is independent of } e_U^+\right] \\
&= \sum_X p(Z \mid X)\,\pi_Z(X) \left[\text{given } X,\ Z \text{ is independent of } e_Y, e_Y^-, e_Q^-\right]
\end{aligned}
$$

where

$$
\begin{aligned}
\pi_Z(X) &= p\left(X \mid e_Y, e_Y^-, e_Q^-, e_U^+\right) = \frac{p\left(e_Y, e_Y^-, e_Q^-, e_U^+, X\right)}{p\left(e_Y, e_Y^-, e_Q^-, e_U^+\right)} \\
&= \frac{p\left(e_Y, e_Y^- \mid e_Q^-, e_U^+, X\right) p\left(e_Q^- \mid e_U^+, X\right) p\left(X \mid e_U^+\right) p\left(e_U^+\right)}{p\left(e_Y, e_Y^-, e_Q^-, e_U^+\right)} \\
&= \alpha\, p\left(e_Y, e_Y^- \mid e_Q^-, e_U^+, X\right) p\left(e_Q^- \mid e_U^+, X\right) p\left(X \mid e_U^+\right) \left[\alpha = \frac{1}{p\left(e_Y, e_Y^-, e_Q^- \mid e_U^+\right)}\right] \\
&= \alpha\, p\left(e_Y, e_Y^- \mid X\right) p\left(e_Q^- \mid e_U^+, X\right) p\left(X \mid e_U^+\right) \left[X \text{ separates } e_Y, e_Y^- \text{ from } e_Q^-, e_U^+\right] \\
&= \alpha\, p\left(e_Y, e_Y^- \mid X\right) p\left(e_Q^- \mid X\right) p\left(X \mid e_U^+\right) \left[X \text{ separates } e_Q^- \text{ from } e_U^+\right] \\
&= \alpha\, \lambda_Y(X) \lambda_Q(X) \pi(X) \\
&= \alpha \frac{\lambda_Y(X) \lambda_Z(X) \lambda_Q(X) \pi(X)}{\lambda_Z(X)} \\
&= \alpha \frac{Bel_{new}(X)}{\lambda_Z(X)}
\end{aligned}
$$

Therefore, the revised $\pi(Z)$ can be computed at the node Z by taking the product of the message $\frac{Bel_{new}(X)}{\lambda_Z(X)}$, sent by its parent X, and the CPT $p(Z \mid X)$. Similarly, X sends $\frac{Bel_{new}(X)}{\lambda_Q(X)}$ to Q to update its belief. Note that $\lambda(Z)$ remains unchanged since e_Z^- is unchanged. The node Z revises its belief as follows:

$$\begin{aligned}Bel_{new}(Z) &= p\left(Z \mid e_Y, e_Y^-, e_Q^-, e_U^+, e_Z^-\right)\\ &= \alpha\, p\left(Z \mid e_Y, e_Y^-, e_Q^-, e_U^+\right) p\left(e_Z^- \mid Z\right) \left[\text{since } Z \text{ separates } e_Z^- \text{ from } e_Y, e_Y^-, e_Q^-, e_U^+\right]\\ &= \alpha \pi_{new}(Z)\lambda(Z)\\ &= \alpha \pi_Z(X)\, p(Z \mid X)\lambda(Z)\end{aligned}$$

Therefore, Z revises its belief by multiplying the message $\pi_Z(X)$, sent by its parent X, with its unchanged λ vector $\lambda(Z)$ and its CPT $p(Z \mid X)$.

6.6.4 Downward Propagation in a Tree Fragment

This case, which is illustrated in Figure 6-22, is similar to the case of downward propagation in a linear fragment presented earlier.

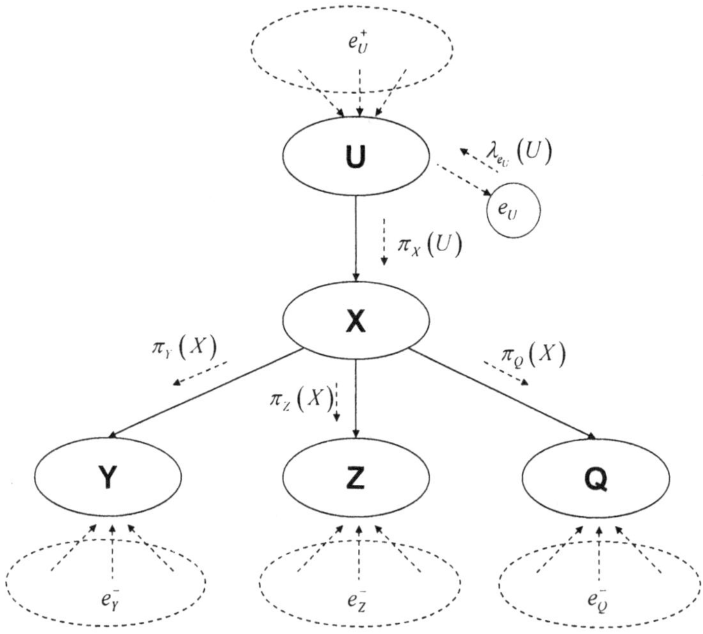

Figure 6-22: Downward propagation of evidence e_U in a tree fragment

6.6.5 Upward Propagation in a Polytree Fragment

This case is illustrated in Figure 6-23, where the node X has three parents U, V, and W, and three children Y, Z, and Q, and the child Y receives evidence e_Y. The node Y updates its belief and sends the message $\lambda_Y(X)$ to X. The node X updates its belief upon receiving the message from Y and, in turn, sends the causal messages $\pi_Z(X)$ and $\pi_Q(X)$ to Z and Q, respectively, to help update their beliefs. The messages $\pi_Z(X)$ and $\pi_Q(X)$ are $\frac{Bel_{new}(X)}{\lambda(Z)}$ and $\frac{Bel_{new}(X)}{\lambda(Q)}$ computed as above in the case of upward propagation in a tree fragment. In the following, we show how U, V, and W update their beliefs upon receiving the messages $\lambda_X(U)$, $\lambda_X(V)$, and $\lambda_X(W)$, respectively, from their common child X.

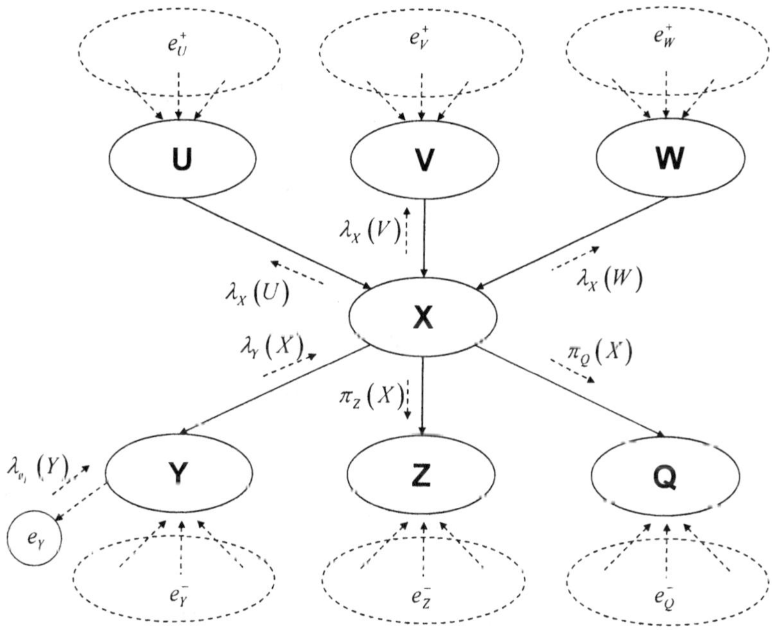

Figure 6-23: Upward propagation of evidence e_Y in a polytree fragment

The revised $\lambda(U)$ can now be computed as follows:

$$\begin{aligned}
\lambda_{new}(U) &= p\left(e_Y, e_Y^-, e_Z^-, e_Q^-, e_V^+, e_W^+ \mid U\right) \\
&= \sum_X \sum_V \sum_W p\left(e_Y, e_Y^-, e_Z^-, e_Q^-, e_V^+, e_W^+ \mid U, V, W, X\right) p(V, W, X \mid U) \\
&= \sum_X \sum_V \sum_W p\left(e_Y, e_Y^-, e_Z^-, e_Q^- \mid e_V^+, e_W^+, U, V, W, X\right) \\
&\qquad\qquad p\left(e_V^+, e_W^+ \mid U, V, W, X\right) p(V, W, X \mid U) \\
&= \sum_X \sum_V \sum_W p\left(e_Y, e_Y^-, e_Z^-, e_Q^- \mid X\right) p\left(e_V^+ \mid e_W^+, U, V, W, X\right) \\
&\qquad\qquad p\left(e_W^+ \mid U, V, W, X\right) p(V, W, X \mid U) \\
&\qquad\qquad \left[\text{since } X \text{ separates } e_Y, e_Y^-, e_Z^-, e_Q^- \text{ from } e_V^+, e_W^+, U, V, W\right] \\
&= \sum_X \sum_V \sum_W \lambda_{new}(X) p\left(e_V^+ \mid e_W^+, V, W\right) p\left(e_W^+ \mid V, W\right) p(V, W, X \mid U) \\
&\qquad\qquad \left[\begin{array}{l}\text{since } V \text{ separates } e_V^+ \text{ from } U, X, \text{ and} \\ W \text{ separates } e_W^+ \text{ from } U, X\end{array}\right] \\
&= \sum_X \lambda_{new}(X) \sum_V \sum_W p\left(e_V^+, e_W^+ \mid V, W\right) p(X \mid V, W, U) p(V, W \mid U)
\end{aligned}$$

Since U, V, and W are marginally independent, $\lambda_{new}(U)$ can further be simplified as follows:

$$\begin{aligned}
\lambda_{new}(U) &= \alpha \sum_X \lambda_{new}(X) \sum_V \sum_W \frac{p\left(V, W \mid e_V^+, e_W^+\right)}{p(V, W)} p(X \mid V, W, U) p(V, W) \\
&= \alpha \sum_X \lambda_{new}(X) \sum_V \sum_W p\left(V \mid W, e_V^+, e_W^+\right) p\left(W \mid e_V^+, e_W^+\right) p(X \mid V, W, U) \\
&= \alpha \sum_X \lambda_{new}(X) \sum_V \sum_W p\left(V \mid e_V^+\right) p\left(W \mid e_W^+\right) p(X \mid V, W, U) \\
&= \alpha \sum_X \lambda_{new}(X) \sum_V \sum_W \pi_X(V) \pi_X(W) p(X \mid V, W, U)
\end{aligned}$$

Therefore, the message $\lambda_X(U)$, which will be sent to U from X as the new λ vector for U, is the above expression computed at X. Note that $p(X \mid V, W, U)$ is the CPT of X stored at the node X. The revised belief of U is obtained by multiplying its unchanged π vector with the above λ vector. The messages that are sent to V and W are the following:

$$\lambda_X(V) = \alpha \sum_X \lambda_{new}(X) \sum_U \sum_W \pi_X(U) \pi_X(W) p(X \mid U, W, V)$$

$$\lambda_X(W) = \alpha \sum_X \lambda_{new}(X) \sum_U \sum_V \pi_X(U) \pi_X(V) p(X \mid U, V, W)$$

Example

Consider the network fragment shown in Figure 6-24 along with the π, λ, and belief vectors, where the two CPTs for the nodes *Mobility* and *Delay* are the same as the two CPTs in Table 6-7 and Table 6-8, respectively, in our main BN example.

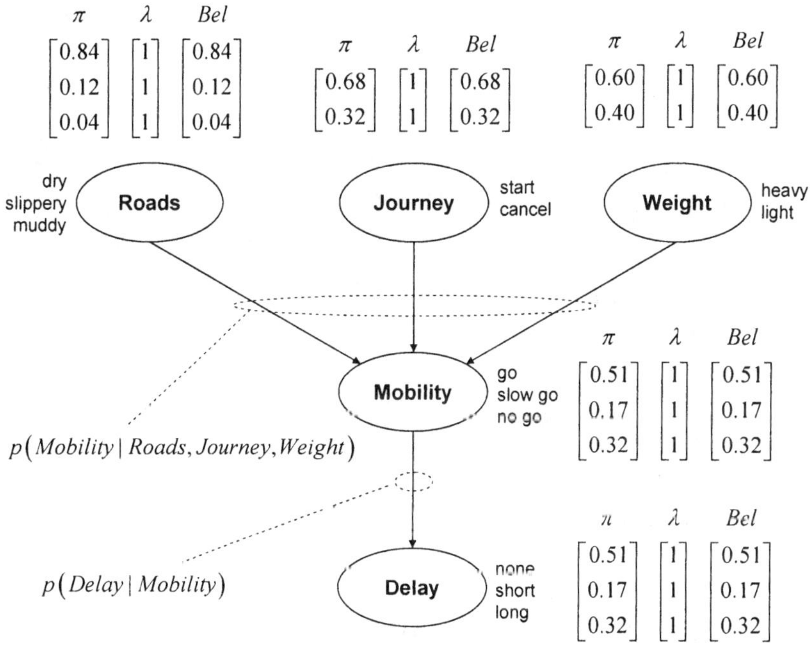

Figure 6-24: Example polytree fragment

The upward propagation of evidence due to the posting of evidence e_{Delay} at the node *Delay* is shown in Figure 6-25.

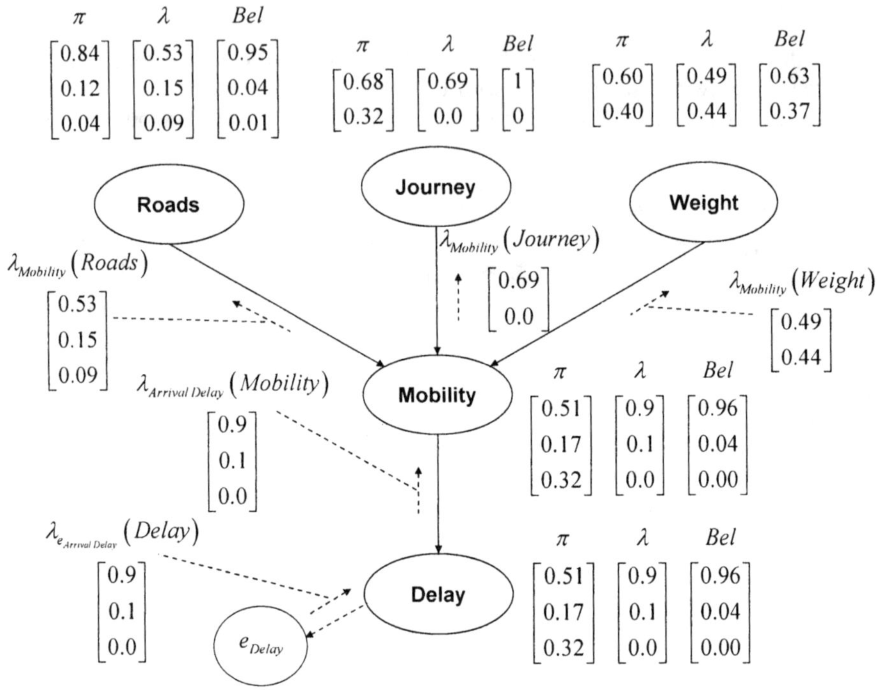

Figure 6-25: Example upward propagation in a polytree fragment

6.6.6 Downward Propagation in a Polytree Fragment

This case is illustrated in Figure 6-26, where the node X has three parents U, V, and W, and three children Y, Z, and Q, and the parent U receives evidence e_U. Node U updates its belief and sends the causal message $\pi_X(U)$ to X. Node X updates its belief upon receiving the message from Y and, in turn, sends the messages $\pi_Y(X)$, $\pi_Z(X)$, and $\pi_Q(X)$ to Y, Z, and Q, respectively, to help update their beliefs. The messages $\pi_Y(X)$, $\pi_Z(X)$, and $\pi_Q(X)$ are $\frac{Bel_{new}(X)}{\lambda(Y)}$, $\frac{Bel_{new}(X)}{\lambda(Z)}$, and $\frac{Bel_{new}(X)}{\lambda(Q)}$, respectively, computed as above in the case of upward propagation in a tree fragment. Next, we show how V and W update their beliefs upon receiving, respectively, the messages $\lambda_X(V)$ and $\lambda_X(W)$ from their common child X.

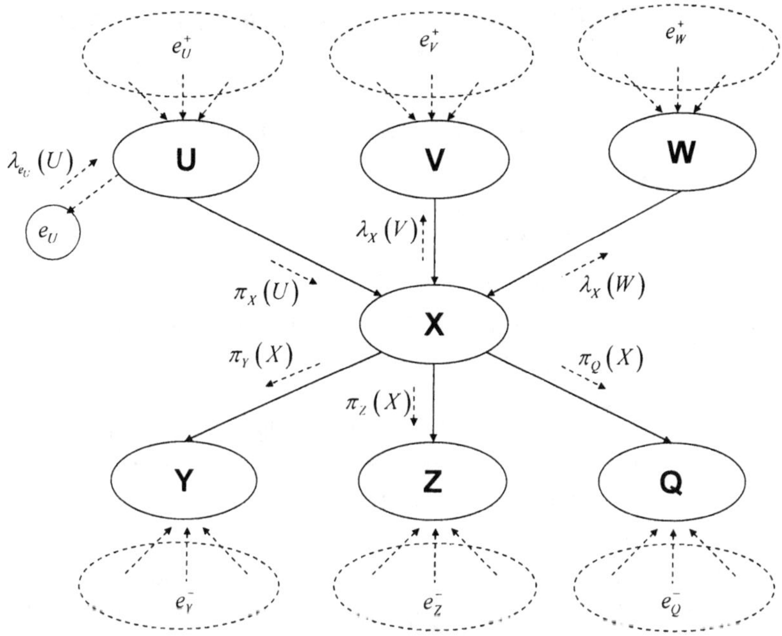

Figure 6-26: Downward propagation of evidence e_U in a polytree fragment

$$
\begin{aligned}
\lambda_{new}(V) &= p\left(e_U, e_U^+, e_W^+, e_Y^-, e_Z^-, e_Q^- \mid V\right) \\
&= \sum_X \sum_U \sum_W p\left(e_U, e_U^+, e_W^+, e_Y^-, e_Z^-, e_Q^- \mid V, U, W, X\right) p\left(U, W, X \mid V\right) \\
&= \sum_X \sum_U \sum_W p\left(e_Y^-, e_Z^-, e_Q^- \mid e_U, e_U^+, e_W^+, V, U, W, X\right) \\
&\qquad p\left(e_U, e_U^+, e_W^+ \mid V, U, W, X\right) p\left(U, W, X \mid V\right) \\
&= \sum_X \sum_U \sum_W p\left(e_Y^-, e_Z^-, e_Q^- \mid X\right) p\left(e_U \mid e_U^+, e_W^+, V, U, W, X\right) \\
&\qquad p\left(e_U^+, e_W^+ \mid V, U, W, X\right) p\left(U, W, X \mid V\right) \\
&\qquad \left[\text{since } X \text{ separates } e_Y^-, e_Z^-, e_Q^- \text{ from } e_U, e_U^+, e_W^+, U, V, W\right]
\end{aligned}
$$

$$=\sum_X\sum_U\sum_W \lambda(X)\,p(e_U^-\mid U)\,p(e_U^+,e_W^+\mid U,W,V)\,p(U,W,X\mid V)$$

$$\begin{bmatrix}\text{since } U \text{ separates } e_U^- \text{ from } e_U^+,e_W^+,X,V,W, \text{and} \\ \text{since } U,\,W \text{ separate } e_U^+,e_W^+ \text{ from } V,X\end{bmatrix}$$

$$=\sum_X\lambda(X)\sum_U\sum_W \lambda_{e_U}(U)\,p(e_U^+\mid U)\,p(e_W^+\mid W)\,p(U,W\mid V)\,p(X\mid U,W,V)$$

$$\left[\text{since } e_U^+,e_W^+ \text{ are independent of each other given } U \text{ or } W\right]$$

$$=\alpha\sum_X\lambda(X)\sum_U\sum_W \lambda_{e_U}(U)\,\pi_X(U)\,\pi_X(W)\,p(X\mid U,W,V)$$

$$\left[\text{since } U \text{ and } W \text{ are marginally independent}\right]$$

Note that if X did not receive any diagnostic evidence from its descendants Y, Z, and Q, then the λ vector for X would still be $(1,1,\ldots,1)$. In this case, the above message sent to V from X due to the evidence on U would still be $(1,1,\ldots,1)$, making no impact on the belief of V. This is consistent with the network marginal independence property, which says that U, V, and W are independent of each other if neither X nor any of its descendants received evidence. This is illustrated below.

Suppose $x_1,\ldots,x_n$ (n states) are all possible instantiations of X. Then, from the CPT of X, U, V, W, we have for any u, v, w:

$$\sum_i p(x_i\mid u,v,w)=1$$

If the variable X did not receive any evidence from its descendants then $\lambda(x_i)$ = 1, for every i. If the variable V has m states and $v_1,\ldots,v_m$ are all possible instantiations of V, then from the derivation of $\lambda_{new}(V)$ above,

$$\begin{aligned}\lambda_X(v_j) &= \alpha\sum_i\lambda(x_i)\sum_U\sum_W \lambda_{e_U}(U)\,\pi_X(U)\,\pi_X(W)\,p(x_i\mid U,W,v_j)\\ &= \alpha\sum_i\sum_U\sum_W \lambda_{e_U}(U)\,\pi_X(U)\,\pi_X(W)\,p(x_i\mid U,W,v_j)\\ &= \alpha\sum_U\sum_W \lambda_{e_U}(U)\,\pi_X(U)\,\pi_X(W)\sum_i p(x_i\mid U,W,v_j)\\ &= \alpha\sum_U\sum_W \lambda_{e_U}(U)\,\pi_X(U)\,\pi_X(W)\end{aligned}$$

Therefore, each $\lambda_X(v_j)$ has the same value, making the vector $\lambda_X(V)$ a unit vector that does not change the belief of V.

Example

Consider the network fragment shown in Figure 6-27 along with the π, λ, and belief vectors, where the two CPTs for the nodes *Mobility* and *Delay* are the same as the two CPTs in Table 6-7 and Table 6-8, respectively, in our main BN example.

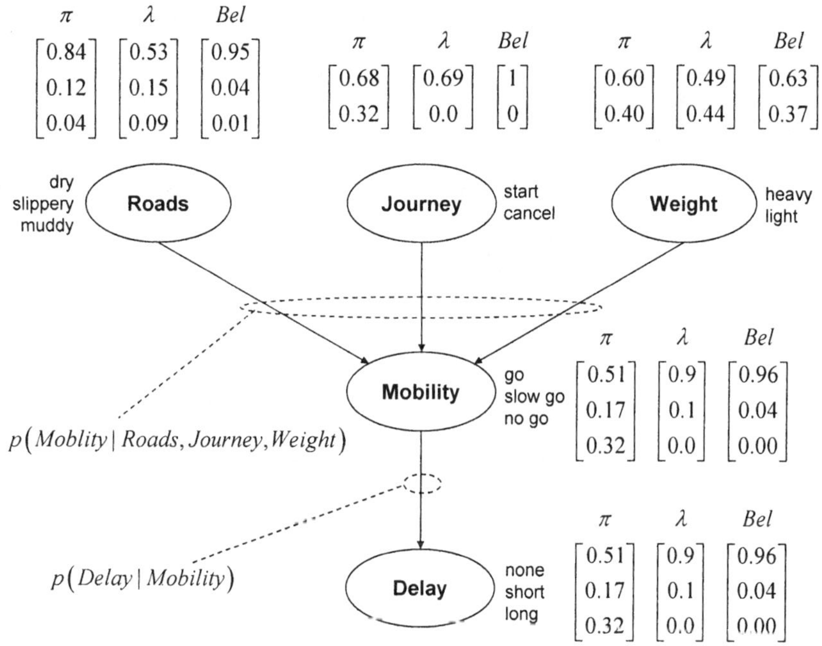

Figure 6-27: Example of a polytree fragment

The downward propagation of evidence due to the posting of evidence e_{Roads} at the node *Roads* is shown in Figure 6-28.

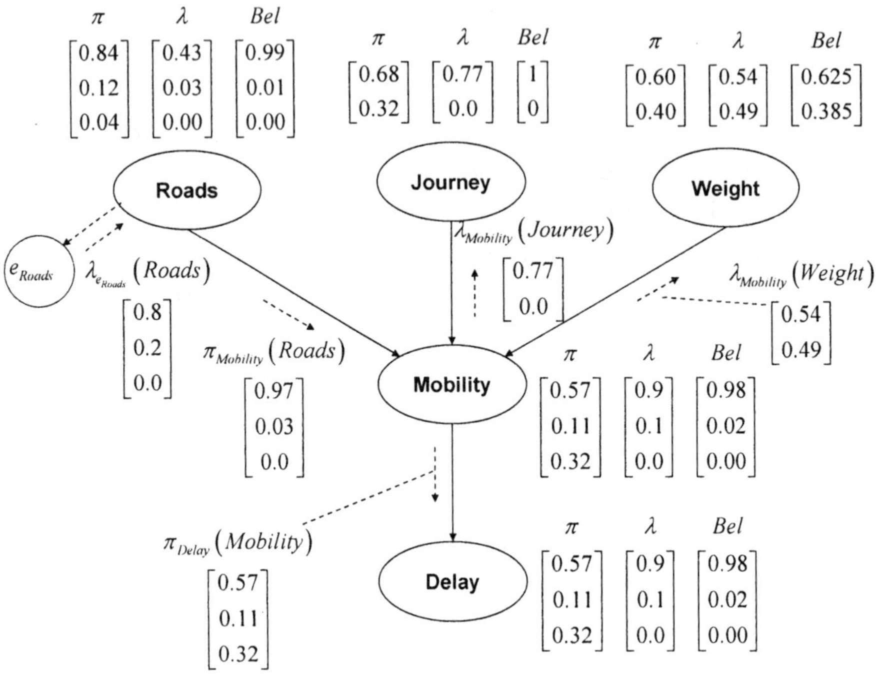

Figure 6-28: Example downward propagation in a polytree fragment

6.6.7 Propagation Algorithm

Now that we have illustrated different belief propagation with a series of examples, we generalize these steps to a concrete algorithm for belief propagation in polytrees.

Input

- BN $\mathcal{N}$ (causal structure with associated CPTs)
- Evidence e_X on the variable X of $\mathcal{N}$

Output

- Revised belief vector for each node, i.e., $p(Y \mid e_X)$, for each node Y

Node Structure

- Each node X in $\mathcal{N}$ with p number of states stores the following information locally in a suitable data structure (as shown in Figure 6-29):

- p-ary π vector $\pi(X)$
- p-ary λ vector $\lambda(X)$
- p-ary belief vector $Bel(X)$
- p-ary evidence vector $\lambda_e(X)$
- $p \times n$ CPT $p(X \mid U_1,...,U_n)$, if X has n parents $U_1, U_2, ..., U_n$
- q-ary parent π vector $\pi_X(U_i)$, for each parent U_i with q number of states
- p-ary child λ vector $\lambda_{Y_i}(X)$, for each child Y_i

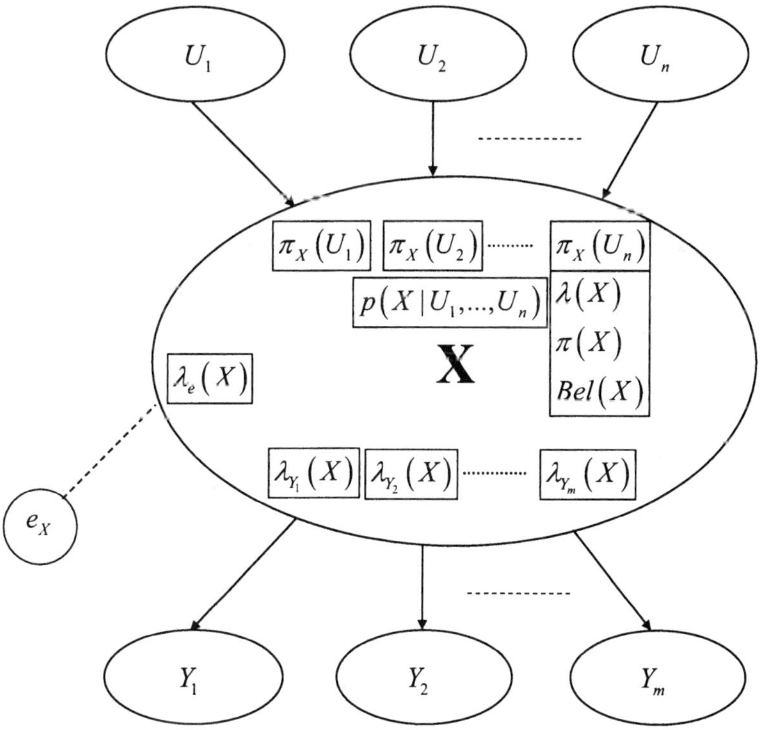

Figure 6-29: Node structure

Initial Probability Computation

- For each node X in $\mathcal{N}$ with p number of states, perform the following:
  - If X is a root node, then set each of its π and belief vectors to its prior probability distribution.
  - Set the evidence vector $\lambda_e(X)$, λ vector $\lambda(X)$, and each child λ vector $\lambda_{Y_i}(X)$ to $(1,...,1)$.
- Mark each root node and then recursively compute the π and belief vectors of the rest of the nodes through the following steps.
- If the parents of a node X are already marked, then perform the following:
  - Set each parent π vector $\pi_X(U_i)$ to the π vector of the parent U_i.
  - Set its π vector $\pi(X)$ to $\sum_{u_1,u_2,...,u_n} p(X \mid u_1,u_2,...,u_n)\prod_{i=1}^{n}\pi_X(u_i)$, if X has parents $U_1,U_2,...,U_n$.
  - Set its belief vector $Bel(X)$ equal to its π vector.

Evidence Propagation

- Set the evidence vector $\lambda_e(X)$ of X to the likelihood vector $p(e_X \mid X)$
- Revise the beliefs of X:
  - Compute the λ vector $\lambda(X)$ of X as the product of all its child λ vectors and the evidence vector.
  - Compute the π vector $\pi(X)$ of X as the product of all its parent π vectors.
  - Compute the belief vector of X as the product of its π and λ vectors.
- Generate messages from X on the update of its evidence vector:
  - Send a message $\alpha\sum_{x}\lambda(x)\sum_{u_1,...,u_{i-1},u_{i+1},...,u_n} p(x \mid u_1,...,u_{i-1},u_{i+1},...,u_n,U_i)\prod_{k\neq i}\pi_X(u_k)$ to each parent U_i of X.
  - Send a message $\alpha\dfrac{Bel(X)}{\lambda_{Y_i}(X)}$ to each child Y_i of X.

- Generate messages from X on the update of the λ_{Y_j} vector:
  - Send a message $\alpha\sum_{x}\lambda(x)\sum_{u_1,\ldots,u_{i-1},u_{i+1}\ldots,u_n} p(x\,|\,u_1,\ldots,u_{i-1},u_{i+1},\ldots,u_n,U_i)\prod_{k\neq i}\pi_X(u_k)$ to each parent U_i of X.
  - Send a message $\alpha\dfrac{Bel(X)}{\lambda_{Y_i}(X)}$ to each child Y_i of X other than the child Y_j.
- Generate messages from X on the update of the $\pi_X(U_j)$ vector:
  - Send a message $\alpha\sum_{x}\lambda(x)\sum_{u_1,\ldots,u_{i-1},u_{i+1}\ldots,u_n} p(x\,|\,u_1,\ldots,u_{i-1},u_{i+1},\ldots,u_n,U_i)\prod_{k\neq i}\pi_X(u_k)$ to each parent U_i of X other than U_j.
  - Send a message $\alpha\dfrac{Bel(X)}{\lambda_{Y_i}(X)}$ to each child Y_i of X.

If a network already received evidence on some of its variables, then only the evidence propagation step is followed, bypassing the initial probability computation step.

6.7 EVIDENCE PROPAGATION IN DIRECTED ACYCLIC GRAPHS

The evidence propagation algorithm presented in the last section cannot usually handle DAGs (such as the one in Figure 6-1) because the evidence propagation on a variable in a DAG that is not a polytree may never terminate due to the recursive computation in the algorithm. This is explained in the context of the DAG shown in Figure 6-30.

Suppose no evidence has been posted to the variable *Weather*, making the two variables *Roads* and *Visibility* marginally independent. Now, if evidence is observed on *Roads*, then *Journey* will receive causal evidence and will send diagnostic evidence to the *Visibility*. The node *Roads* then receives evidence from *Visibility* via *Weather*. This cyclic process of evidence propagation continues unless a stopping criterion is put in place based on repeated evidence and changes in node belief. A similar cyclic process continues in the other direction when *Roads* send diagnostic evidence to *Weather* upon receiving evidence on itself.

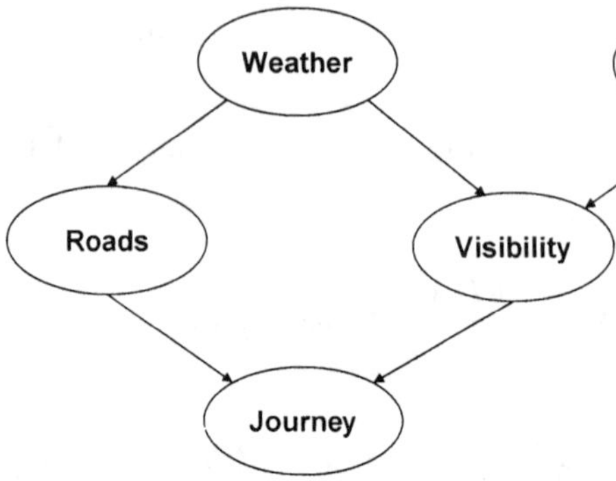

Figure 6-30: Example of a DAG that is not a polytree

A solution to this problem is to deal with the joint probability distribution of the variables in the network. Although BN technology reduces the joint probability distribution of a network to merely the products of a few CPTs, we have ruled out this approach due to the large number of entries that need to be computed for a belief revision. Even for the small network in Figure 6-30, there are 72 entries in its joint probability distribution table that need to be computed using the following formula each time a node receives evidence (only the initial letter has been used from each variable name).

For example, *Roads* receives evidence e_{Roads}:

$$p(W,E,R,V,J \mid e_{Roads}) = p(W)\,p(E)\,p(R \mid W)\,p(V \mid W,E)\,p(J \mid R,V)\,p(e_{Roads} \mid R)$$

To make use of the simple and elegant message passing propagation algorithm while avoiding large joint probability distributions, one approach is to transform a DAG into a polytree by identifying and clustering loops into single nodes. Thus a node in a transformed cluster network is a set of nodes (also called a *clique*) instead of a single node, and joint probability distributions are computed in smaller chunks locally at the network nodes. For example, the network in Figure 6-30 is transformed into the network (undirected) in Figure 6-31 with two cliques.

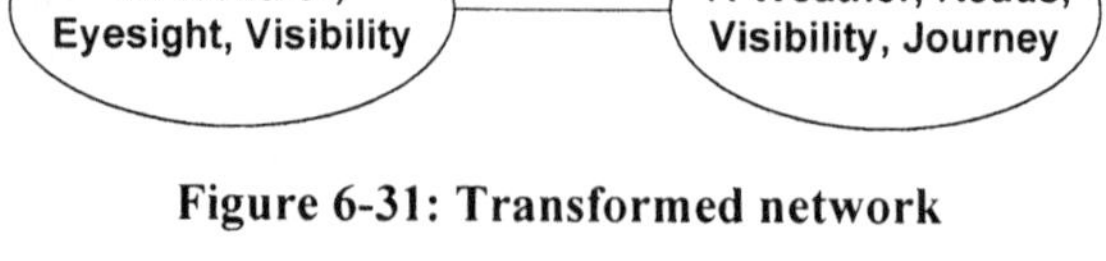

Figure 6-31: Transformed network

Evidence on a variable in the original network is posted to the nodes in the transformed network containing the variable. The joint probability distributions are computed at the nodes, where evidence is posted and messages are passed to neighboring nodes. For example, if evidence on the variable *Weather* is observed in the cluster network in Figure 6-31, then the joint probability distribution $p(W,R,V,J)$ is computed at the node *Y* using some of the CPTs. *Visibility* and *Weather* are the only variables in the intersection of *X* and *Y*. Thus *Y* sends the change in the joint probability distribution of *Visibility* and *Weather* as a message to the variable *X*, which then computes the joint probability distribution of its three variables and derives the new belief of the variable *Eyesight* via marginalization.

But in a complex network, a cycle could be very large and therefore local computation of joint probability distribution is still impractical. A solution is to make smaller size clusters. Since the joint probability distribution of a network is derived from the CPTs and the CPT of a variable involves only its immediate neighbors, clusters around variables can be formed using only their neighboring nodes. One such cluster network constructed from the network in Figure 6-30 is shown in Figure 6-32.

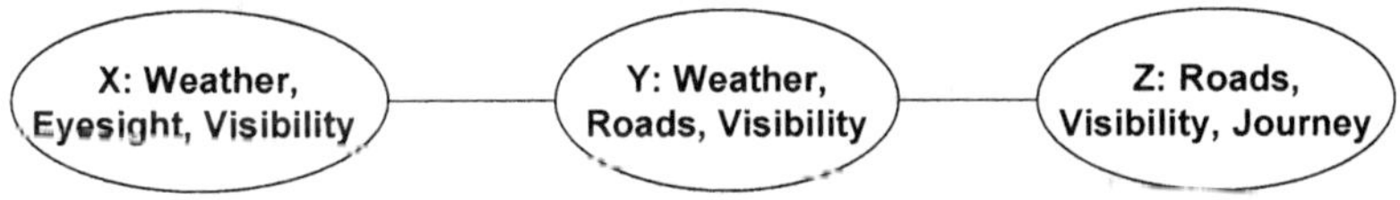

Figure 6-32: Cluster network

As before, if evidence on a variable, such as *Roads*, is observed, then it is posted to one of two nodes *Y* and *Z* (say, *Y*) containing the variable *Roads*. Each of these nodes then computes the joint distribution locally to find the beliefs of the individual variables. The node *Y* then passes the revised joint probability distributions of the variables *Visibility* and *Weather* to *X* (respectively, *Visibility* and *Roads* to *Z*) to help compute the revised belief for the node *Eyesight* (respectively, *Journey*) given the evidence on *Roads*.

In the following subsections, we present the *junction tree* algorithm for dealing with DAGs in general. This algorithm systematically constructs a cluster network from a BN called a *junction tree* (or *clique tree* or *join tree*). The stages of the algorithm are shown on the left side of Figure 6-33.

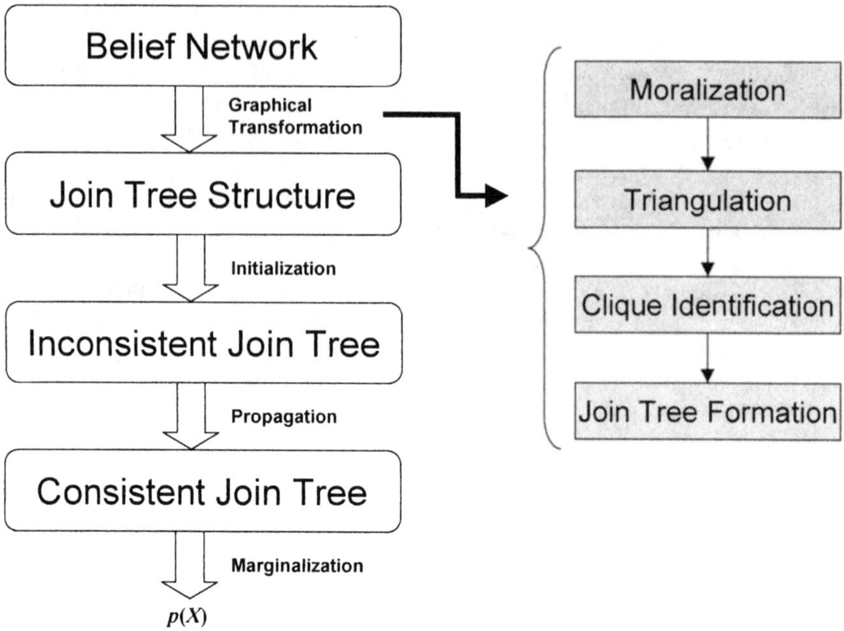

Figure 6-33: Steps for constructing join trees

The first stage, which is the graphical transformation stage, is expanded on the right side of the figure. This stage consists of four steps: moralization, triangulation, clique identification, and join tree formation. These steps construct a join tree from a BN. The subsequent stages of the algorithm then compute prior beliefs of the variables in the network via an initialization of the join tree structure, followed by propagation and marginalization. The final subsection discusses how to handle evidence by computing the posterior beliefs of the variables in the network.

6.7.1 Graphical Transformation

The *moral graph* of a DAG is obtained by adding a link between any pair of variables with a common child, and dropping the directions of the original links in the DAG.

Example

The moral graph of the network in Figure 6-1 is shown in Figure 6-34. The dotted lines in the network in Figure 6-34 are the links added to the original network. For example, the nodes *Roads* and *Weight* have a common child *Mobility*, and therefore are linked with a dotted line.

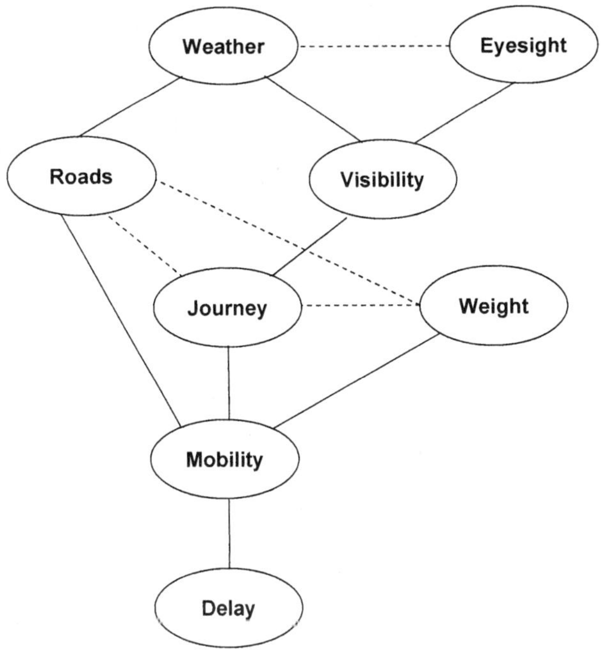

Figure 6-34: The moral graph of the network in Figure 6-1 (added links are indicated by dotted lines)

An undirected graph is *triangulated* if any cycle of length greater than 3 has a chord, that is, an edge joining two nonconsecutive nodes along the cycle.

Example

The triangulated graph of the network in Figure 6-34 is shown in Figure 6-35. The only link added by triangulation is the link between the nodes *Roads* and *Visibility*. These two nodes are two nonconsecutive nodes along the cycle *Weather – Roads – Journey – Visibility* of length 4.

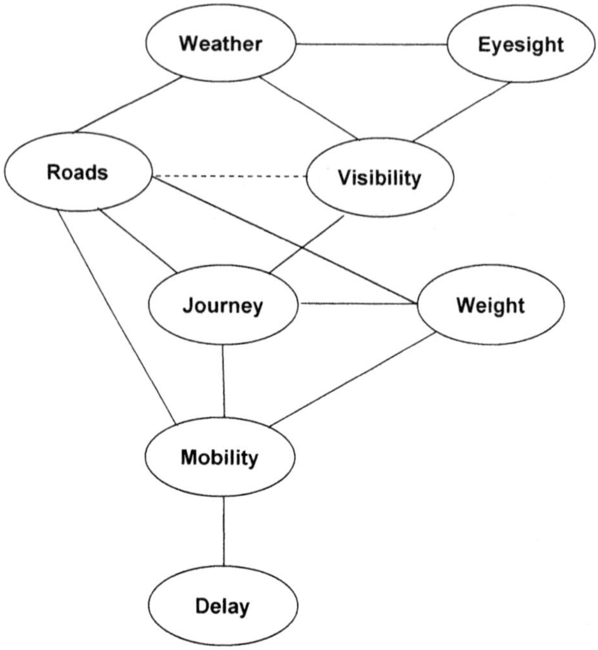

Figure 6-35: A triangulated graph of the network in Figure 6-34 (new link is indicated by dotted lines)

In general there are multiple ways to triangulate a given graph. In the case of the graph in Figure 6-34, we could have added a link between the nodes *Weather* and *Journey*, instead of between the nodes *Roads* and *Visibility* along the cycle *Weather – Roads – Journey – Visibility*, yielding a different triangulation.

The nodes of a join tree for a graph are the *cliques* in the graph (maximal sets of variables that are all pairwise linked).

Example

The five cliques in the graph in Figure 6-35 are listed below:

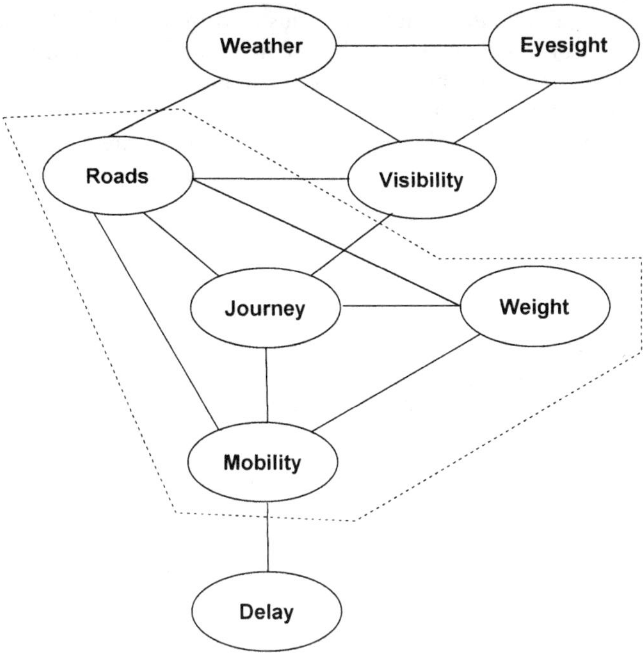

Figure 6-36: An example clique (surrounded by dotted lines) in the network in Figure 6-35

$\mathbf{C}_1$: {*Weather*, *Roads*, *Visibility*}
$\mathbf{C}_2$: {*Weather*, *Eyesight*, *Visibility*}
$\mathbf{C}_3$: {*Roads*, *Visibility*, *Journey*}
$\mathbf{C}_4$: {*Roads*, *Journey*, *Weight*, *Mobility*}
$\mathbf{C}_5$: {*Mobility*, *Delay*}

The clique $\mathbf{C}_4$ is shown in Figure 6-36, surrounded by dotted lines. Note that the set {*Weather*, *Eyesight*, *Roads*, *Visibility*} is not a clique because its nodes are not all pairwise connected since *Roads* and *Eyesight* are not connected. Though the nodes in the set {*Roads*, *Journey*, *Weight*} are pairwise connected, it is not a clique because the set is contained in another clique {*Roads*, *Journey*, *Weight*, *Mobility*}, and it is therefore not maximal.

The triangulation and clique identification techniques described above are combined into a formal two-step recursive algorithm for systematically identifying cliques from an arbitrary undirected moral graph:

- Select a node X from the network $\mathcal{N}$ and make the cluster consisting of the node and its neighbor complete by adding the necessary edges. Choose the node that causes the least number of edges to be added. Break ties by choosing the node that induces the cluster with the smallest weight, where the *weight* of a node is the number of states, and the weight of a cluster is the product of the weights of its constituent nodes.

- Remove X along with its edges (forming a clique) and repeat the previous step if there are still nodes left. Otherwise, if there are no nodes left, $\mathcal{N}$ is now triangulated. Note that a clique is formed in this stage only if it is not a subset of a subsequently induced clique.

Example

The graph in Figure 6-37 shows the triangulated graph obtained from the moral graph in Figure 6-34. The table in the figure shows the variable selection ordering in the first step of the algorithm. In each selection, the weight of the variable is computed and the induced cluster or clique is identified if it is bold and shaded.

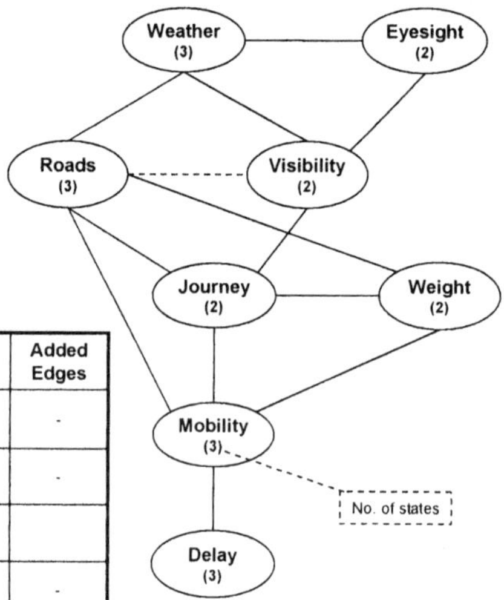

Selected Variable	Weight	Induced Cluster	Added Edges
Delay	9	**{Mobility, Delay}**	-
Eyesight	12	**{Weather, Eyesight, Visibility}**	-
Mobility	36	**{Roads, Journey, Weight, Mobility}**	
Weight	12	{Roads, Journey, Weight}	-
Journey	12	**{Roads, Visibility, Journey}**	(Roads, Visibility)
Visibility	18	**{Weather, Roads, Visibility}**	-
Weather	9	{Weather, Roads}	-
Roads	3	{Roads}	-

Figure 6-37: Triangulation and clique identification

The candidate variables to be selected first are *Eyesight*, *Delay*, and *Weight* since, for each of these three variables, the cluster consisting of the variable and its neighbors is already complete without the requirements for adding any edge. For example, the cluster consisting of the variable *Eyesight* and its two neighbors

(*Weather* and *Visibility*) is complete, but the cluster consisting of the variable *Weather* and its two neighbors (*Roads* and *Visibility*) is not complete because an edge between *Roads* and *Visibility* needs to be added. The variable *Delay* is chosen among the three candidate variables because its weight 9 is the least among the three weights 12, 9, and 36 for the three candidate variables, respectively. The variable *Delay* is then removed along with the edge from it leading to *Mobility*. The process then continues with the remainder of the network and the variable *Eyesight* is chosen next. When the variable *Weight* is chosen, the corresponding induced cluster is not identified as a clique because it is a subset of a clique that is already identified.

The network is now left with the four variables *Weather*, *Roads*, *Visibility*, and *Journey*, and the selection of any of these will require an edge to be added. The two candidate variables are *Visibility* and *Journey* as each of these has weight 12, and each of the remaining two variables *Weather* and *Roads* has weight 18. The selection of the variable *Visibility* needs the edge from *Weather* to *Journey* to be added, and the selection of the variable *Journey* needs the edge from *Roads* to *Visibility* to be added.

Once we have identified cliques, we rank the nodes by assigning numbers to systematically construct a join tree out of cliques. Nodes are numbered from 1 to *n* in increasing order by assigning the next number to the node with the largest set of previously numbered neighbors. For example, an ordering of the nodes of the graph in Figure 6-1 is given below:

1 – *Weather*
2 – *Eyesight*
3 – *Visibility*
4 – *Roads*
5 – *Journey*
6 – *Weight*
7 – *Mobility*
8 – *Delay*

The choice between the two nodes *Weather* and *Eyesight* as a starting node is arbitrary to break the tie. The node *Visibility* comes before the node *Roads* because *Visibility* has two neighbors, *Weather* and *Eyesight*, which have already been numbered. On the other hand, only one neighbor *Weather* of the node *Roads* has been numbered. The rest of the sequence is numbered in a similar manner.

To form a join tree based on the above ranking scheme of graph nodes, first order the cliques of the graph by rank of the highest vertex of each clique. For example, the set of five cliques $\mathbf{C}_1$, $\mathbf{C}_2$, $\mathbf{C}_3$, $\mathbf{C}_4$, and $\mathbf{C}_5$ in Figure 6-35 is as follows:

$\mathbf{C}_5$: {*Mobility*, *Delay*}
$\mathbf{C}_4$: {*Roads*, *Journey*, *Weight*, *Mobility*}
$\mathbf{C}_3$: {*Roads*, *Visibility*, *Journey*}
$\mathbf{C}_1$: {*Weather*, *Roads*, *Visibility*}
$\mathbf{C}_2$: {*Weather*, *Eyesight*, *Visibility*}

For example, the clique $\mathbf{C}_5$ comes before the clique $\mathbf{C}_4$ because $\mathbf{C}_5$ contains the node *Mobility* whose rank is higher than the rank of any node in $\mathbf{C}_4$.

The join tree from an ordered set of cliques $\mathbf{C}_1, \mathbf{C}_2, \ldots, \mathbf{C}_n$ formed by connecting each $\mathbf{C}_i$ to a predecessor $\mathbf{C}_j$ sharing the highest number of vertices with $\mathbf{C}_i$. For example, a join tree for the above ordered set of cliques $\mathbf{C}_5$, $\mathbf{C}_4$, $\mathbf{C}_3$, $\mathbf{C}_1$, $\mathbf{C}_2$ is shown in Figure 6-38. Each edge in the tree is labeled with the intersection of the adjacent cliques, which is called a *separator set* or a *sepset.*

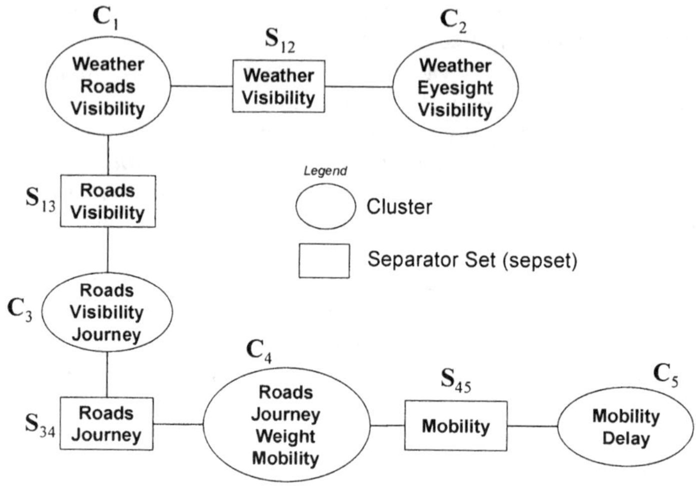

Figure 6-38: Join tree for the graph in Figure 6-35

The following algorithm helps systematically construct a join tree as shown in Figure 6-38. The algorithm is optimal with respect to the mass and cost criteria adopted during the selection of sepsets.

- Create a forest of n distinct trees, where each tree consists of only one node made out of the set of n cliques produced by the triangulation and clique identification procedure above. Also, create a set $\mathbf{S}$ of $n(n-1)$ distinct sepsets obtained by intersecting all possible pairs of distinct elements in the set of cliques.
- Repeat the following steps $n-1$ times:
  - Select a sepset $\mathbf{S}_{XY}$ (that is, $\mathbf{C}_X \cap \mathbf{C}_Y$) from $\mathbf{S}$ that has the largest mass, where mass of a sepset $\mathbf{S}_{XY}$ is the number of variables it contains. If two or more sepsets of equal mass can be chosen then choose the candidate sepset with the smallest cost, where the *cost of a sepset* $\mathbf{S}_{XY}$ is the sum of the product of the number of states of the

variables in **X** and product of the number of states of the variables in **Y**.

- Insert the selected sepset $\mathbf{S}_{XY}$ between the cliques **X** and **Y** only if **X** and **Y** are on different trees in the forest.

In the construction of the join tree in Figure 6-38, first the forest is formed containing all five cliques $\mathbf{C}_1$, $\mathbf{C}_2$, $\mathbf{C}_3$, $\mathbf{C}_4$, and $\mathbf{C}_5$. Each of the sepsets $\mathbf{S}_{12}$, $\mathbf{S}_{13}$, and $\mathbf{S}_{34}$ has a mass 2 and weight 6, and is therefore inserted first into the join tree. Note that the sepset {*Visibility*} was not inserted between $\mathbf{C}_2$ and $\mathbf{C}_3$ before $\mathbf{S}_{12}$ or $\mathbf{S}_{13}$ as {*Field*} has lower mass than each of $\mathbf{S}_{12}$ or $\mathbf{S}_{13}$.

6.7.2 Join Tree Initialization

A join tree maintains a joint probability distribution at each node, cluster, or sepset in terms of a belief *potential*, which is a function that maps each instantiation of the set of variables in the node into a real number. The belief potential of a set **X** of variables will be denoted as $\varphi_{\mathbf{X}}$, and $\varphi_{\mathbf{X}}(x)$ is the number that the potential maps x onto. The probability distribution of a set **X** of variables is just the special case of a potential whose elements add up to 1. In other words,

$$\sum_{x \in \mathbf{X}} \varphi_{\mathbf{X}}(x) = \sum_{x \in \mathbf{X}} p(x) = 1$$

The marginalization and multiplication operations on potentials are defined in a manner similar to the same operations on probability distributions.

Belief potentials encode the joint distribution $p(\mathbf{X})$ of the BN according to the following:

$$p(\mathbf{X}) = \frac{\prod_i \phi_{\mathbf{C}_i}}{\prod_j \phi_{\mathbf{S}_j}}$$

where $\varphi_{\mathbf{C}_i}$ and $\varphi_{\mathbf{S}_j}$ are the cluster and sepset potentials, respectively. We have the following joint distribution for the join tree in Figure 6-38 (only the initial letter has been used for each variable name except *Wt* stands for *Weight*):

$$p(W, E, R, V, J, Wt, M, D) = \frac{\phi_{\mathbf{C}_1}\phi_{\mathbf{C}_2}\phi_{\mathbf{C}_3}\phi_{\mathbf{C}_4}\phi_{\mathbf{C}_5}}{\phi_{\mathbf{S}_{12}}\phi_{\mathbf{S}_{13}}\phi_{\mathbf{S}_{34}}\phi_{\mathbf{S}_{45}}}$$

$$= \frac{\phi_{WRV}\phi_{WEV}\phi_{RVJ}\phi_{RJWtM}\phi_{MD}}{\phi_{WV}\phi_{RV}\phi_{RJ}\phi_{M}}$$

It is imperative that a cluster potential agrees on the variables in common with its neighboring sepsets up to marginalization. This imperative is formalized by the concept of local consistency. A join tree is *locally consistent* if for each cluster **C** and neighboring sepset S, the following holds:

$$\sum_{\mathbf{C} \setminus \mathbf{S}} \phi_{\mathbf{C}} = \phi_{\mathbf{S}}$$

To start initialization, for each cluster **C** and sepset **S**, set the following:

$$\phi_C \leftarrow 1, \quad \phi_S \leftarrow 1$$

Then assign each variable X to a cluster **C** that contains X and its parents $pa(X)$. Then set the following:

$$\phi_C \leftarrow \phi_C p(X \mid pa(X))$$

Example

To illustrate the initialization process, consider the join tree in Figure 6-38. The allocation of prior and conditional probability tables is shown in Figure 6-39. The variable *Visibility* has been assigned to cluster $\mathbf{C}_2$ as it contains the variable and its parents *Weather* and *Eyesight*. The probability table *p*(*Weather*) could have been assigned to any of $\mathbf{C}_1$ and $\mathbf{C}_2$, but $\mathbf{C}_1$ is chosen arbitrarily.

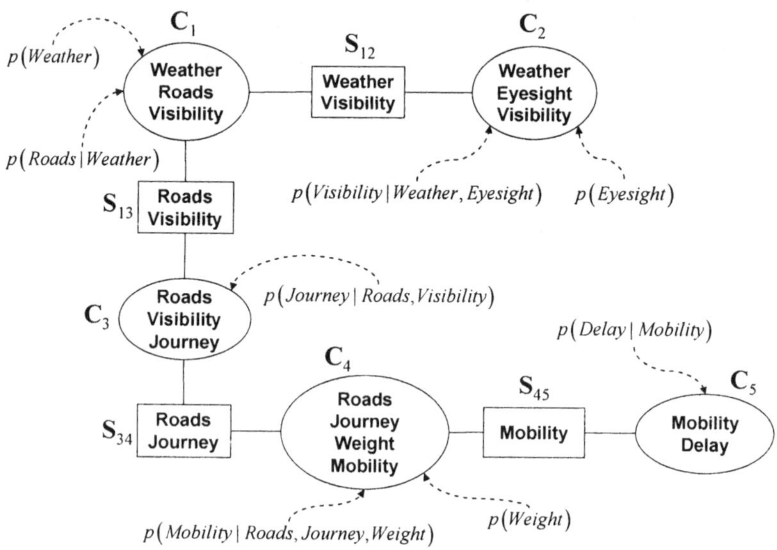

Figure 6-39: Allocation of prior and conditional probability tables for join tree initialization

As an example, Figure 6-40 shows the computation of potential for the clique $\mathbf{C}_1$ by multiplying $p(Weather)$ and $p(Roads \mid Weather)$.

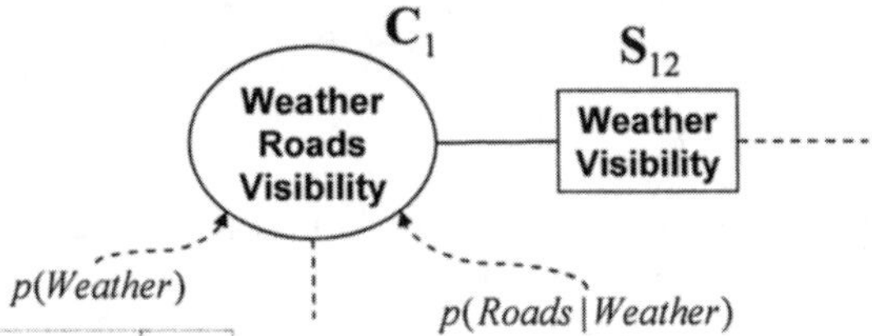

p(**Weather** = normal)	0.55
p(**Weather** = rainy)	0.15
p(**Weather** = foggy)	0.30

Weather ⇒	normal	rainy	foggy
p(**Roads** = dry)	0.90	0.70	0.80
p(**Roads** = slippery)	0.09	0.20	0.15
p(**Roads** = muddy)	0.01	0.10	0.05

$\phi_{S_{12}}$

Weather	Visibility	Initial Values
normal	clear	1
normal	poor	1
rainy	clear	1
rainy	poor	1
foggy	clear	1
foggy	poor	1

ϕ_{C_1}

Weather	Roads	Visibility	Initial Values
normal	dry	clear	0.90 × 0.55 × 1 = 0.4950
normal	dry	poor	0.90 × 0.55 × 1 = 0.4950
normal	slippery	clear	0.09 × 0.55 × 1 = 0.0495
normal	slippery	poor	0.09 × 0.55 × 1 = 0.0495
normal	muddy	clear	0.01 × 0.55 × 1 = 0.0055
normal	muddy	poor	0.01 × 0.55 × 1 = 0.0055
rainy	dry	clear	0.70 × 0.15 × 1 = 0.1050
rainy	dry	poor	0.70 × 0.15 × 1 = 0.1050
rainy	slippery	clear	0.20 × 0.15 × 1 = 0.0300
rainy	slippery	poor	0.20 × 0.15 × 1 = 0.0300
rainy	muddy	clear	0.10 × 0.15 × 1 = 0.0150
rainy	muddy	poor	0.10 × 0.15 × 1 = 0.0150
foggy	dry	clear	0.80 × 0.30 × 1 = 0.2400
foggy	dry	poor	0.80 × 0.30 × 1 = 0.2400
foggy	slippery	clear	0.15 × 0.30 × 1 = 0.0450
foggy	slippery	poor	0.15 × 0.30 × 1 = 0.0450
foggy	muddy	clear	0.05 × 0.30 × 1 = 0.0150
foggy	muddy	poor	0.05 × 0.30 × 1 = 0.0150

Figure 6-40: Computation of potential

6.7.3 Propagation in Join Tree and Marginalization

The join tree thus formed is not locally consistent as, for example, $\sum_{C_1 \setminus S_{12}} \phi_{C_1} \neq \phi_{S_{12}}$.

An inconsistent join tree can be made consistent using a *global propagation.* The message passing mechanism is at the heart of global propagation.

Consider two adjacent clusters $\mathbf{C}_1$ and $\mathbf{C}_2$ with sepset $\mathbf{S}$. A message pass from $\mathbf{C}_1$ to $\mathbf{C}_2$ consists of the following two steps:

- Projection: $\phi_S^{old} \leftarrow \phi_S \qquad \phi_S \leftarrow \sum_{C_1 \setminus S} \phi_{C_1}$

- Absorption: $\phi_{C_2} \leftarrow \phi_{C_2} \dfrac{\phi_S}{\phi_S^{old}}$

It can be easily verified that any number of messages passing as shown above encodes the joint distribution $p(\mathbf{X})$ of the BN. Global propagation is a systematic collection of message passing via the following two recursive procedures: *Collect Evidence* and *Distribute Evidence*.

- Choose an arbitrary cluster **C**.
- Unmark all clusters and call the following three steps of *Collect Evidence*(**C**):
  - Mark **C**.
  - Recursively call *Collect Evidence* on each unmarked neighboring cluster of **C**.
  - Pass a message from **C** to the cluster that invoked *Collect Evidence*(**C**).
- Unmark all clusters and call the following three steps of *Distribute Evidence*(**C**):
  - Mark **C**.
  - Pass a message from **C** to each of its unmarked neighboring clusters.
  - Recursively call *Distribute Evidence* on each unmarked neighboring cluster of **C**.

Example

Figure 6-41 shows the message flow order when the cluster $\mathbf{C}_3$ is chosen as the starting cluster in the above algorithm. First, *Collect Evidence* is called on cluster $\mathbf{C}_3$, which causes two calls of *Collect Evidence* on each of $\mathbf{C}_1$ and $\mathbf{C}_4$. The first call of these two calls on $\mathbf{C}_1$ triggers a call of *Collect Evidence* on $\mathbf{C}_2$. The node then passes message 1 to $\mathbf{C}_1$. The process continues, yielding a total of eight messages as shown in Figure 6-41.

Once the join tree has been made consistent, prior probabilities of variables can be obtained using marginalization. First, identify a cluster **C** that contains the variable X of interest. Then compute $p(X)$ by marginalizing ϕ_C as $p(X) = \sum_{C \setminus \{X\}} \phi_C$.

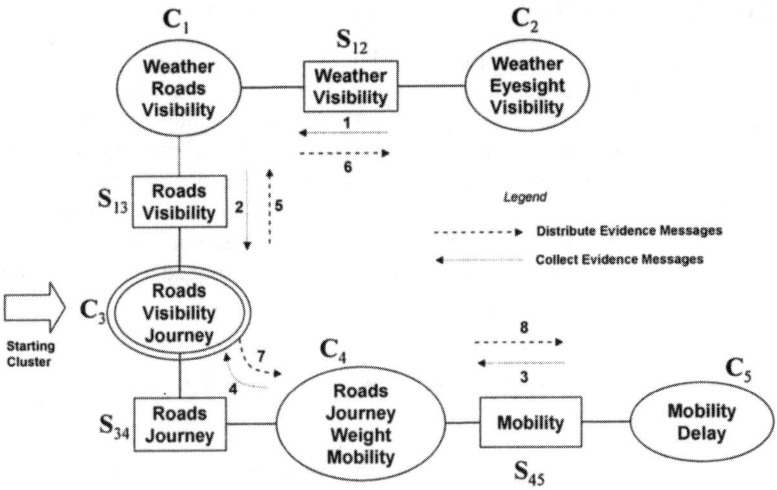

Figure 6-41: Message flow in global propagation

An example of this computation for the variable *Visibility* from cluster $\mathbf{C}_2$ is shown in Figure 6-42.

Clique $\mathbf{C}_2 = \{Weather, Eyesight, Visibility\}$

Weather	Eyesight	Visibility	Potential ϕ_{C_2}
normal	good	clear	0.8
normal	good	poor	0.2
normal	bad	clear	0.99
normal	bad	poor	0.01
rainy	good	clear	0.1
rainy	good	poor	0.9
rainy	bad	clear	0.6
rainy	bad	poor	0.4
foggy	good	clear	0.01
foggy	good	poor	0.99
foggy	bad	clear	0.15
foggy	bad	poor	0.85

Sum (Shaded Rows): 2.65
Sum (Clear Rows): 3.35

p(**Visibility** = clear)	0.44 = 2.65 / (2.65 + 3.35)
p(**Visibility** = poor)	0.56 = 3.35 / (2.65 + 3.35)

Figure 6-42: Marginalization from potential and normalization

6.7.4 Handling Evidence

Figure 6-43 shows the overall flow for using evidence in join trees to compute the variables' posterior probabilities. Compare this figure with Figure 6-33, which shows the flow for computing only prior probabilities. When new evidence on a variable is entered into the tree, it becomes inconsistent and requires a global propagation to make it consistent. The posterior probabilities can be computed via marginalization and normalization from the global propagation. If evidence on a variable is updated, then the tree requires initialization. Next, we present initialization, normalization, and marginalization procedures for handling evidence.

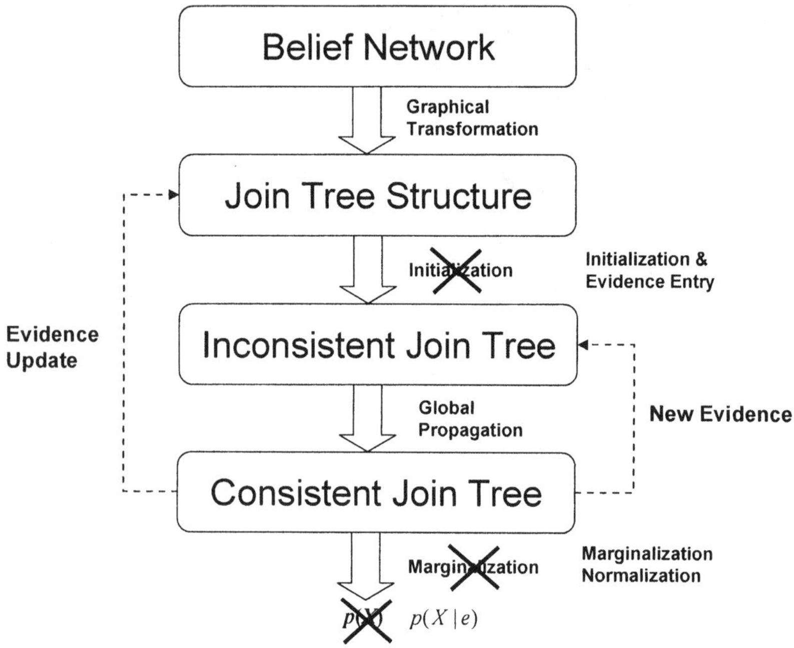

Figure 6-43: Steps for handling evidence in join trees

As before, to start initialization, for each cluster **C** and sepset **S**, set the following:

$$\phi_C \leftarrow 1, \quad \phi_S \leftarrow 1$$

Then assign each variable X to a cluster **C** that contains X and its parents $pa(X)$, and then set the following:

$$\phi_C \leftarrow \phi_C \, p(X \mid pa(X))$$
$$\lambda_X \leftarrow 1$$

where λ_X is the likelihood vector for the variable X. Now, perform the following steps for each piece of evidence on a variable X:

- Encode the evidence on the variable as a likelihood λ_X^{new}.
- Identify a cluster **C** that contains X (e.g., one containing the variable and its parents).
- Update as follows:

$$\phi_C \leftarrow \phi_C \frac{\lambda_X^{new}}{\lambda_X}$$

$$\lambda_X \leftarrow \lambda_X^{new}$$

Now perform a global propagation using the *Collect Evidence* and *Distribute Evidence* procedures. Note that if the belief potential of one cluster **C** is modified, then it is sufficient to unmark all clusters and call only *Distribute Evidence*(**C**).

The potential φ_C for each cluster **C** is now $p(\mathbf{C}, e)$, where e denotes evidence incorporated into the tree. Now marginalize C into the variable as

$$p(X, e) = \sum_{\mathrm{C} \setminus \{X\}} \phi_{\mathrm{C}}$$

Compute posterior $p(X \mid e)$ as follows:

$$p(X \mid e) = \frac{p(X,e)}{p(e)} = \frac{p(X,e)}{\sum_X p(X,e)}$$

To update evidence, for each variable X on which evidence has been obtained, update its likelihood vector. Then initialize the join tree by incorporating the observations. Finally, perform global propagation, marginalization, etc.

6.8 COMPLEXITY OF INFERENCE ALGORITHMS

Probabilistic inference using BNs is computationally intractable, that is, NP-hard (Cooper, 1990). Informally, this means if there exists an algorithm that solves our problems in polynomial time, then the polynomial-time algorithm would exist for practically all discrete problems, such as the propositional satisfiability problem. The fact that inferencing in BNs is not tractable does not mean it can never be applied; it simply means that there are cases when its inferencing time will take too long for this algorithm to be practical.

The computational complexity of Pearl's message passing algorithm for acyclic networks can be shown to be $O(n \times d \times 2^d)$, where n is the number of vertices in the network and d is the network's maximal indegree. The computational complexity of Lauritzen and Spiegelhalter's junction tree algorithm equals $O(n \times 2^c)$, where n is the number of vertices in the network and c is the number of vertices in the largest clique in the clique tree that is constructed from the network. Note that the algorithm's complexity is exponential in the size of the largest clique.

If the clique sizes in the junction tree algorithm are bounded by a constant, then the algorithm takes linear time. Since the computational complexity of the junction tree algorithm relates exponentially to clique size, the best clique tree to use in practical applications is a tree inducing the smallest state space. The problem of finding such a clique tree is known to be NP-hard (Wen, 1990). Various efficient heuristic algorithms are available for finding a clique tree for an acyclic network. However, these algorithms do not exhibit any optimality properties.

Therefore, it seems unlikely that an exact algorithm can be developed to perform probabilistic inference efficiently over all classes of BNs. This result suggests that research should be directed away from the search for a general, efficient, probabilistic inference algorithm, and towards the design of efficient special-case (for example, tree structure or inherently modular network), average-case, and approximation algorithms.

6.9 ACQUISITION OF PROBABILITIES

The acquisition of probabilities for BN structures involves eliciting conditional probabilities from subject matter experts along causal directions. These probabilities are "causal" conditional probabilities of the form $p(Report \mid Rain, Sensor)$, indicating the chance of obtaining a sensor report from the appropriate sensor when it is raining. This chance is related to the sensor functionality in the presence of rain, which may be best estimated by the sensor designers. Similarly, a causal probability of the form $p(Journey \mid Roads, Visibility)$ indicates the chance of starting or canceling a journey given the road and visibility conditions. The logistics analysts, together with weather and terrain analysts, can best estimate this probability. On the other hand, the "diagnostic" conditional probabilities in the belief context are probabilities of the form $p(Visibility \mid Journey)$, indicating various visibility conditions given a vehicle journey. An experienced logistics analyst may best estimate this probability from previous experience. Both causal and diagnostic probabilities can be used to compute joint probability distributions.

The major issue related to eliciting probabilities from subject matter experts is how to phrase questions to experts so as to accurately and efficiently determine relevant prior and conditional probabilities (Druzdzel and van der Gaag, 1995).

Example

Consider, for example, the fragment of the network as shown in Figure 6-44, which says that bad weather (*BW*), terrorist threat (*TT*), and routine maintenance

(*RM*) together cause road closures (*RC*). Each variable *X* (*BW*, *TT*, *RM*, and *RC*) in the network is binary with two states X and $\neg X$. If we are eliciting causal probabilities, then experts will be asked questions for determining the priors $p(BW)$, $p(TT)$, and $p(RM)$, and the conditional $p(RC \mid BW, TT, RM)$.

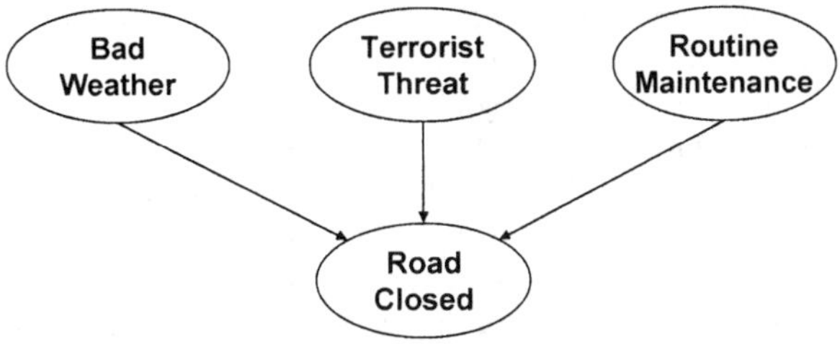

Figure 6-44: BN illustrating probability acquisition

If prior probability $p(RC)$ is available, then diagnostic conditional probabilities are elicited by employing the arc reversal approach (Shachter and Heckerman, 1987). But even if the diagnostic conditional probabilities $p(BW \mid RC)$, $p(TT \mid RC)$, and $p(RM \mid RC)$ are known, it is not possible to uniquely determine the causal probability $p(RC \mid BW, TT, RM)$.

Our experience suggests that subject matter experts, scientists, and knowledge engineers are often comfortable drawing arrows in the causal direction once the term "causality" is explained. The directionality has great impact on the resultant ease of knowledge elicitation. For example, if all three arrows in the network in Figure 6-44 are reversed, then the resultant network is equally effective in determining whether the road is going to be closed or not, but now the a priori probability $p(RC)$ is required, as are the conditional probabilities $p(BW \mid RC)$, $p(TT \mid RC)$, and $p(RM \mid RC)$.

The number of conditional probabilities required for a node to complete its conditional probability table, like $p(RC \mid BW, TT, RM)$ for *RC*, grows exponentially with the number of parents. But we can employ the *noisy-or* technique to avoid building large probability tables, provided certain conditions are met (Pearl, 1988). Informally, the noisy-or technique states that any members of a set of independent conditions are likely to cause a certain event and that this chance is not reduced when several of these conditions occur simultaneously. In the case of the node *RC*, chances for the road to be closed can only increase when several of the three conditions, bad weather, terrorist threat, and routine maintenance, occur simultaneously. Given this knowledge, we can generate the CPT $p(RC \mid BW, TT, RM)$ of eight independent entries from only three values. Formally, the noisy-or is a BN node representing an event (e.g., *Road Closed*) at which the following assumptions are made:

- The *accountability* condition requires that we explicitly list as parent nodes all the conditions likely to influence the event. In the example, this assumption requires that we explicitly list all the conditions likely to cause a road to be closed. In other words, the condition states that an event is presumed false if all conditions listed as causes of the event are false.
- The *exception independence* condition requires that whatever inhibits each parent node from implying the event is independent of whatever inhibits other parent nodes from implying the event. For example, assume that the only factor inhibiting the road to be closed when the weather is bad is a paved road, and the only factor inhibiting the road to be closed when there is a terrorist threat is tight security. The exception independence condition holds since the two inhibitory factors can be assumed independent of each other. On the other hand, if we had considered a variable representing only maintenance due to bad weather, instead of routine maintenance, then good weather would have inhibited both the parent nodes for bad weather and maintenance from implying the event, thus it would have violated the exception independence condition.

Example

An incomplete CPT, like the one in Table 6-9, can be completed by deriving the missing probabilities through the noisy-or technique. Suppose only the following entries in the CPT are known:

$$p(RC \mid BW, \neg TT, \neg RM) = 0.7$$
$$p(RC \mid \neg BW, TT, \neg RM) = 0.6$$
$$p(RC \mid \neg BW, \neg TT, RM) = 0.9$$

Bad Weather (BW)		BW				¬BW			
Terrorist Threat (TT)		TT		¬TT		TT		¬TT	
Routine Maintenance (RM)		RM	¬RM	RM	¬RM	RM	¬RM	RM	¬RM
Road Closed (TS)	RC	?	?	?	0.7	?	0.6	0.9	?
	¬RC	?	?	?	0.3	?	0.4	0.1	?

Table 6-9: An incomplete CPT to be completed by the noisy-or technique

Therefore, we have the following:

$$p(\neg RC \mid BW, \neg TT, \neg RM) = 0.3$$
$$p(\neg RC \mid \neg BW, TT, \neg RM) = 0.4$$
$$p(\neg RC \mid \neg BW, \neg TT, RM) = 0.1$$

Now,

$$p(RC \mid BW, TT, \neg RM) = 1 - p(\neg RC \mid BW, TT, \neg RM)$$

The accountability condition states that *TS* is false if all conditions listed as causes of *RC* are false. Therefore,

$$p(\neg RC \mid BW, TT, \neg RM) = p(\neg TS \mid BW, TT)$$

Thus,

$$p(RC \mid BW, TT, \neg RM) = 1 - p(\neg RC \mid BW, TT)$$

Now, the exception independence condition states that the inhibitory conditions *BW* and *TT* for *RC* are independent of each other. Therefore,

$$p(\neg RC \mid BW, TT) = p(\neg RC \mid BW)\, p(\neg RC \mid TT)$$

Thus,

$$\begin{aligned} p(RC \mid BW, TT, \neg RM) &= 1 - p(\neg RC \mid BW)\, p(\neg RC \mid TT) \\ &= 1 - (1 - p(RC \mid BW))(1 - p(RC \mid TT)) \\ &= 1 - (1 - 0.7)(1 - 0.6) \\ &= 0.88 \end{aligned}$$

The rest of the entries in the CPT can be computed in a similar manner.

6.10 ADVANTAGES AND DISADVANTAGES OF BELIEF NETWORKS

Like any other computational formalism, BN technology offers certain advantages and disadvantages. Advantages of BNs include:

- *Sound theoretical foundation*: The computation of beliefs using probability estimates is guaranteed to be consistent with the probability theory. This advantage stems from the Bayesian update procedure's strict derivation from the axioms of probability.
- *Graphical models*: BNs graphically depict the interdependencies that exist between related pieces of domain knowledge, enhancing understanding of the domain. The structure of a BN captures the cause-effect relationships that exist amongst the variables of the domain. The ease of causal interpretation in BN models typically makes them easier to construct than other models, minimizing the knowledge engineering costs and making them easier to modify.
- *Predictive and diagnostic reasoning*: BNs combine both deductive/predictive and abductive/diagnostic reasoning. Interdependencies among variables in a network are accurately captured and speculative if-then-type computation can be performed.
- *Computational tractability*: BNs are computationally tractable for most practical applications. This efficiency stems principally from the exploitation of conditional independence relationships over the domain. We have presented an efficient single-pass evidence propagation algorithm for networks without loops.

- *Evidence handling*: Evidence can be posted to any node in a BN. This means that subjective evidence can be posted at an intermediate node representing an abstract concept.

A major disadvantage of BN technology is the high level of effort required to build network models. Although it is relatively easy to build a BN structure with the help of subject matter experts, the model will require a significant amount of probability data as the number of nodes and links in the structure increases. The size of a CPT corresponding to a node with multiple parents can potentially be huge. For example, the number of independent entries in the CPT of a binary node (a node with two states) with 8 binary parent variables is 128.

BNs are also poor at handling continuous variables. Current software handles continuous variables in a very restrictive manner (for example, they must be Gaussian and can only be children). Lerner et al. (2001) developed an inference algorithm for static hybrid BNs, which are Conditional Linear Gaussian models, where the conditional distribution of the continuous variables assigned to the discrete variables is a multivariate Gaussian. Cobb and Shenoy (2003) developed an inference algorithm in hybrid BNs using Mixtures of Truncated Potentials. But these techniques are yet to be incorporated in commercial software.

6.11 THEATER MISSILE DEFENSE APPLICATION

In this section, we use BNs to model the problem of determining types of detected sites (missile reload vs. logistic site), suspected missile warhead types (nuclear vs. chemical/biological), and their vector capabilities (short range vs. long range). The detected missiles pose threats to the deployed troops at the theater, cities, and ports before, during, and after an operation. This assessment supports a decision to attack, avoid, or defend against the detected missile. We discuss here a sequence of three BNs illustrating three components used in the process of analyzing N-B-C (Nuclear, Biological, Chemical) missile threats:

- *Classification of suspected reload/logistic sites*: Reload and logistic sites are critical installations needed for missile attacks. Their presence is a strong indicator of the intent to prepare a missile offensive.
- *Distinction between nuclear and biological/chemical reload sites*: Once a suspected site has been identified as a reload site, it becomes important to know whether the installation is intended for a nuclear, biological, or chemical attack.
- *Classification of missile launch sites*: Missile launch sites are mobile offensive installations. The presence of a logistic or reload site, together with additional site-specific information, allows the analyst to characterize the vector capabilities of the installed missiles as short-range, long-range, or dummy missiles.

The BN in Figure 6-45 allows the analyst to classify a suspected site into one of the following three: N-B-C reload or logistic site, conventional reload or logistic site, and nonsite. The BN consists of two types of knowledge: strategic doctrine knowledge and physical doctrine knowledge.

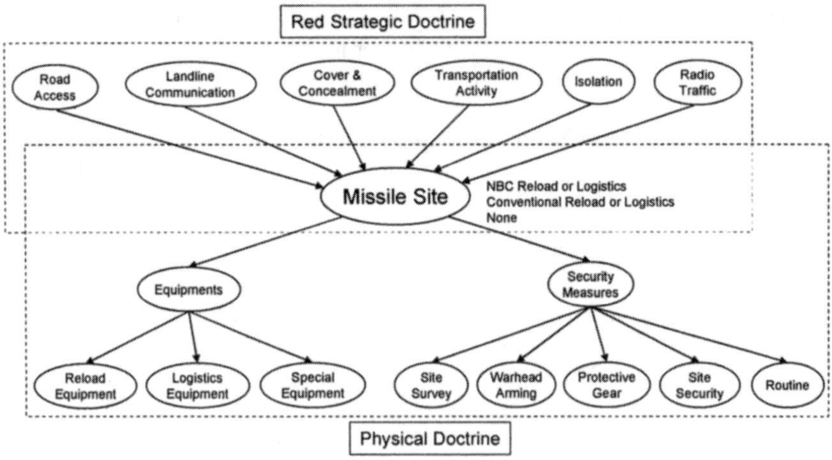

Figure 6-45: BN for classifying suspected reload/logistic sites

The strategic doctrine knowledge nodes are located in the upper half of the BN in Figure 6-45. They encode information to allow reasoning for the areas where the enemy would place reload or logistic sites according to its strategic doctrine. The strategic doctrine represents background knowledge that makes an event or a situation likely whenever a set of general criteria is satisfied. The strategic doctrine does not include observations or intelligence reports about the site per se.

For example, the BN in Figure 6-45 indicates that an area characterized by good access to multiple roads, high density of landline communications, and a transportation activity that follows a heterogeneous pattern is adequate for the placement of an N-B-C logistic site. An area with reduced landline communications, reduced density of enemy units, and where encrypted radio communications are frequent indicates a site that is more likely to be used for missile reload purposes.

The lower part of the BN in Figure 6-45 includes physical doctrine nodes. The nodes correspond to physical observations representing confirming evidence for one of the possible values of the Missile Site node. The evidence for the physical doctrine nodes originates in intelligence reports about the suspected site received from various sources. For example, evidence indicating the presence of reload equipment, combined with evidence of radioactive traces received from RADINT sources, and with observations of warhead arming activities are a strong indication that the observed site is likely to be a nuclear weapons reload site.

It is important to notice that some of the nodes combine evidence from sub-nodes. Thus, the Equipment node combines evidence on the type of equipment at the suspected site received from various imagery sources. Such techniques are designed to take into account varying reliability of the information, before including it into inference processes at a higher level.

Assuming that a suspected site has been classified as an N-B-C reload site, the analyst needs to infer with more accuracy whether the preparations associated with this site are intended to support a nuclear missile attack or a biological/chemical attack. The BN in Figure 6-46 provides precisely this type of analysis.

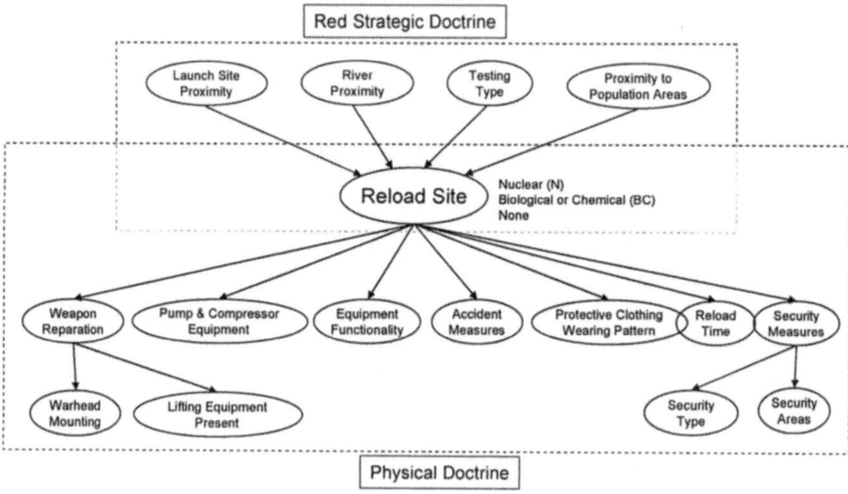

Figure 6-46: BN for classifying reload sites into nuclear, biological, and chemical sites

The nodes in the strategic doctrine category include criteria such as the proximity to launch sites, the proximity of rivers, the nature of the testing activities carried out by the enemy, and the proximity of populated areas. To exemplify, in a context where the enemy is known to have carried out underground testing activities, a reload site that is situated in river proximity and that is close to populated areas leads to the hypothesis that the site is of the nuclear type.

The criteria falling in the physical doctrine category take into account observations about the weapon preparations, the type of pump and compressor equipment, the equipment functionality, the accident measures, the pattern of wearing protective clothing, the reload time, and the security measures. Collected evidence showing the presence of pump and compressor equipment usable for decontamination purposes, the use of heavy lifting equipment, a temporary pattern of wearing protective clothing, and a reload activity pattern that spans the entire day is conclusive for a reload site for nuclear-type weapons. Again, like in Figure 6-45, some of the nodes combine higher-level evidence from lower-level cues. For example, the type of weapon preparation procedures is given by the combination of evidence on warhead mounting activities and the temporal pattern of the presence of the lifting equipment. A one-step warhead procedure combined with a permanent presence of the lifting equipment suggests weapon preparation procedures of the nuclear type.

Finally, the BN in Figure 6-47 supports the classification of missile launch sites into sites for short-range missiles, long-range missiles, and dummy/no threat missiles.

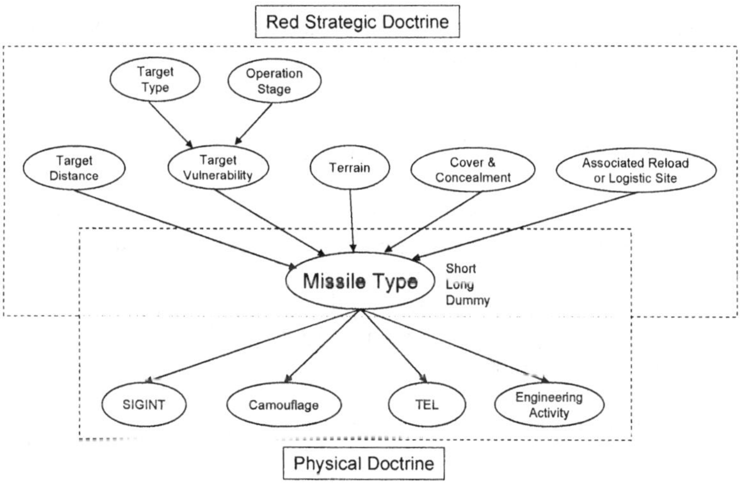

Figure 6-47: BN for missile-type assessment

The strategic doctrine part of the BN is based on the enemy's strategic doctrine and missile performance capabilities. It can be expected that theater missile target requirements will evolve as the political and military situation changes. Initially, the operational objective of enemy theater missile forces will be to slow the friendly forces. Thus, at an entry stage of the operation the enemy missiles are typically used against high-value targets such as ports. On the other hand, during an operation, cities and command posts are more vulnerable to enemy missiles than ports. Thus, given a target (city, port, or command post) and an operational stage (entry, operation, and postconflict), the degree of target vulnerability is determined as low or high. This is how the nodes Target Type and Operation Stage drive the node Target Vulnerability.

The figure also shows a second layer of nodes in the strategic doctrine part, which directly drive the beliefs associated with Missile Type:

- Target Distance will help determine Missile Type because of the missile's range capabilities.
- Target Vulnerability is also a factor influencing the location of a missile within launch range.
- Terrain is a key driver, since Transportable Erector Launcher (TEL) mobility strongly constrains TEL location.
- Cover and Concealment also contribute to missile location, since natural cover will preclude early detection of the TEL.
- Finally, the Associated Reload and Logistic Sites will influence the location, since it is less likely that a TEL will be located too far from these sites.

The physical doctrine evidence nodes for the BN in Figure 6-47 include information from SIGINT sources, the presence of camouflage, the existence of TELs, and the existence of engineering activities.

6.12 BELIEF NETWORK TOOLS

Various free and commercial software tools and packages are currently available for manipulating BNs incorporating some of the above functionalities. Of these, HUGIN (www.hugin.com) is the most widely used commercial package that contains a flexible, user friendly, and comprehensive graphical user interface. The package allows modeling decision-making problems via influence diagrams and handles continuous variables with some restrictions. Other popular tools and packages include Netica (http://www.norsys.com/), BayesiaLab (http://www.bayesia.com/), and BNet™ (http://www.cra.com/).

6.13 FURTHER READING

The book by Pearl (1988) is still the most comprehensive account on BNs, and more generally on using probabilistic reasoning to handle uncertainty. Various cases of the evidence propagation algorithms in polytrees presented here closely follow Pearl's book. Though Pearl himself developed an exact inference algorithm for DAGs, called loop cutset conditioning (Pearl, 1986), the junction tree algorithm of Lauritzen and Spiegelhalter (1988), as refined by Jensen et al. (1990) in HUGIN, is more general and the most popular inference algorithm for general BNs. A good comprehensive procedural account of the algorithm can be found in Huang and Darwiche (1996). Jensen's books (1996, 2002) are also useful guides in this field. See Das et al. (2002) for an application of BNs for conventional battlefield SA.

Chapter 7

Modeling Time for Situation Assessment

This chapter presents temporal model-based approaches to Situation Assessment (SA) in dynamic environments. Temporal approaches, as opposed to static- or snapshot-based approaches, handle time formally in that random variables in a temporal model change their values as time changes. Temporal models are therefore appropriate for modeling the stochastic nature of a dynamic environment. We present the Markov model, Hidden Markov Model (HMM)-based, Hierarchical Hidden Markov Model (HHMM)-based, and Dynamic Bayesian Network (DBN)-based approaches to temporal modeling for SA. HMMs are also useful for sequence recognition, such as understanding semantics of word sequences in text data. DBNs generalize HMMs, and DBNs are temporal extensions of Bayesian Belief Networks (BNs) presented in Chapter 6. Readers are therefore advised to understand the BN technology before reading the section on DBNs in this chapter. We conclude the chapter by presenting an application of DBNs to unit life status estimation.

7.1 MARKOV MODELS

A simple Markov model of a stochastic process is specified as a triplet $\langle S, A, \Pi \rangle$ defined as follows:

- A set $S = \{s_1, ..., s_n\}$ of n distinct *states*, one of which is assumed to be the state of the process at time step t. The process undergoes a change of state at regularly spaced discrete time steps $t = 1, 2, ...$
- A set $A = \{a_{ij}\}_{i,j=1,...,n}$ of *state transition probabilities*, where each a_{ij} represents the probability of transitioning from state s_i to state s_j. Symbolically, $p(X_t = s_j \mid X_{t-1} = s_i) = a_{ij}$, where X_t is the random variable representing the state at time step t.

- An initial probability distribution $\Pi = \{\pi_1, ..., \pi_n\}$, respectively, over the states $S = \{s_1, ..., s_n\}$, where π_i is the probability that the model will start in state s_i.

The total sum of initial distribution probabilities must be equal to 1:

$$\sum_{i=1}^{n} \pi_i = 1$$

The total sum of the outgoing transition probabilities from a given state must also sum to 1:

$$\sum_{j=1}^{n} a_{ij} = 1$$

The above Markov model can be represented as a probabilistic graphical model with states representing nodes and directed arcs representing transitions. Figure 7-1 represents a Markov model with four states along with its transition matrix shown on the right. Note that the transitions with the zero probability are not shown.

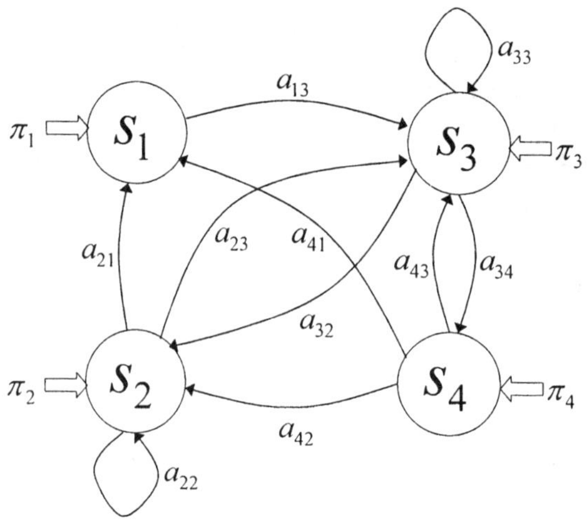

$$A = \begin{bmatrix} 0 & 0 & a_{13} & 0 \\ a_{21} & a_{22} & a_{23} & 0 \\ 0 & a_{32} & a_{33} & a_{34} \\ a_{41} & a_{42} & a_{43} & 0 \end{bmatrix}$$

Figure 7-1: Markov model and transition matrix with four states

In a *first-order Markov model*, the probability of a particular state is dependent only on the previous state:

$$p(X_t = s_i \mid X_{t-1} = s_j, X_{t-2} = s_k, ...) = p(X_t = s_i \mid X_{t-1} = s_j)$$

The model above will be called observable if each state corresponds to a physical observable event. Given a model as above, we can then ask questions, for example,

What is the probability that the observation sequence $\{s_3, s_4, s_1, s_3, s_2\}$ of states corresponding to $t = 1,2,3,4,5$ will be observed? This probability is evaluated as follows:

$$\begin{aligned} p(\{s_3,s_4,s_1,s_3,s_2\}) &= p(s_3)\cdot p(s_4 \mid s_3)\cdot p(s_1 \mid \{s_3,s_4\})\cdot \\ &\qquad p(s_3 \mid \{s_3,s_4,s_1\})\cdot p(s_2 \mid \{s_3,s_4,s_1,s_3\}) \\ &= p(s_3)\cdot p(s_4 \mid s_3)\cdot p(s_1 \mid s_4)\cdot p(s_3 \mid s_1)\cdot p(s_2 \mid s_3) \\ &= \pi_3 \cdot a_{43} \cdot a_{14} \cdot a_{31} \cdot a_{23} \end{aligned}$$

Here is a concrete instantiation of the Markov model in Figure 7-1.

Example

Consider a four-state Markov model for selecting appropriate motion models during target tracking. We assume that once in every 10 seconds, the state of the target vehicle is observed as being one of the following:

$$\begin{aligned} s_1 &= \text{Stopped}\,(stop) \\ s_2 &= \text{In acceleration}\,(accl) \\ s_3 &= \text{Traveling at a constant speed}\,(cons) \\ s_4 &= \text{In deceleration}\,(decl) \end{aligned}$$

The model and the transition probability matrix A are given in Figure 7-2:

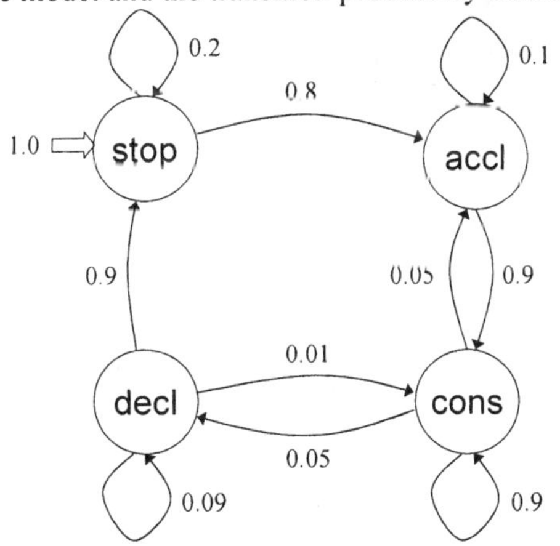

$$A = \begin{bmatrix} 0.2 & 0.8 & 0 & 0 \\ 0 & 0.1 & 0.9 & 0 \\ 0 & 0.05 & 0.9 & 0.05 \\ 0.9 & 0 & 0.1 & 0.09 \end{bmatrix}$$

Figure 7-2: Example Markov model and transition matrix with four states

So the probability of, for example, the observation sequence $\{stop, accl, cons, cons, decl\}$ can be computed as below:

$$p(\{stop, accl, cons, cons, decl\}) = p(stop) \cdot p(accl \mid stop) \cdot p(cons \mid accl) \cdot p(cons \mid cons) \cdot p(dccl \mid cons)$$
$$= 1.0 \cdot 0.8 \cdot 0.9 \cdot 0.9 \cdot 0.05$$
$$= 0.0324$$

7.2 HIDDEN MARKOV MODELS (HMM)

In a Markov model, each state corresponds to an observable event. However, the events we are interested in are not observable in many applications. For example, the exact motion model of a vehicle may not be directly observable but returns can be obtained from remote sensors about the vehicle's position, velocity, etc. In this case, the motion models are *hidden* because they are not observed. The notion of the Markov model is extended to the Hidden Markov Model (HMM) to encode both hidden and observed events. Formally, an HMM is a quintuplet $\langle S, O, A, B, \Pi \rangle$ defined as follows:

- A set $S = \{s_1, ..., s_n\}$ of n distinct *states*, one of which is assumed to be the state of the process at time step t.
- A set $O = \{o_1, ..., o_m\}$ of m distinct *observation symbols* correspond to the physical output of the system or process being modeled.
- A set $A = \{a_{ij}\}_{i,j=1,...,n}$ of *state transition probabilities*, where each a_{ij} represents the probability of transitioning from state s_i to state s_j. Symbolically, $p(X_t = s_j \mid X_{t-1} = s_i) = a_{ij}$, where X_t is the random variable representing the state at time point step t.
- A set $B = \{b_i(o_j)\}_{i=1,...,n;j=1,...,m}$ of *observation likelihoods*, where each $b_i(o_j)$ represents the probability of observing o_j at state s_i. Symbolically, $p(Y_t = o_j \mid X_t = s_i) = b_{ij}$, where Y_t is the random variable representing the observation at time step t.
- An initial probability distribution $\Pi = \{\pi_1, ..., \pi_n\}$, respectively, over the states $S = \{s_1, ..., s_n\}$, where π_i is the probability that the model will start in state s_i.

As before, we have the following constraints for a Markov model:

$$\sum_{i=1}^{n} \pi_i = 1, \sum_{j=1}^{n} a_{ij} = 1$$

In a *first-order HMM*, as in the case of a first-order Markov model, the probability of a particular state is dependent only on the previous state:

$$p(X_t = s_i \mid X_{t-1} = s_j, X_{t-2} = s_k, ...) = p(X_t = s_i \mid X_{t-1} = s_j)$$

Finally, in a first-order HMM, the probability of an output observation is dependent only on the state that produced the observation. Symbolically,

$$p(Y_t = o_i \mid X_i = s_j, \ldots, X_t = s_k, \ldots) = p(Y_t = o_i \mid X_t = s_k)$$

Figure 7-3 represents an HMM, extended from the simple Markov model shown in Figure 7-1 with four states, along with its observation matrix shown on the right. The transition is as before in Figure 7-1. Note that the observations with the zero probability are not shown.

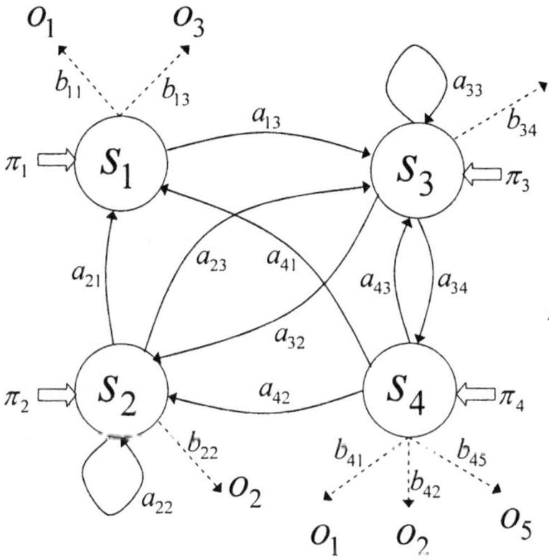

$$B = \begin{bmatrix} b_{11} & 0 & b_{13} & 0 & 0 \\ 0 & b_{22} & 0 & 0 & 0 \\ 0 & 0 & 0 & b_{34} & 0 \\ b_{41} & b_{42} & 0 & 0 & b_{45} \end{bmatrix}$$

Figure 7-3: HMM with four states along with its observation matrix

An HMM $\langle S, O, A, B, \Pi \rangle$ can be used as a generator to give an observation sequence $Y_1 Y_2 \ldots Y_t \ldots$ as follows:

1) Choose an initial state X_1 from S according to the initial state distribution Π.
2) Set $t = 1$.
3) Choose Y_t from O according to the observation likelihood in state X_t.
4) Transit to a new state X_{t+1} from S according to the state transition distribution for state X_t.
5) Set $t = t + 1$ and return to Step 3.

Figure 7-4 shows the evolution of this generator around the time step t.

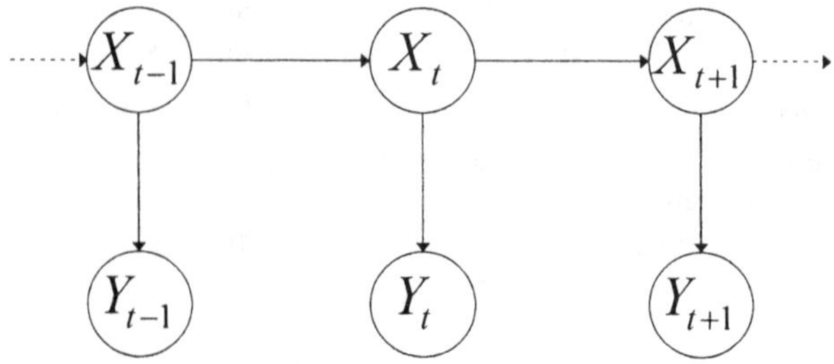

Figure 7-4: Evolution of the HMM

Example

The output of the process for analyzing trafficability results in a product known as the Modified Combined Obstacles Overlay (MCOO). The MCOO classifies terrain into one of three coarse categories: *go*, *slow go*, and *no go*. Roads often become a slow go or a no go after heavy rains, causing increased soil moisture content and heavy fog resulting in decreased visibility, or citizens' protests causing blockades. Naturally, vehicle speed varies with the traficability condition. The observable here is the average number (0, 1, 2, or 3) of cars that are passing through a check post every five minutes. We consider a trafficability HMM with $\{go, no\text{-}go, slow\text{-}go\}$ as the set of states. The prior and transition probabilities and observable likelihoods are shown in Figure 7-5.

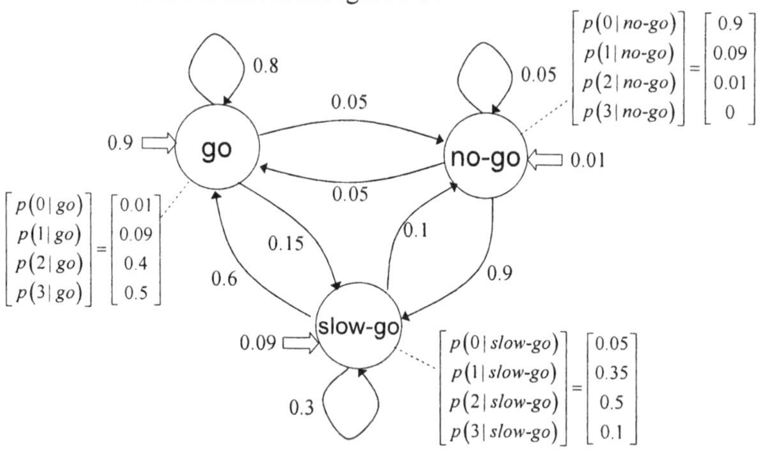

Figure 7-5: Example HMM with observable likelihoods

Note that the HMM in Figure 7-5 is *fully connected* in the sense that the transition probability between any two states is nonzero.

Three natural problems typically arise in applications that use HMM:

1) Computing the likelihood of an observation sequence (*the forward algorithm*): Find the probability of a given sequence of observations to be generated by the model.
2) Computing the most probable state sequence (*the Viterbi algorithm*): Find the most probable sequence of states that is likely to generate a given observation sequence.
3) Estimating the parameters of a model (*the Baum-Welch algorithm*): Find the most probable parameter set of a model given its structure and one or more observation sequences.

The forward and Viterbi algorithms are detailed in the following two subsections. The Baum-Welch algorithm, because of its machine learning slant, will be described in Chapter 10 on model learning.

7.2.1 The Forward Algorithm

Given an HMM $\langle S, O, A, B, \Pi \rangle$ and an observation sequence $Y_1 ... Y_T$, we want to compute the probability $p(Y_1 ... Y_T)$, which can be obtained by summing over all possible state sequences of length T as follows:

$$p(Y_1...Y_T) = \sum_{X_1...X_T} p(Y_1...Y_T \mid X_1...X_T) p(X_1...X_T)$$

Since the observations are independent,

$$p(Y_1...Y_T \mid X_1...X_T) = \prod_{t=1}^{T} p(Y_t \mid X_1...X_T)$$

Since an output observation is dependent only on the state that produced the observation, for every t,

$$\begin{aligned} p(Y_t \mid X_1...X_T) &= p(Y_t \mid X_t) \\ &= b_{X_t}(Y_t) \end{aligned}$$

Moreover,

$$\begin{aligned} p(X_1...X_T) &= p(X_1) p(X_2 \mid X_1) p(X_3 \mid X_1, X_2) ... p(X_T \mid X_1, ..., X_{T-1}) \\ &= p(X_1) p(X_2 \mid X_1) p(X_3 \mid X_2) ... p(X_T \mid X_{T-1}) \\ &= \pi_{X_1} a_{X_1 X_2} a_{X_2 X_3} ... a_{X_{T-1} X_T} \\ &= \pi_{X_1} \prod_{t=1}^{T-1} a_{X_T X_{T+1}} \end{aligned}$$

Therefore,

$$\begin{aligned} p(Y_1...Y_T) &= \sum_{X_1...X_T} \left(\prod_{t=1}^{T} b_{X_t}(Y_t) \right) \left(\pi_{X_1} \prod_{t=1}^{T-1} a_{X_t X_{t+1}} \right) \\ &= \sum_{X_1...X_T} \pi_{X_1} b_{X_1}(Y_1) a_{X_1 X_2} b_{X_2}(Y_2) ... a_{X_{T-1} X_T} b_{X_T}(Y_T) \end{aligned}$$

The above probability computation for each sequence $X_1...X_T$ of states can be interpreted as follows. First at time 1 in the process we are in state X_1 with probability π_{X_1}. The symbol Y_1 is generated at this state with probability $b_{X_1}(Y_1)$. When the time changes, a transition takes place from state X_1 to state X_2 with probability $a_{X_1X_2}$. The symbol Y_2 at this state is generated with probability $b_{X_2}(Y_2)$. The process continues until the final transition takes place from state X_{t-1} to state X_t with probability $a_{X_{T-1}X_T}$, and the symbol Y_t is generated with probability $b_{X_T}(Y_T)$.

The simplest way of computing $p(Y_1...Y_T)$ is to enumerate all possible state sequences and then, for each such sequence $X_1...X_T$, compute the term in the summation by following the above interpretation.

Example

Consider an evolution in Figure 7-6 of the HMM in Figure 7-5. This evolution provides a term in the summation for computing $p(122)$:

$$\begin{aligned}&\pi_{slow\text{-}go}b_{slow\text{-}go}(1)\,a_{slow\text{-}go\,go}b_{go}(2)\,a_{go\,no\text{-}go}b_{no\text{-}go}(2)\\&=0.09\times0.35\times0.6\times0.4\times0.05\times0.01\\&=0.000000378\end{aligned}$$

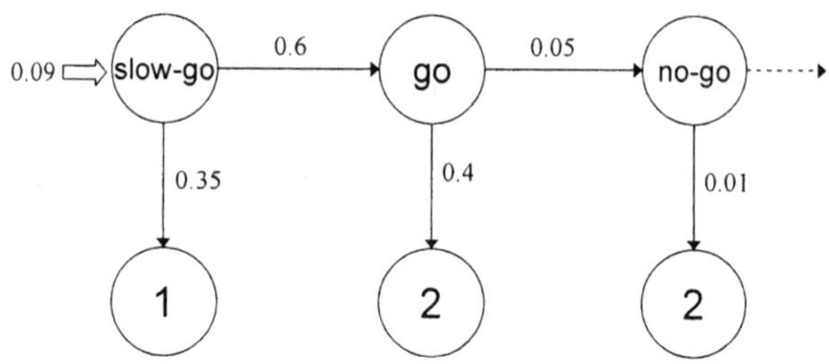

Figure 7-6: An evolution of the HMM in Figure 7-5

But the total number of sequences to be handled in this way can be very large (e.g., 4^{100} for four states and 100 time steps). The forward algorithm below is a more efficient implementation for computing $p(Y_1...Y_T)$.

The algorithm considers the forward variable $\alpha_t(s_i)$ as defined below:

$$\alpha_t(s_i)=p(Y_1...Y_t,X_t=s_i)$$

The variable represents the joint probability of the partial observation sequence $Y_1...Y_t$ and state s_i at time t. The three-step forward algorithm is developed via an inductive computation of $\alpha_t(s_i)$.

Step 1: Initialization

$$\alpha_1(s_i) = \pi_i b_i(Y_1), 1 \le i \le n$$

Step 2: Induction

$$\alpha_{t+1}(s_j) = \left(\sum_{i=1}^{n} \alpha_t(s_i) a_{ij}\right) b_j(Y_{t+1}), \quad 1 \le t \le T-1; 1 \le j \le n$$

Step 3: Termination

$$p(Y_1...Y_T) = \sum_{i=1}^{n} \alpha_T(s_i)$$

Figure 7-7 generically illustrates the computation of a forward variable.

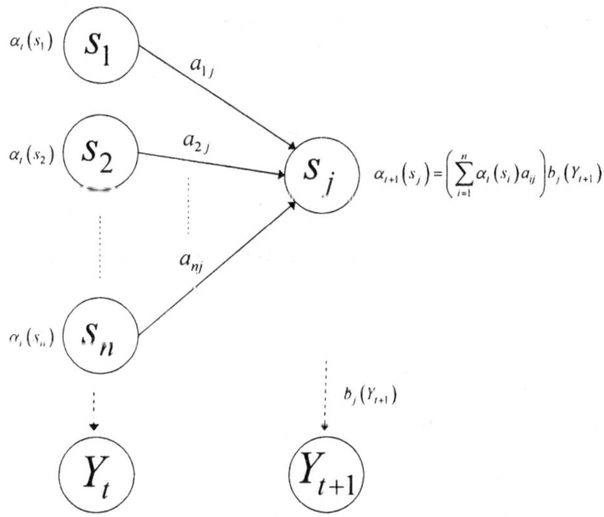

Figure 7-7: Illustration of the computation of a forward variable

Example

Consider the HMM in Figure 7-5. Figure 7-8 illustrates the forward algorithm with the following computation of values of α-variables at various stages (g – *go*; ng – *no-go*; sg – *slow-go*).

$$\alpha_1(g) = \pi_g p(1 \mid g) = 0.9 \cdot 0.09 = 0.081$$

$$\alpha_1(ng) = \pi_{ng} p(1 \mid ng) = 0.01 \cdot 0.09 = 0.0009$$

$$\alpha_1(sg) = \pi_{sg} p(1 \mid sg) = 0.09 \cdot 0.35 = 0.0315$$

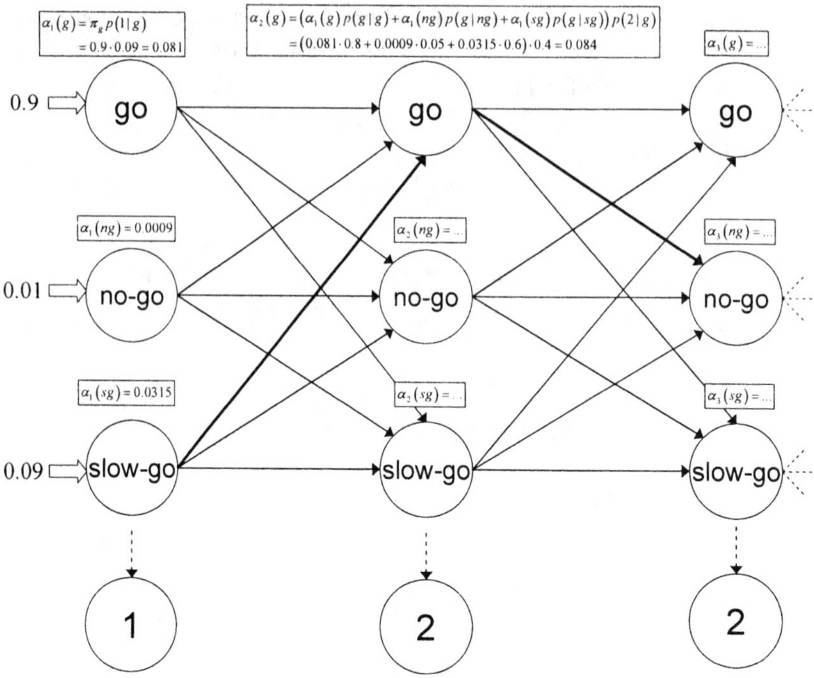

Figure 7-8: Illustration of the forward algorithm (*g* – *go*; *ng* – *no-go*; *sg* – *slow-go*)

$$\alpha_2(g)=(\alpha_1(g)p(g\mid g)+\alpha_1(ng)p(g\mid ng)+\alpha_1(sg)p(g\mid sg))p(2\mid g)$$
$$=(0.081\cdot0.8+0.0009\cdot0.05+0.0315\cdot0.6)\cdot0.4=0.0335$$
$$\alpha_2(ng)=(\alpha_1(g)p(ng\mid g)+\alpha_1(ng)p(ng\mid ng)+\alpha_1(sg)p(ng\mid sg))p(2\mid ng)$$
$$=(0.081\cdot0.05+0.0009\cdot0.05+0.0315\cdot0.1)\cdot0.01=0.0008$$
$$\alpha_2(g)=(\alpha_1(g)p(sg\mid g)+\alpha_1(ng)p(sg\mid ng)+\alpha_1(sg)p(sg\mid sg))p(2\mid sg)$$
$$=(0.081\cdot0.15+0.0009\cdot0.9+0.0315\cdot0.3)\cdot0.5=0.0419$$

7.2.2 The Viterbi Algorithm

Given an HMM $\langle S,O,A,B,\Pi\rangle$ and an observation sequence $Y_1...Y_T$, we want to find the most probable sequence $X_1...X_T$ of states that is likely to generate the observation sequence. In other words, we have the following maximization problem at hand:

$$\max_{X_1...X_T} p(X_1...X_T \mid Y_1...Y_T)$$

which is equivalent to the maximization problem

$$\max_{X_1 \ldots X_T} p(X_1 \ldots X_T, Y_1 \ldots Y_T)$$

To compute this, let us define the following inductive variable:

$$\delta_t(s_i) = \max_{X_1 \ldots X_t} p(X_1 \ldots X_t = s_i, Y_1 \ldots Y_t)$$

which is the highest probability along a single path that ends in state s_i. The induction principle provides the following:

$$\delta_{t+1}(s_j) = \left(\max_i \delta_t(s_i) a_{ij}\right) b_j(Y_{t+1})$$

We want to keep track of the argument that maximized the above at each time point t via the variable $\psi_t(s_j)$. The Viterbi algorithm steps are given below:

Step 1: Initialization

$$\delta_1(s_i) = \pi_i b_i(Y_1) \qquad 1 \le i \le n$$

$$\psi_1(i) = start \qquad 1 \le i \le n$$

Step 2: Recursion

$$\delta_t(s_j) = \left(\max_{1 \le i \le n} \delta_{t-1}(s_i) a_{ij}\right) b_j(Y_t) \qquad 1 \le j \le n; 2 \le t \le T$$

$$\psi_t(s_j) = \left(\arg\max_{1 \le i \le n} \delta_{t-1}(s_i) a_{ij}\right) \qquad 1 \le j \le n; 2 \le t \le T$$

Step 3: Termination

Best score $P^* = \max_{1 \le i \le n} \delta_T(s_i)$

Path backtracking start $X_T^* = \arg\max_{1 \le i \le n} \delta_T(s_i)$

Step 4: State sequence

$$X_t^* = \psi_{t+1}(X_{t+1}^*) \qquad t = T-1, T-2, \ldots, 1$$

The Viterbi algorithm is similar to the forward algorithm except that it maintains the ψ-variable for backtracking and employs maximization in Step 2 instead of a summation.

Example

Consider the HMM in Figure 7-5. Figure 7-9 illustrates the Viterbi algorithm with the computation of values of δ- and ψ-variables at various stages.

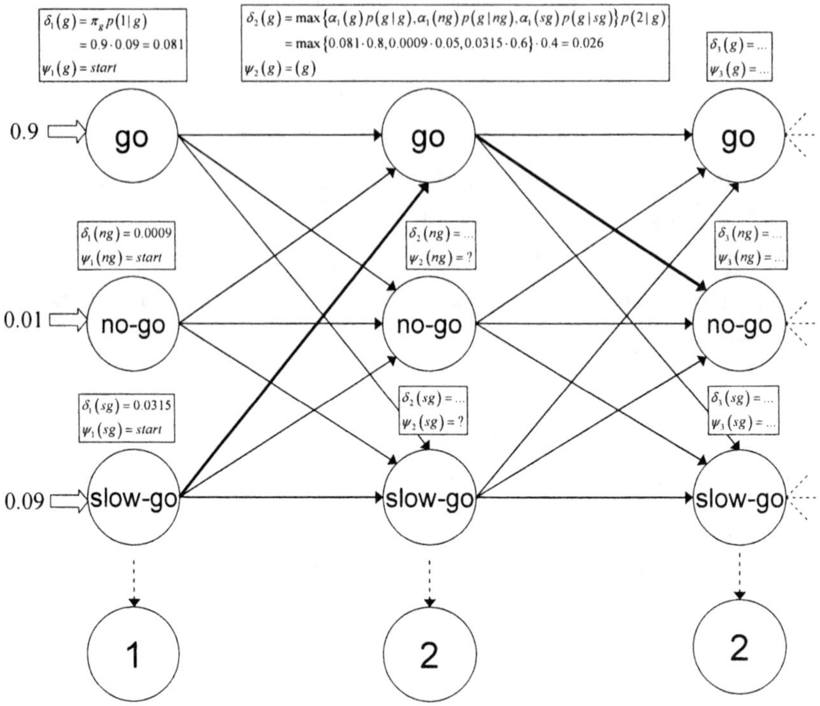

Figure 7-9: Illustration of a Viterbi algorithm

7.3 HIERARCHICAL HIDDEN MARKOV MODELS (HHMM)

Hierarchical Hidden Markov Models (HHMMs) generalize HMMs by making each of the hidden states an HHMM as well. Therefore, an HHMM state emits sequences of symbols, rather than a single symbol, by a recursive activation of one of the sub-states of a state. A hidden state that does not emit an observable symbol directly is called an *internal state*. An internal state makes a vertical transition to a lower-level state, which, in turn, can make a transition to another lower-level state or a horizontal transition to a state in the same level. Thus the set of states induces a tree structure: the root state is the node at the top, leaves are the *production states*, and the rest of the intermediate nodes are internal states.

Example

Figure 7-10 shows a set of HHMMs at different levels. The process at the first level consists of only a single node s_1^1 representing the start of the process. The

HHMM H^2 is a process at Level 2. The state s_2^2 of H^2 is itself the HHMM H^3. Both states s_1^3 and s_2^3 of H^3 are themselves HHMMs H_1^4 and H_2^4, respectively.

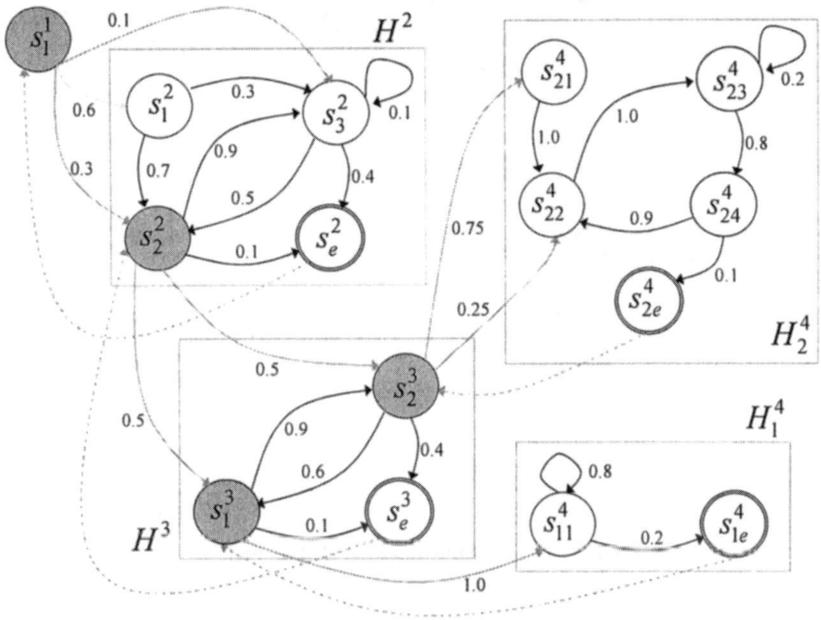

Figure 7-10: An example HHMM

The tree structure in Figure 7-11 is an evolution of the above HHMM generating the observation sequence $Y_1 ... Y_9$. The levels are indicated on the left. The shaded circles represent internal nodes and the clear circles represent production nodes. An HMM at a certain level will return its control to a level higher via its end state upon the completion of a vertical transition. For example, internal state s_1^3 makes a vertical transition from Level 3 to state s_{11}^4 of H_1^4 at Level 4. After the generation of the observation sequence $Y_2 Y_3 Y_4$ by H_1^4, its end state s_{1e}^4 then returns the control to s_1^3.

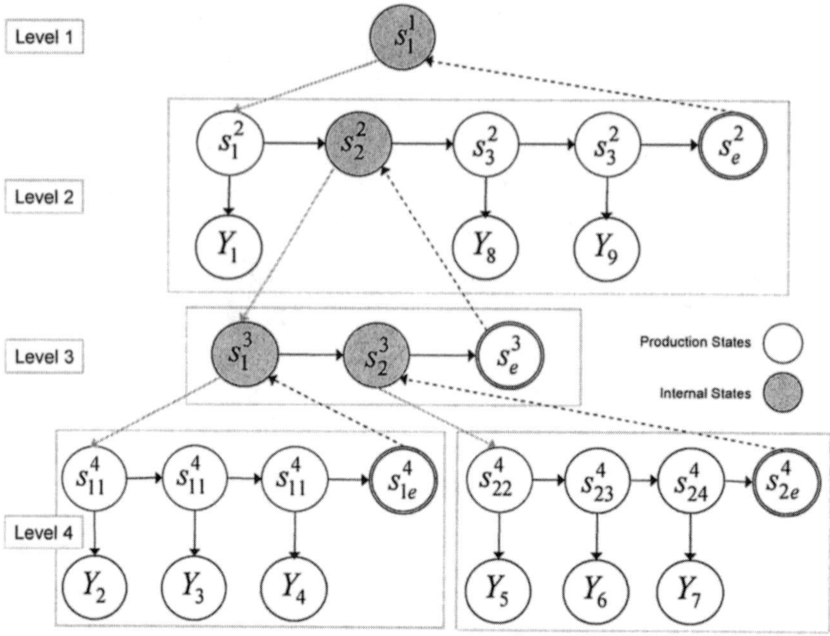

Figure 7-11: An evolution of the HHMM in Figure 7-10

Formally, an HHMM with the depth of hierarchy D is a quintuplet $\langle S, O, A, B, \Pi \rangle$ defined as follows:

- For each level or hierarchy index $d\,(1 \le d \le D)$, there is a set $S^d = \left(s_i^d\right)_i$ of distinct *states*, one of which is assumed to be the state of the process at time step t.
- A set $O = \{o_1, ..., o_m\}$ of m distinct observation symbols correspond to the physical output of the system or process being modeled.
- For each internal state $s_k^d\,(1 \le d \le D-1)$ at depth d, a set $A^{dk} = \left(a_{ij}^{dk}\right)_{ij}$ of *state transition probabilities*, where each a_{ij}^{dk} represents the probability of making a horizontal transition at depth $d+1$ from state s_i^{d+1} to state s_j^{d+1}, both of which are sub-states of state s_k^d. Symbolically, $p\left(X_t = s_j^{d+1} \mid X_{t-1} = s_i^{d+1}\right) = a_{ij}^{dk}$, where X_t is the random variable representing the state at time step t.
- For each production state s_i^D at depth D, a set $B^{Di} = \left(b^{Di}\left(o_j\right)\right)_j$ of *observation likelihoods*, where each $b^{Di}\left(o_j\right)$ represents the probability of

observing o_j at state s_i^D. Symbolically, $p\left(Y_t^D = o_j \mid X_t^D = s_i^D\right) = b^{Di}\left(o_j\right)$, where Y_t^D is the random variable representing the observation at level D at time step t.

- For each internal state $s_k^d \left(1 \le d \le D-1\right)$ at depth d, a set $\Pi^{dk} = \left(\pi_i^{dk}\right)_i$ of an initial probability distribution, respectively, over the substates $\left(s_i^{d+1}\right)_i$ of s_k^d, where π_i^{dk} is the probability that s_k^d will initially activate s_i^{d+1}.

7.3.1 The Forward Algorithm for HHMM

Given an HHMM $\langle S, O, A, B, \Pi \rangle$ and an observation sequence $Y_1 ... Y_T$, we want to compute the probability $p\left(Y_1 ... Y_T\right)$. Let $s_{i_1}^2 ... s_{i_l}^2$ be the sequence of the states at the second level visited during the generation of $Y_1 ... Y_T$. Thus, $s_{i_l}^2 = s_e^2$. Let $t_1, ..., t_l$ be the temporal positions of the first symbols generated, respectively, by states $s_{i_1}^2, ..., s_{i_l}^2$. Therefore, $t_1 = 1$ and $t_l = T$, and $s_{i_1}^2$ were activated by s_1^1. Thus,

$$
\begin{aligned}
& p\left(Y_1 ... Y_T \mid t_1, ..., t_l, s_1^1, s_{i_1}^2 ... s_{i_l}^2\right) = \\
& \qquad \pi_{i_1}^{11} p\left(Y_1 ... Y_{t_1 - 1} \mid s_1^1, s_{i_1}^2\right) a_{i_1 i_2}^{11} p\left(Y_{t_1} ... Y_{t_2 - 1} \mid s_1^1, s_{i_2}^2\right) a_{i_2 i_3}^{11} ... \\
& \qquad ... p\left(Y_{t_{l-2}} ... Y_{t_{l-1} - 1} \mid s_1^1, s_{i_{l-2}}^2\right) a_{i_{l-2} i_{l-1}}^{11} p\left(Y_{t_{l-1}} ... Y_T \mid s_1^1, s_{i_{l-1}}^2\right) a_{i_{l-1} e}^{11}
\end{aligned}
$$

To compute each of the probability terms in their generic form $p\left(Y_t ... Y_{t_k} \mid s_k^{d-1}, s_i^d\right)$, we define the following generalized forward variable:

$\alpha\left(t, t+k, s_k^{d-1}, s_i^d\right) =$ Probability that the partial observation sequence $Y_t ... Y_{t+k}$ was generated by state s_k^{d-1} and that state s_i^d was the last state activated by s_k^{d-1} during the generation of the sequence

$$
= p\left(Y_t ... Y_{t+k}, X_{t+k}^d = s_i^d \mid X_t^{d-1} = s_k^{d-1}\right)
$$

Now, $\alpha(.)$ can be recursively computed as below:

$$
\alpha\left(t, t, s_k^{D-1}, s_i^D\right) = \pi_i^{(D-1)k} b^{Di}\left(Y_t\right)
$$

$$
\alpha\left(t, t+\tau, s_k^{D-1}, s_i^D\right) = \left(\sum_{j=1}^{\left|s_k^{D-1}\right|} \alpha\left(t, t+\tau-1, s_k^{D-1}, s_j^D\right) a_{ji}^{(D-1)k} \right) b^{(D-1)i}\left(Y_{t+\tau}\right)
$$

$$
\alpha\left(t, t, s_k^{d-1}, s_i^d\right) = \pi_i^{(d-1)k} \left(\sum_{j=1}^{\left|s_i^d\right|} \alpha\left(t, t, s_i^d, s_j^{d+1}\right) a_{je}^{di} \right)
$$

$$\alpha\left(t,t+\tau,s_k^{d-1},s_i^d\right)=\sum_{l=0}^{\tau-1}\left(\sum_{j=1}^{\left|s_k^{d-1}\right|}\alpha\left(t,t+l,s_k^{d-1},s_j^d\right)a_{ji}^{(d-1)k}\right)$$

$$\left(\sum_{j=1}^{\left|s_i^d\right|}\alpha\left(t+l+1,t+\tau,s_i^d,s_j^{d+1}\right)a_{je}^{di}\right)$$

$$+\pi_i^{(d-1)k}\left(\sum_{j=1}^{\left|s_i^d\right|}\alpha\left(t,t+\tau,s_i^d,s_j^{d+1}\right)a_{je}^{di}\right)$$

In general, $Y_t...Y_{t+k}$ can be a prefix of a longer sequence generated by s_k^{d-1}. Therefore, to calculate the probability that $Y_t...Y_{t+k}$ is generated by s_k^{d-1}, we need to sum over all possible states at Level d ending at s_e^{d-1}. Thus,

$$p\left(Y_t...Y_{t+k}\mid X_t^{d-1}=s_k^{d-1}\right)=\sum_{i=1}^{\left|s_k^{d-1}\right|}\alpha\left(t,t+k,s_k^{d-1},s_i^d\right)a_{ie}^{(d-1)k}$$

Finally, the likelihood of $Y_1...Y_T$ is obtained by summing over all possible starting states:

$$p\left(Y_1...Y_T\right)=\sum_{i=1}^{\left|s_1^1\right|}\alpha\left(1,T,s_1^1,s_i^2\right)$$

The α-variables are evaluated recursively bottom-up in the sense that the α-value of an internal state is evaluated based on the α-values of the substates of the state.

7.3.2 The Viterbi Algorithm for HHMM

We generalize the definitions of the inductive variable δ and ψ variables as follows, along with the definition of a supporting time variable to help generate the most probable state sequence via backtracking:

- $\delta\left(t,t+k,s_k^{d-1},s_i^d\right)$ is the likelihood of the most probable state sequence generating $Y_t...Y_{t+k}$. A recursive activation from state s_k^{d-1} at time step t started the generation of this sequence and ended at s_i^d, which returned to s_k^{d-1} at time step $t+k$.
- $\psi\left(t,t+k,s_k^{d-1},s_i^d\right)$ is the index of the most probable state to be activated by s_k^{d-1} before s_i^d. This index is zero if $Y_t...Y_{t+k}$ is solely generated by s_i^d.
- $\tau\left(t,t+k,s_k^{d-1},s_i^d\right)$ is the time step at which the state s_i^d was most probable to be called by s_k^{d-1}. This time step is t if $Y_t...Y_{t+k}$ is solely generated by s_i^d.

The algorithm computes the δ, ψ, and τ variables in a bottom-up manner starting from the production states.

Step 1: Production State – Initialization

$$\delta\left(t,t,s_k^{D-1},s_i^D\right)=\pi_i^{(D-1)k}b^{Di}\left(Y_t\right)$$

$$\psi\left(t,t,s_k^{D-1},s_i^D\right)=0$$

$$\tau\left(t,t,s_k^{D-1},s_i^D\right)=0$$

Step 2: Production State – Recursion

$$\text{Set } \Delta_j=\delta\left(t,t+k-1,s_k^{D-1},s_j^D\right)a_{ji}^{(D-1)k}b^{Di}\left(Y_{t+k}\right)$$

$$\delta\left(t,t+k,s_k^{D-1},s_i^D\right)=\max_{1\le j\le\left|s_k^{D-1}\right|}\left(\Delta_j\right)$$

$$\psi\left(t,t+k,s_k^{D-1},s_i^D\right)=\arg\max_{1\le j\le\left|s_k^{D-1}\right|}\left(\Delta_j\right)$$

$$\tau\left(t,t+k,s_k^{D-1},s_i^D\right)=t+k$$

Step 3: Internal State – Initialization

$$\delta\left(t,t,s_k^{d-1},s_i^d\right)=\max_{1\le j\le\left|s_i^d\right|}\left(\pi_i^{(d-1)k}\delta\left(t,t,s_i^d,s_j^{d+1}\right)a_{je}^{di}\right)$$

$$\psi\left(t,t,s_k^{d-1},s_i^d\right)=0$$

$$\tau\left(t,t,s_k^{d-1},s_i^d\right)=0$$

Step 4: Internal State – Recursion

$$\text{Set } \Delta_i\left(t'\right)=\max_{1\le l\le\left|s_i^d\right|}\delta\left(t',t+k,s_i^d,s_l^{d+1}\right)a_{le}^{di}$$

$$\text{Set } \Delta_{ji}\left(t'\right)=\delta\left(t,t'-1,s_k^{d-1},s_j^d\right)a_{ji}^{(d-1)k}\Delta_i\left(t'\right)$$

$$\begin{aligned}\Lambda\left(t'\right)&=\pi_{di}^{(d-1)k}\Delta_{ji}\left(t'\right), \qquad t'=t\\&=\max_{1\le j\le\left|s_k^{d-1}\right|}\left(\delta\left(t,t'-1,s_k^{d-1},s_j^d\right)a_{ji}^{(d-1)k}\Delta_{ji}\left(t'\right)\right), \quad t'=t+1,\ldots,t+k\end{aligned}$$

$$\begin{aligned}\psi\left(t'\right)&=0, \qquad t'=t\\&=\arg\max_{1\le j\le\left|s_k^{d-1}\right|}\left(\delta\left(t,t'-1,s_k^{d-1},s_j^d\right)a_{ji}^{(d-1)k}\Delta_{ji}\left(t'\right)\right),t'=t+1,\ldots,t+k\end{aligned}$$

The values of the variables δ, ψ, and τ are the following:

$$\delta\left(t,t+k,s_k^{d-1},s_i^d\right)=\max_{t\le t'\le t+k}\Lambda\left(t'\right)$$

$$\tau\left(t,t+k,s_k^{d-1},s_i^d\right)=\arg\max_{t\le t'\le t+k}\Lambda\left(t'\right)$$

$$\psi\left(t,t+k,s_k^{d-1},s_i^d\right)=\Psi\left(\tau\left(t,t+k,s_k^{d-1},s_i^d\right)\right)$$

Therefore, the probability of the most probable state sequence is $\max_{1\le i\le\left|s_1^1\right|}\delta\left(1,T,s_1^1,s_i^2\right)$, and the most probable state sequence itself can be obtained by scanning the lists ψ and τ, starting from $\tau\left(1,T,s_1^1,s_{last}^2\right)$ and $\psi\left(1,T,s_1^1,s_{last}^2\right)$, where $s_{last}^2 = \underset{1\le i\le |s_1^1|}{\arg\max}\, \delta\left(1,T,s_1^1,s_i^2\right)$.

7.4 MARKOV MODELS FOR TEXT ANALYSES

The order of words that appear in a sentence is relevant to understanding its semantics. HMMs can be used to recognize such sequential data. The HMM on the left side of Figure 7-12 recognizes names that appear in text fragments or sentences.

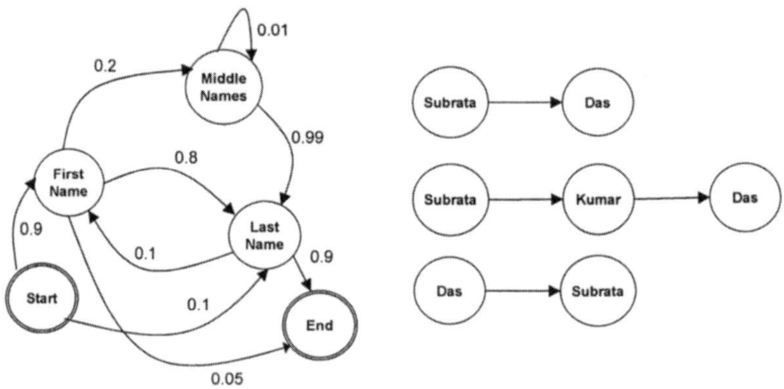

Figure 7-12: (*left*) Example HMM to recognize names; (*right*) some example evolutions of the HMM

Some example evolutions of the HMMs are shown on the right of the figure. The words "Subrata," "Kumar," and "Das" will be parsed and interpreted, respectively, as first, middle, and last names with the help of a database. Clearly, the probability of the first observation sequence will be the highest among all three.

The HHMM shown in Figure 7-13 understands the semantics of a sentence about a meeting between two people at some place on some date. It can also understand the semantics of two consecutive sentences; the first describes the two names and the second describes the place and date.

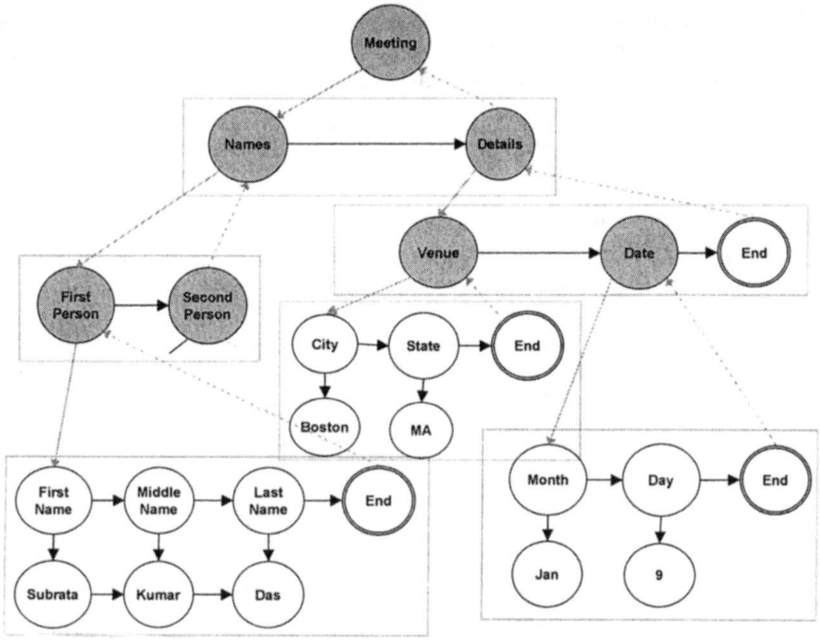

Figure 7-13: HHMM to recognize consecutive text fragments

7.5 HMM WITH EXPLICIT STATE DURATION

The inherent probability density $p_i(d)$, associated with state s_i with self-transition probability a_{ii} as illustrated in Figure 7-14, is the probability of consecutive observations in state s_i. It has the form

$$p_i(d) = (1 - a_{ii}) a_{ii}^{d-1}$$

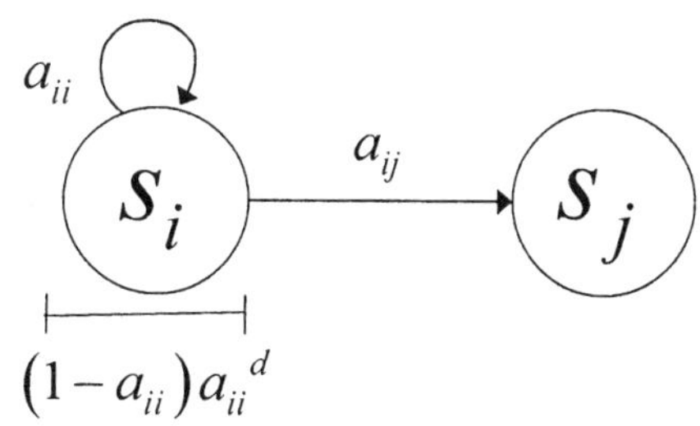

Figure 7-14: Ordinary HMM with exponential state duration

In some applications, it is preferable to have explicit state durations instead of self-transition, which can be specified via some density function. This kind of variable duration HMM, called Hidden Semi-Markov Model (HSMM), can be specified as follows:

- A duration d_k $(1 \le d_k \le D_i)$ is chosen, upon entering into a state s_i, according to the state duration density $p_i(d_k)$, where D_i is the maximum duration within the state s_i.
- Observations are assumed to be independent of each other within a state, and a sequence $Y_1...Y_{d_k}$ is chosen according to the joint observation density

$$b_i(Y_1...Y_{d_k}) = \prod_{t=1}^{d_k} b_i(Y_{d_k})$$

So a state transition in an HSMM from state s_i to state s_j, as shown in Figure 7-15, occurs as usual according to the state transition probability a_{ij}, but with constraint $a_{ij} = 0$.

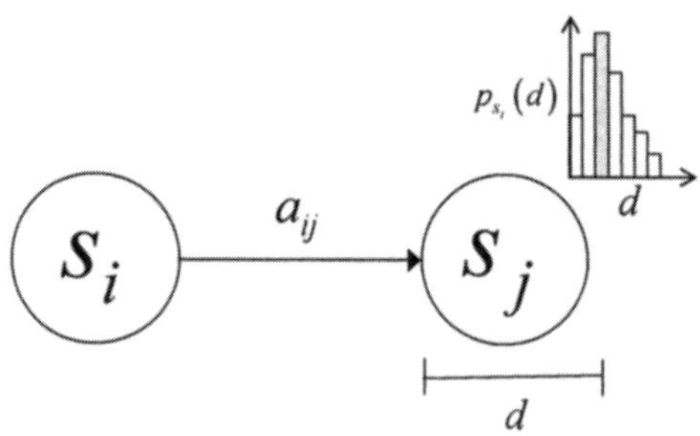

Figure 7-15: HSMM with explicit state duration probability distribution

The forward algorithm computes $p(Y_1...Y_T)$ for a given sequence $Y_1...Y_T$ of observations. As in the case of ordinary HMMs, we define the forward variable α as

$$\alpha_t(s_i) = p(Y_1...Y_t, X_t = s_i)$$

Let $S_1,...,S_m$ be the first m states visited during the sequence $Y_1...Y_t$ of observations and that the associated durations are, respectively, $d_1,...d_m$. Thus, $S_m = X_t = s_i$ and

$$\sum_{j=1}^{m} d_j = t$$

Therefore, the forward variable can be rewritten as

$$\alpha_t(s_i) = \sum_{S_1...S_m} \sum_{d_1...d_m} \pi_1 \cdot p_1(d_1) \cdot b_i(Y_1...Y_{d_1}) \cdot$$
$$a_{12} \cdot p_2(d_2) \cdot b_2(Y_{d_1+1}...Y_{d_2}) \cdot ... \cdot$$
$$a_{(m-1)m} \cdot p_m(d_m) \cdot b_m(Y_{d_1+...+d_{m-1}+1}...Y_{d_m})$$

where S_is are random variables representing the states. The above expression can be inductively written as follows:

$$\alpha_1(s_i) = \pi_i \cdot p_i(1) \cdot b_i(Y_1)$$

$$\alpha_2(s_i) = \pi_i \cdot p_i(2) \cdot \prod_{j=1}^{2} b_i(Y_j) + \sum_{j=1, j\neq i}^{N} \alpha_1(s_j) \cdot a_{ji} \cdot p_i(1) \cdot b_i(Y_2)$$

$$\alpha_3(s_i) = \pi_i \cdot p_i(3) \cdot \prod_{j=1}^{3} b_i(Y_j) + \sum_{j=1, j\neq i}^{N} \alpha_1(s_j) \cdot a_{ji} \cdot p_i(2) \cdot \prod_{k=2}^{3} b_i(Y_k)$$
$$+ \sum_{j=1, j\neq i}^{N} \alpha_2(s_j) \cdot a_{ji} \cdot p_i(1) \cdot b_i(Y_3)$$
$$= \pi_i \cdot p_i(3) \cdot \prod_{j=1}^{3} b_i(Y_j) + \sum_{d=1}^{2} \sum_{j=1, j\neq i}^{N} \alpha_{3-d}(s_j) \cdot a_{ji} \cdot p_i(d) \cdot \prod_{k=4-d}^{3} b_i(Y_k)$$

...

$$\alpha_{D_i}(s_i) = \pi_i \cdot p_i(D) \cdot \prod_{j=1}^{D} b_i(Y_j) + \sum_{d=1}^{D-1} \sum_{j=1, j\neq i}^{N} \alpha_{D-d}(s_j) \cdot a_{ji} \cdot p_i(d) \cdot \prod_{k=D-d+1}^{D} b_i(Y_k)$$

In general, for every $t > D_i$, where D is the maximum duration within the state s_i, $\alpha_t(s_i)$ can be computed using the following formula:

$$\alpha_t(s_i) = \sum_{d=1}^{D_i} \sum_{j=1, j\neq i}^{N} \alpha_{t-d}(s_j) \cdot a_{ji} \cdot p_i(d) \cdot \prod_{k=t-d+1}^{t} b_i(Y_k)$$

As in the case of ordinary HMMs, the desired probability of the observation sequence $Y_1 ... Y_T$ is

$$p(Y_1 ... Y_t) = \sum_{i=1}^{N} \alpha_T(s_i)$$

Example

Figure 7-16 is an example HSMM based on the HMM in Figure 7-5. The probability of the observation sequence 12333320 can be computed using the formula above.

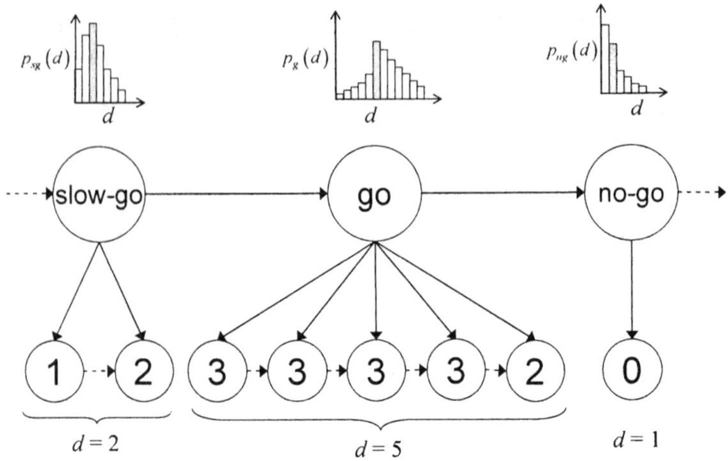

Figure 7-16: An example HSMM

In some applications, transition from a state depends on the amount of time already spent on the state itself. For example, once we enter into the state representing rain, to transition to a sunny state will depend on how long it has been raining. This can be captured in a Nonstationary Hidden Semi-Markov Model (NHSMM) by modifying the state transition probabilities to be dependent on the duration as follows.

- $A = (a_{ij}(\delta))$, where $a_{ij}(\delta)$, as shown in Figure 7-17, is the transition probability from state s_i to state s_j after having remained in state s_i exactly $\delta - 1$ length duration, and

$$\sum_{j=1}^{N} \sum_{\delta=1}^{D_i} a_{ij}(\delta) = 1, \qquad i = 1, \dots, N$$

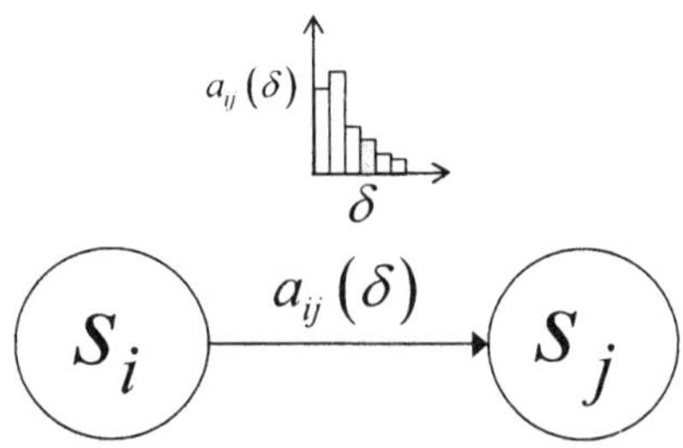

Figure 7-17: NHSMM with state transition probability distribution

We then define the forward probability α as the following:

$$\alpha_t(s_i) = \sum_{d=1}^{D_i} \sum_{j=1,j}^{N} \sum_{\delta=1}^{D_j} \alpha_{t-d-\delta+1}(s_j) \cdot a_{ji}(\delta) \cdot p_i(d) \cdot \prod_{k=t-d+1}^{t} b_i(Y_k)$$

The above recursive computation can be initiated as in the case of HSMM using $\alpha_t(s_i)$, for $t \le D_i$.

7.6 DYNAMIC BAYESIAN NETWORKS (DBNs)

One limitation of BNs presented in Chapter 6 is that they do not explicitly or adequately model temporal information that may be required when modeling a dynamic operational environment. This limitation can reduce overall assessment accuracy and thereby reduce operational utility. To illustrate, consider the BN shown in Figure 7-18.

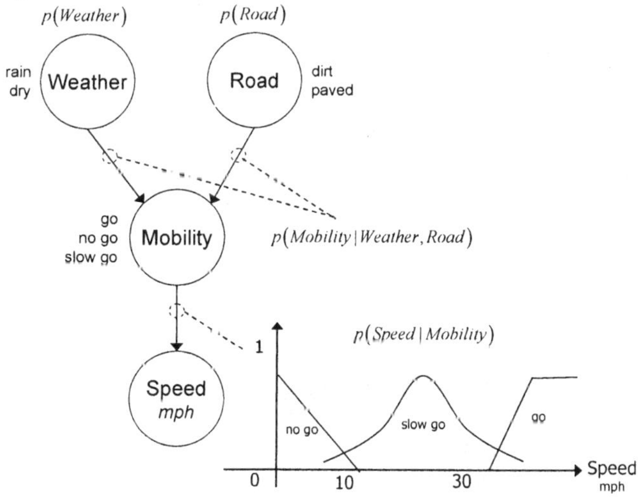

Figure 7-18: An example static BN

The type of a road and the weather at the current time period together affect the mobility of the road at this period of time, which is one of go, no go, and slow go. The mobility then determines the average speed of vehicles along the road defined by the three distributions corresponding to the three mobility states. The fact that continuous rain over several time periods may affect the road mobility differently than a single period of time is not modeled in this BN.

To address this problem, we introduce Dynamic Bayesian Networks (DBNs) (Murphy, 2002; Ghahramani, 2001), which are simply BNs for modeling time series data. There are two key ideas behind the extension of a BN to a DBN:

- All nodes of the BN are associated with particular time steps, simply by indexing the nodes with a time step value.
- Some BN nodes for a given time step may have causal dependencies on nodes from earlier time steps (in addition to the usual causal dependencies on nodes from their own time step); such dependencies are called temporal dependencies.

The result is a DBN. Figure 7-19 illustrates the general case of time indexing and temporal dependency: variable X_t from time step t has causal dependencies on some variables X_t^i from its own time step t; it also has causal dependencies on X_{t-j}^i from earlier time steps. When dependencies between nodes are only from the immediately prior time step, we say we have a Markovian model.

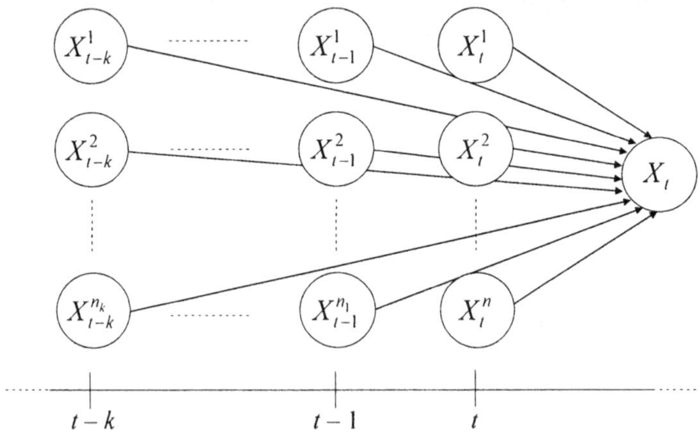

Figure 7-19: General temporal dependency in DBNs

As a concrete illustration, consider a temporal extension of the static BN of Figure 7-18 to the DBN in Figure 7-20. The weather node at time step t depends on the weather node at the previous time point. The mobility node at time step t not only depends on the weather and road nodes at t, but also on the weather at one and two time periods before. Therefore, the CPT of the weather node at time step t is $p(W_t \mid W_{t-1})$, and the CPT of the mobility node at time step t is $p(M_t \mid R_t, W_t, W_{t-1}, W_{t-2})$, where M and W represent mobility and weather, respectively. In the case of a Markovian model, the CPT of the mobility node is $p(M_t \mid R_t, W_t, W_{t-1})$. Note that the exact length of the time period is defined as needed by the specific application at hand, which could be, for example, a minute, an hour, or a day.

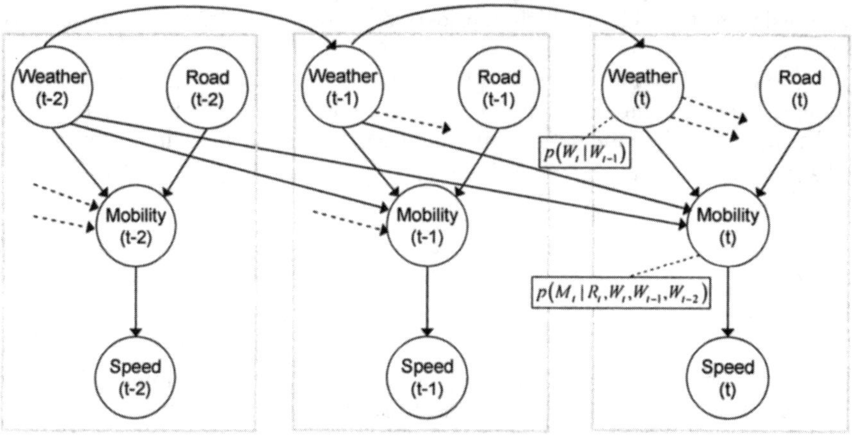

Figure 7-20: An example DBN based on the BN in Figure 7-18

We assumed that the current and some past weather conditions affect mobility. An alternative modeling strategy would be to account for all the past weather effects via the mobility node in the immediate past, thus yielding a Markovian model as shown in Figure 7-21. In other words, an overall effect of the past weather on the mobility node at time step t is summarized by the mobility node at time step $t-1$.

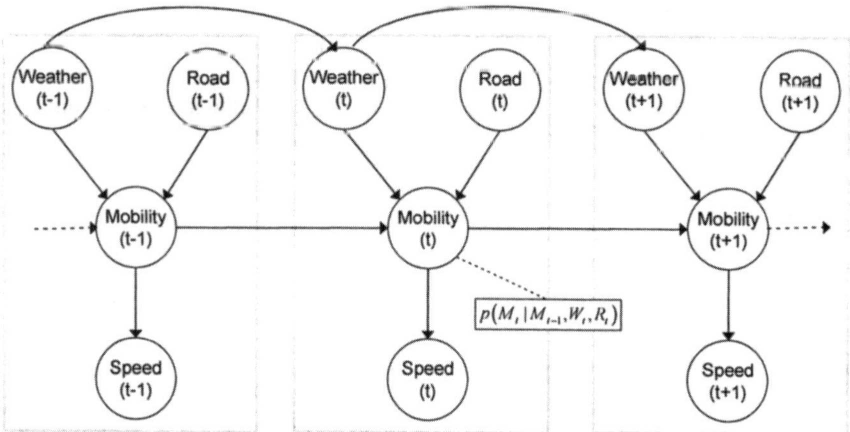

Figure 7-21: A Markovian version of the DBN in Figure 7-20

A DBN is assumed *stationary* in the sense that the node relationships within a time step t and the transition probabilities from time step t to time step $t+1$ (both expressed in CPTs) do not depend on t. Therefore, we can carry out inference in a

DBN using just a 2TBN, which considers only the nodes from two consecutive time steps.

A fragment of the evolution of an HMM can be considered a DBN. For example, the fragment in Figure 7-4 is a DBN with six nodes, where the members of the sets *S* and *O* in the definition of the HMM form the mutually exclusive and exhaustive set of states of nodes X_is and Y_js respectively. The CPTs $p(X_0)$, $p(X_i \mid X_{i-1})$, and $p(Y_i \mid X_i)$ can be constructed, respectively, from Π, *A*, and *B* in the definition of the HMM. But DBNs model problems at a finer level than HMMs. The concept of a state in a DBN is a possible combination of the states of its variables. So, an example state for the DBN in Figure 7-21 is

$$\langle Weather = rain, Road = dirt, Mobility = go, Visibility = Fog, Speed = 36.5 \rangle$$

There will be an infinite number of such states if one or more variables is continuous in nature (e.g., *Speed*); otherwise, the number of states that can be composed from a DBN consisting of only categorical variables is finite. For example, there will be 108 states if the variable *Speed* is considered categorical with its states *high*, *medium*, and *low*, as shown later in Figure 7-23. These states then constitute the states of an HMM equivalent to the DBN. The transition probabilities $p(X_t \mid X_{t-1})$ in this HMM can be computed from the CPTs of the DBN.

7.6.1 Inference Algorithms for DBNs

A naïve approach to inference in a DBN is to evolve it for a desired number of time steps and treat it as a static BN. Figure 7-22 shows an unrolled version of the DBN in Figure 7-21 up to time step 5. Evidence can be posted at any time step to obtain the necessary prediction. So if the current time step is 3 and evidence has been posted then querying at time step 4 or 5 provides prediction. Smoothing can be done by reposting any evidence in the past time step 1 or 2.

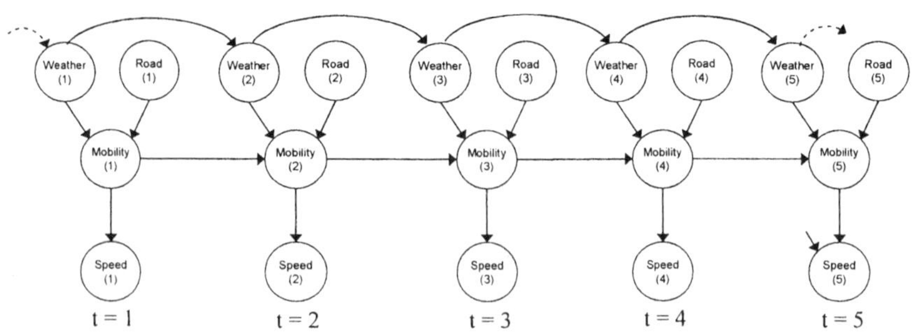

Figure 7-22: Unrolled up to time step 5 for the DBN in Figure 7-21

Though a naïve approach to inference in DBNs is simple to implement, the size of an unrolled DBN becomes large with many time steps. What we need is a process to maintain just a 2TBN by advancing forward one step at a time and dynamically trimming past BN slices. The following two algorithms leverage the junction tree algorithm on the 2TBN:

- Frontier algorithm (Zweig, 1998), and
- Interface algorithm (Murphy, 2002).

Here we present the interface algorithm as it is more efficient than the frontier algorithm. The algorithm uses the interface to *d*-separate the past from the present during the evolution of a DBN process. Here we consider a simplified version of the algorithm that does one-step prediction and filtering.

Inference in hybrid DBNs involving both discrete and continuous variables requires the particle filtering technique to be presented in the next section. Here we consider the BN in Figure 7-23, a discrete and extended version of the BN in Figure 7-18, to illustrate an exact inference algorithm for DBNs involving only discrete variables.

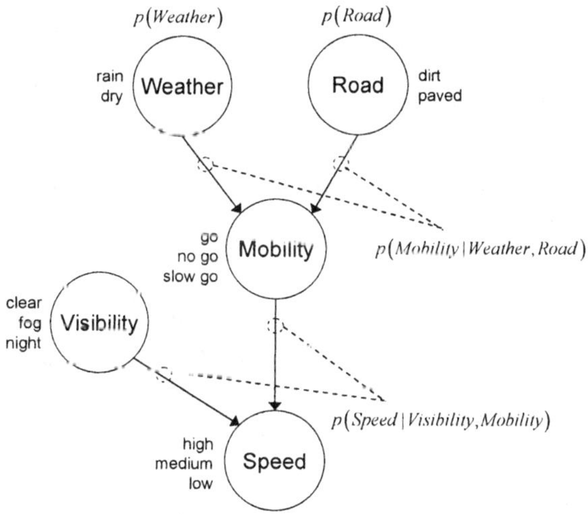

Figure 7-23: An example BN to illustrate a DBN inference algorithm

Figure 7-24 is a 2TBN of the BN in Figure 7-23. The shaded nodes are the outgoing interface $\mathbf{I}_t = \{Weather(W), Mobility(M), Visibility(V)\}$, that is, the set of nodes in the current time step t with children in time step $t+1$. Interface $\mathbf{I}_t$ *d*-separates past from future and encapsulates all the historical information necessary to carry out prediction and filtering.

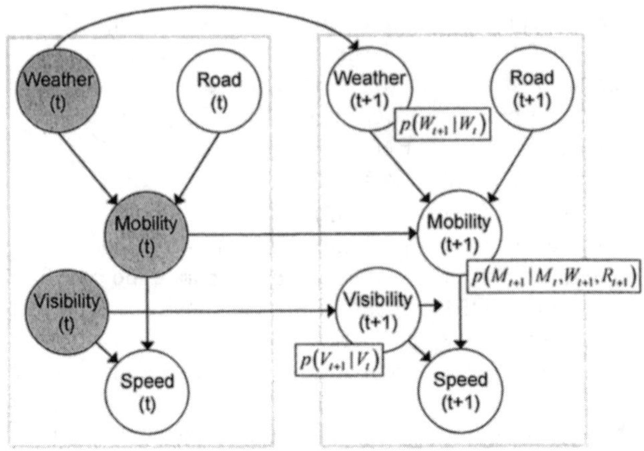

Figure 7-24: 2TBN based on the BN in Figure 7-23

As part of the initialization step of the interface algorithm, create the following two junction trees $\mathcal{J}_1$ and $\mathcal{J}_t$;

- $\mathcal{J}_1$ is the junction tree created from time slice 1 of the 2TBN and used for inferencing at time step $t = 1$.
- $\mathcal{J}_t$ is the junction tree for each subsequent time step and is created from time slice 2 of the 2TBN and the outgoing interface of time slice 1.

The steps in creating $\mathcal{J}_1$ are same as the steps for building junction trees from static BNs except necessary edges are added during the moralization stage to guarantee that the interface $\mathbf{I}_1$ is complete to fall within a single clique. For example, Figure 7-25 shows the steps involved in creating junction tree $\mathcal{J}_1$ from time slice 1 of 2TBN in Figure 7-24. Figure 7-25 shows that an additional edge has been added between the two nodes *Weather* and *Visibility*. The clique that contains $\mathbf{I}_1$ is called *in-clique*, and also *out-clique* in $\mathcal{J}_1$.

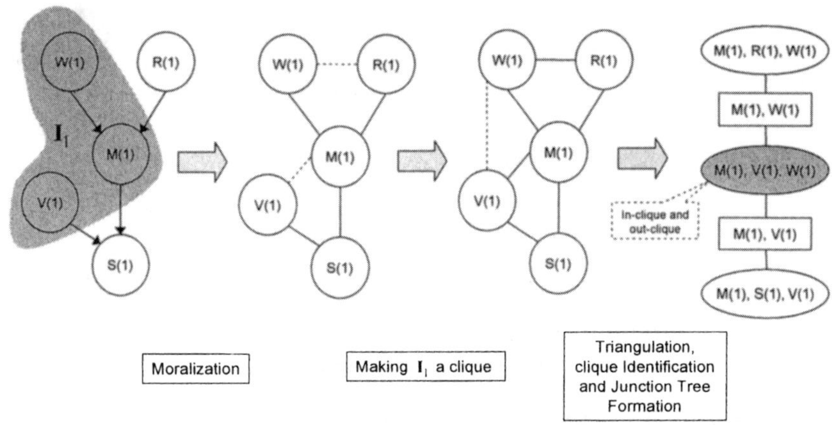

Figure 7-25: Junction tree for the slice at $t = 1$ of the 2TBN in Figure 7-24 (W – *Weather*; R – *Road*; V – *Visibility*; S – *Speed*; M – *Mobility*)

To form $\mathcal{J}_t$, start with the whole 2TBN, and then identify nodes $\mathbf{I}_t$ and $\mathbf{I}_{t+1}$ in the outgoing interfaces of time slices t and $t+1$, respectively. The resulting 2TBN is then converted to 1.5TBN by removing noninterface nodes in time slice t. Figure 7-26 shows these two steps starting from the 2TBN in Figure 7-24.

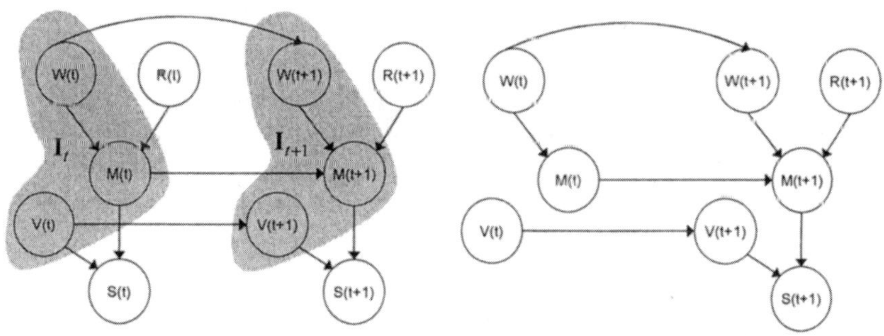

Figure 7-26: (*left*) Identification of outgoing interfaces in 2TBN, and (*right*) construction of 1.5TBN from 2TBN

The next step is to moralize the resulting 1.5TBN the same way as in static BNs, but by adding necessary edges in $\mathbf{I}_t$ and $\mathbf{I}_{t+1}$ to make them complete. Figure 7-27 shows these two steps applied to the 1.5TBN from Figure 7-26.

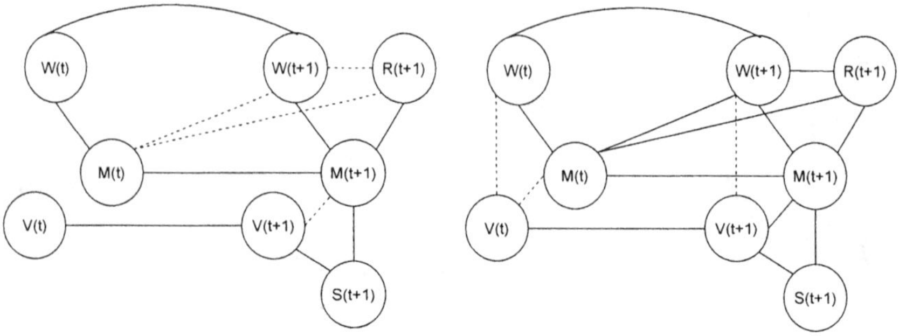

Figure 7-27: (*left*) Moralization of 1.5TBN in Figure 7-26, and (*right*) making outgoing interfaces complete

The moralized 1.5TBN is then triangulated the same as the way in static BNs, and cliques are identified to form the junction tree $\mathcal{J}_t$. The cliques in the junction tree containing outgoing interface $\mathbf{I}_t$ and $\mathbf{I}_{t+1}$ are called in-clique and out-clique, respectively. The DBN on the left side of Figure 7-28 is the triangulation of the resulting DBN of Figure 7-27, and the corresponding junction tree is shown on the right of Figure 7-28.

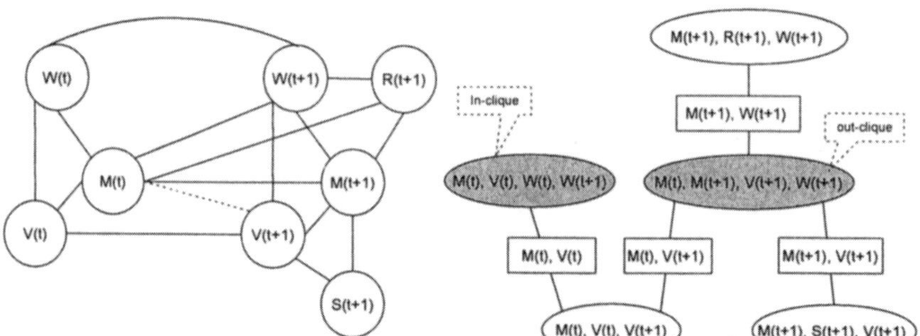

Figure 7-28: (*left*) Triangulation of the DBN on the right of Figure 7-27, and (*right*) clique identification from the triangulated DBN

In the beginning ($t = 0$), $\mathcal{J}_1$ is initialized to obtain clique potentials by multiplying CPTs same as the way in static BNs and the potential for its in-clique and out-clique is determined. Get the potential ϕ_1 of interface $\mathbf{I}_1$ by marginalizing the out-clique potential in $\mathcal{J}_1$. Then the time is forwarded ($t = 1$) as shown, for example, in Figure 7-29 for our running example.

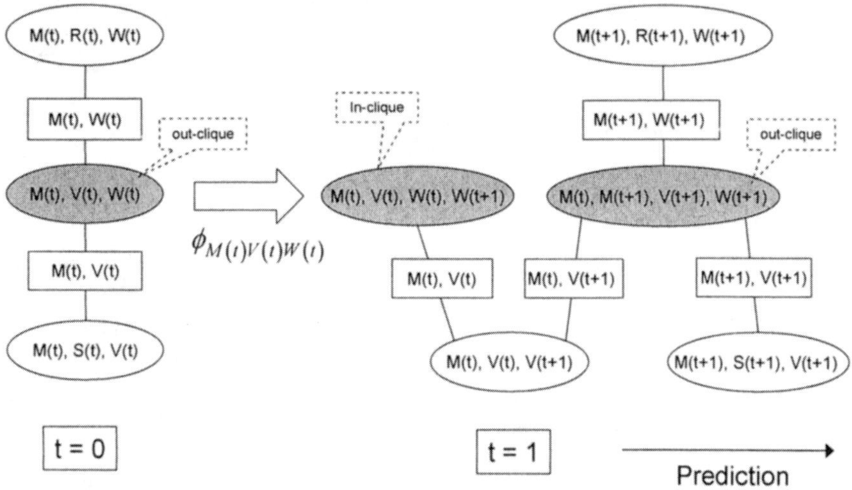

Figure 7-29: Advancing DBN inference form $t = 0$ to $t = 1$

$\mathcal{J}_2$ is initialized the same as the way in static BNs except multiplying only second time slice nodes' CPTs from a clique onto that clique's potential. Considering, for example, the DBN in Figure 7-28, none of the CPTs for the nodes $M(t)$, $V(t)$, and $W(t)$ is multiplied. Potential ϕ_1 is multiplied onto the in-clique of $\mathcal{J}_2$. Any evidence at $t = 1$ can be posted to $\mathcal{J}_2$, which can then be used to make a one-step prediction.

Time is then forwarded ($t = 2$) as shown, for example, in Figure 7-30 for our running example. The interface potential ϕ_2 is computed from the out-clique potential of $\mathcal{J}_2$ and multiplied with the in-clique potential of $\mathcal{J}_3$. The filtering and prediction from $\mathcal{J}_3$ are done as before, and the process continues.

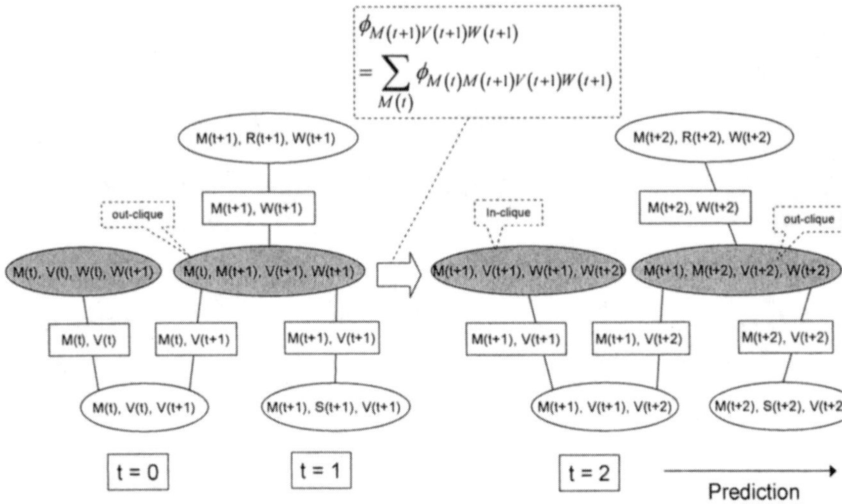

Figure 7-30: Advancing the DBN inference form $t = 1$ to $t = 2$

7.7 DBN APPLICATION FOR LIFE STATUS ESTIMATION

The life status estimation task involves making a simple determination as to whether or not a unit is alive on the basis of remote sensor data. In this task, overcoming possible sensor failures and determining where such failures lie are far more difficult than the clinical assessment itself. We developed the Probabilistic Unit Life States Estimation (PULSE) architecture (Das et al., 2004). The PULSE system applies DBNs to produce dynamic clinical assessments. As a by-product of the clinical assessment, the PULSE system also produces an assessment of reliability of the underlying hardware, and supports fault isolation analysis techniques.

We have developed a modular architecture for the development and testing of DBNs for life status estimation. A schematic of this architecture is shown in Figure 7-31.

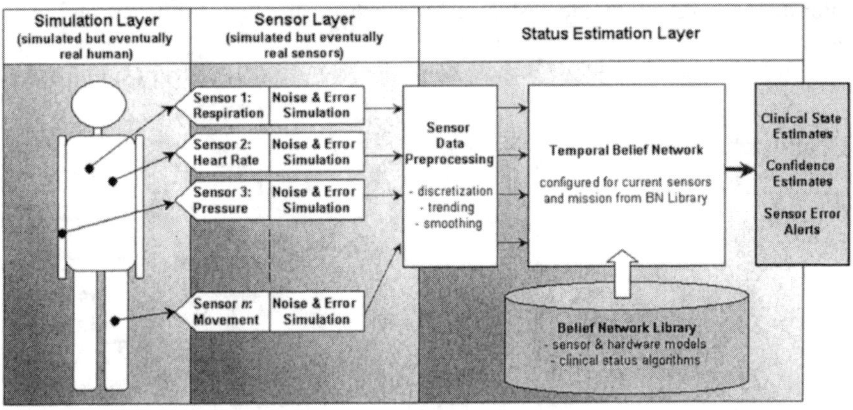

Figure 7-31: PULSE architecture

As seen in the diagram, there are three major functional components to the PULSE system:

- *Simulation Layer*: Creates and/or provides the various biometric measurements to be used;
- *Sensor Layer*: Retrieves biometric measurements from the simulation layer, and adds noise depending on the type of sensor failure to be simulated. These parameters can be modified at run time;
- *Status Estimation Layer*: Takes sensor readings and performs assessments using DBNs to deliver clinical status estimates along with confidence estimates and sensor error alerts.

In order to develop an assessment network for mortality assessment, knowledge elicitation sessions were held with medical experts. In these elicitation sessions, we sought to accumulate two types of information.

- Physiological patterns indicating likely trauma, thus providing information for assessment;
- Physiological patterns that were unlikely to appear, providing information for sensor validation.

The elicited information was transcribed into a Markovian DBN that uses two sensors: systolic blood pressure, and pulse. Figure 7-32 is a screenshot of the Mortality Assessment Network in three time slices. The network covers three distinct time slices, and has two history nodes that accumulate state in the network, and four input nodes that are posted with trend and absolute evidence from the sensors. The network spans 90 seconds (i.e., there are 30 seconds between two time steps) including history nodes, as this is a useful time span to keep in view when trying to detect the physiological patterns that accompany serious trauma (discussed farther below). The output is read from the Confidence and Alive nodes. The current implementation of PULSE makes use of a history node-based inference algorithm developed in-house. In this algorithm, the interface, discussed in the last section, consists of the Confidence and Alive nodes.

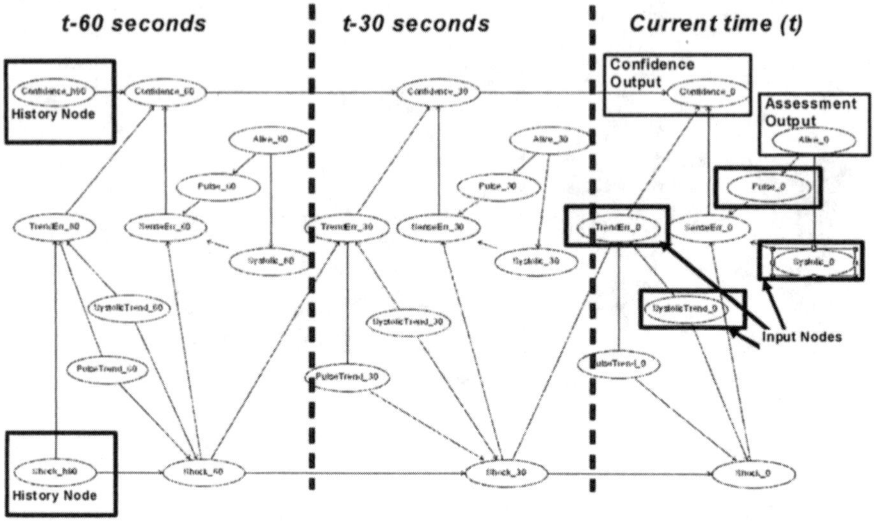

Figure 7-32: The assessment network

Figure 7-33 provides a conceptual breakdown of a single slice of the above network. There are three "modules" within a time slice. These modules are:

- *Physiological Module*: Provides a confidence component according to how well-observed metrics match known physiological (in this case, shock or normal physiological states) trends;
- *Hardware Module*: Provides a confidence given the operating parameters of the sensor hardware (maximum and minimum values allowed for each sensor);
- *Clinical Module*: Provides an assessment of a subject's mortality on the basis of current readings from the sensor hardware.

Note that there is no direct connection between the physiological module and the clinical module in this network; thus, the goodness of fit between the known physiological patterns and the observed sensor readings contributes to the confidence in the hardware, but does not influence the assessment directly. This separation was useful in the development of this network, but is not a necessary feature.

The mortality assessment is made by detecting whether or not the monitored unit has both a pulse and viable blood pressure reading. Illegal states (impossible values on either heart rate or blood pressure sensors) or states that are unlikely given the current stage of shock register an error at the "SenseErr" node, and this also reduces confidence in the hardware.

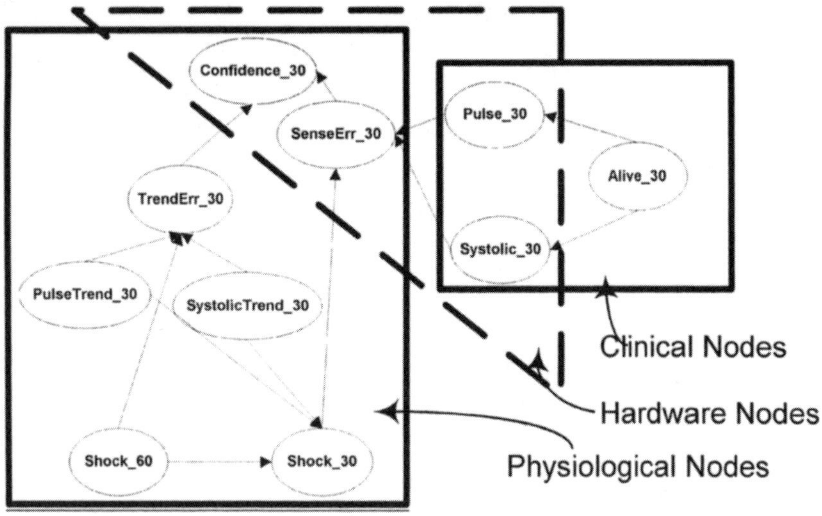

Figure 7-33: Conceptual breakdown of mortality network

As shown in Figure 7-33, the network has two outputs, a clinical assessment (the Alive node), and a hardware confidence (the Confidence node). These are to be interpreted as follows:

- Alive_0 – The current best estimate as to whether or not the unit is still alive. In general, this value should be either "yes" or "no" (with some very small deviance from certainty). Occasionally, when the clinical state is not recognizable (e.g., no blood pressure yet continued heart beat, which is possible due to sensor error) each state will have a nonnegligible probability. This may be interpreted as an "indeterminate" state, in which an assessment is not really possible.
- Confidence_0 – The current estimate as to the reliability of the underlying hardware. The interpretation of this node is fairly straightforward.

Ultimately, the output from these nodes should be used to drive a "user-friendly" layer that interprets the output so as to provide a well-tuned decision support tool. Our network performed very well for simulated data over a variety of sensor conditions.

7.8 FURTHER READING

A seminal paper containing good background information on HMMs is Rabiner (1989). The discussion on HHMMs in this chapter largely follows Fine et al. (1998). Murphy (2002) provides a good introduction on DBNs and inference algorithms.

Chapter 8

Handling Nonlinear and Hybrid Models

This chapter presents inference algorithms for handling nonlinear and hybrid transition and observation functions. We introduce the Extended Kalman Filter (EKF), the Unscented Kalman Filter (UKF), and the particle filter (PF) and its various extensions. The PF is especially effective for handling hybrid Dynamic Bayesian Networks (DBNs) containing both continuous and discrete variables.

8.1 EXTENDED KALMAN FILTER (EKF)

The Kalman Filter (KF), as presented in Chapter 4, is a Bayesian optimal Minimum Mean Square Error (MMSE) estimator for linear systems with Gaussian noise. The Extended Kalman Filter (EKF) replaces the nonlinear transition and observation equation

$$X_k = f(X_{k-1}, U_{k-1}) + W_{k-1}$$

$$Z_k = h(X_k) + V_k$$

with an approximate linear system obtained via the Taylor series expansion of the transition equation about mean $\bar{X}$ of the random variable X:

$$\begin{aligned} f(X) &= f(\bar{X} + \delta X) \\ &= f(\bar{X}) + \nabla f \delta X + \frac{1}{2}\nabla^2 f \delta X^2 + \frac{1}{3!}\nabla^3 f \delta X^3 + \frac{1}{4!}\nabla^4 f \delta X^4 + \ldots \end{aligned}$$

where δX is a zero-mean Gaussain variable with covariance P_{XX}, and ∇f is the Jacobian of f evaluated at $\bar{X}$. So, if f has the form

$$f(x, y, z) = \begin{bmatrix} f_1(x, y, z) \\ f_2(x, y, z) \end{bmatrix}$$

then its *Jacobian* has the form

$$\nabla f = \begin{bmatrix} \frac{\partial f_1}{\partial x} & \frac{\partial f_1}{\partial y} & \frac{\partial f_1}{\partial z} \\ \frac{\partial f_2}{\partial x} & \frac{\partial f_2}{\partial y} & \frac{\partial f_2}{\partial z} \end{bmatrix}$$

Taking expectations of the equation of $f(X)$ on both sides, it can be shown that the transformed mean and covariance are as follows:

$$\overline{Y} = f(\overline{X}) + \frac{1}{2}\nabla^2 f P_{XX} + \frac{1}{2}\nabla^4 f E\left[\delta_X{}^4\right] + \ldots$$

$$P_{YY} = \nabla F P_{XX} (\nabla f)^T + \frac{1}{2\times 4!}\nabla^2 f\left(E\left[\delta_X{}^4\right] - \ldots\right)(\nabla^2 f)^T + \frac{1}{3!}\nabla^3 f E\left[\delta_X{}^4\right](\nabla^3 f)^T + \ldots$$

If the moments and derivatives can be evaluated correctly up to the n-th order, the mean and covariance are correct up to the n-th order as well. Each term in the series is scaled by a progressively smaller and smaller term due to an increasing power of δX. Therefore, the prediction procedure should be concentrated on evaluating the lower-order terms. From this point of view, linearization assumes the second and higher terms of δX can be neglected to obtain

$$\overline{Y} = f(\overline{X})$$

$$P_{YY} = \nabla f P_{XX} (\nabla f)^T$$

Considering, $\hat{X}_{k-1|k-1}$ for $\overline{X}$, $\hat{X}_{k|k-1}$ for $\overline{Y}$, $P_{k-1|k-1}$ for P_{XX}, and $P_{k|k-1}$ for P_{YY}, we formulate the following recursive estimation for EKF:

Prediction:

$$\hat{X}_{k|k-1} = f\left(\hat{X}_{k-1|k-1}, U_{k-1}\right) + W_{k-1}$$

$$P_{k|k-1} = \tilde{F}_{k-1} P_{k-1|k-1} \tilde{F}_{k-1}^T + Q_{k-1}$$

Observation Update:

$$K_k = P_{k|k-1}\tilde{H}_k{}^T\left(\tilde{H}_k P_{k|k-1}\tilde{H}_k{}^T + R_k\right)^{-1}$$

$$\hat{X}_{k|k} = \hat{X}_{k|k-1} + K_k\left(Z_k - H_k\hat{X}_{k|k-1}\right)$$

$$P_{k|k} = \left(1 - K_k\tilde{H}_k\right)P_{k|k-1}\left(1 - K_k\tilde{H}_k\right)^T + K_k R_k K_k{}^T$$

where $\tilde{F}_{k-1} = \nabla f\left(\hat{X}_{k-1|k-1}\right)$ and $\tilde{H}_k = \nabla h\left(\hat{X}_{k|k-1}\right)$. The following example illustrates an estimation via EKF.

Example

Consider the state vector of a 2-dimensional Tactical Ballistic Missile (TBM) in the re-entry phase (see Chang (1980) and Bhowmik and Sadhukhan (2007) for details):

$$X_k = \begin{bmatrix} x_k \\ y_k \\ v_{xk} \\ v_{yk} \end{bmatrix}$$

where x_k and y_k are measured, respectively, along the x and y axes at time step k centering at the radar site, and v_{xk} and v_{yk} are the lateral velocities of the missile, respectively, along the x and y directions. The lateral accelerations a_{xk} and a_{yk}, respectively, along the x and y axes, generated by a maneuver are related in successive time steps as follows:

$$a_{x(k+1)} = a_{xk} - \frac{\rho\left(v_{xk}^{2} + v_{yk}^{2}\right)}{2\beta} g\cos\gamma$$

$$a_{y(k+1)} = a_{yk} + \frac{\rho\left(v_{xk}^{2} + v_{yk}^{2}\right)}{2\beta} g\sin\gamma$$

where ρ is the air density, β is the ballistic coefficient defined in terms of reference area and zero lift drag coefficient, g is the gravitational acceleration, and γ is the flight path angle defined as $\tan^{-1}\left(-\frac{v_x}{v_y}\right)$. The nonlinear transition equation is then the following:

$$\begin{aligned} X_{k+1} &= f(X_k) + U_k + W_k \\ &= \begin{bmatrix} f_1(X_k) \\ f_2(X_k) \\ f_3(X_k) \\ f_4(X_k) \end{bmatrix} + \begin{bmatrix} u_{1k} \\ u_{2k} \\ u_{3k} \\ u_{4k} \end{bmatrix} + \begin{bmatrix} w_{1k} \\ w_{2k} \\ w_{3k} \\ w_{4k} \end{bmatrix} \\ &= \begin{bmatrix} v_{xk} \\ v_{yk} \\ -\frac{\rho\left(v_{xk}^{2} + v_{yk}^{2}\right)}{2\beta} g\cos\gamma \\ \frac{\rho\left(v_{xk}^{2} + v_{yk}^{2}\right)}{2\beta} g\sin\gamma - g \end{bmatrix} + \begin{bmatrix} 0 \\ 0 \\ a_{xk} \\ a_{yk} \end{bmatrix} + \begin{bmatrix} w_{1k} \\ w_{2k} \\ w_{3k} \\ w_{4k} \end{bmatrix} \end{aligned}$$

Therefore,

$$\tilde{F}_k = \nabla g\left(\hat{X}_{k|k}\right) = \left.\frac{\partial g\left(X_{k|k}\right)}{\partial X_k}\right|_{X_k = \hat{X}_{k|k}}$$

$$= \begin{bmatrix} \frac{\partial f_1}{\partial x_k} & \frac{\partial f_1}{\partial y_k} & \frac{\partial f_1}{\partial v_{xk}} & \frac{\partial f_1}{\partial v_{yk}} \\ \frac{\partial f_2}{\partial x_k} & \frac{\partial f_2}{\partial y_k} & \frac{\partial f_2}{\partial v_{xk}} & \frac{\partial f_2}{\partial v_{yk}} \\ \frac{\partial f_3}{\partial x_k} & \frac{\partial f_3}{\partial y_k} & \frac{\partial f_3}{\partial v_{xk}} & \frac{\partial f_3}{\partial v_{yk}} \\ \frac{\partial f_4}{\partial x_k} & \frac{\partial f_4}{\partial y_k} & \frac{\partial f_4}{\partial v_{xk}} & \frac{\partial f_4}{\partial v_{yk}} \end{bmatrix} = \begin{bmatrix} 0 & 0 & 1 & 0 \\ 0 & 0 & 0 & 1 \\ 0 & 0 & f_{33} & f_{34} \\ 0 & 0 & f_{43} & f_{44} \end{bmatrix}$$

where ρ is considered constant and

$$f_{33} = -\frac{\rho}{2\beta} g\left(2v_{xk}\cos\gamma - v_{yk}\sin\gamma\right), f_{34} = -\frac{\rho}{2\beta} g\left(2v_{yk}\cos\gamma + v_{xk}\sin\gamma\right)$$

$$f_{43} = -\frac{\rho}{2\beta} g\left(2v_{xk}\sin\gamma + v_{yk}\cos\gamma\right), f_{44} = -\frac{\rho}{2\beta} g\left(2v_{yk}\sin\gamma - v_{xk}\cos\gamma\right)$$

The measurement vector is defined as

$$Z_k = \begin{bmatrix} r \\ \dot{r} \\ \theta \end{bmatrix}$$

where r, $\dot{r}$, and θ are, respectively, range, range-rate, and bearing of the target with respect to the observing sensor, and are defined below:

$$r = \sqrt{x_k^{\,2} + y_k^{\,2}}, \dot{r} = \frac{x_k v_{xk} + y_k v_{yk}}{r}, \theta = \tan^{-1}\frac{y_k}{x_k}$$

The measurement equation is

$$Z_k = h\left(X_k\right) + V_k$$

$$= \begin{bmatrix} h_1\left(X_k\right) \\ h_2\left(X_k\right) \\ h_3\left(X_k\right) \end{bmatrix} + \begin{bmatrix} v_{1k} \\ v_{2k} \\ v_{3k} \end{bmatrix} = \begin{bmatrix} \sqrt{x_k^{\,2} + y_k^{\,2}} \\ \frac{\left(x_k v_{xk} + y_k v_{yk}\right)}{r} \\ \tan^{-1}\left(\frac{y_k}{x_k}\right) \end{bmatrix} + \begin{bmatrix} v_{1k} \\ v_{2k} \\ v_{3k} \end{bmatrix}$$

$$\tilde{H}_k = \nabla h\left(\hat{X}_{k|k}\right) = \left.\frac{\partial h\left(X_{k|k}\right)}{\partial X_k}\right|_{X_k = \hat{X}_{k|k}}$$

$$= \begin{bmatrix} \frac{\partial h_1}{\partial x_k} & \frac{\partial h_1}{\partial y_k} & \frac{\partial h_1}{\partial v_{xk}} & \frac{\partial h_1}{\partial v_{yk}} \\ \frac{\partial h_2}{\partial x_k} & \frac{\partial h_2}{\partial y_k} & \frac{\partial h_2}{\partial v_{xk}} & \frac{\partial h_2}{\partial v_{yk}} \\ \frac{\partial h_3}{\partial x_k} & \frac{\partial h_3}{\partial y_k} & \frac{\partial h_3}{\partial v_{xk}} & \frac{\partial h_3}{\partial v_{yk}} \end{bmatrix} = \begin{bmatrix} h_{11} & h_{12} & 0 & 0 \\ h_{21} & h_{22} & 0 & 0 \\ h_{31} & h_{32} & x_k & y_k \end{bmatrix}$$

$$h_{11} = \frac{x_k}{\sqrt{x_k^{\ 2} + y_k^{\ 2}}}, h_{12} = \frac{y_k}{\sqrt{x_k^{\ 2} + y_k^{\ 2}}}, \text{ and so on}$$

8.2 UNSCENTED KALMAN FILTER (UKF)

The Unscented Kalman Filter (UKF) uses a sampling approach to capture the mean and covariance estimates (Julier and Uhlmann, 1997). The UKF is founded on the intuition that "it is easier to approximate a Gaussian distribution than it is to approximate an arbitrary nonlinear function or transformation." A set of "sigma" points are chosen with sample mean and covariance of the nonlinear function. Samples are drawn deterministically. The nonlinear function is applied to each point in turn to yield a cloud of transformed points, and new mean and covariance are the statistics of the transformed points. The steps of the filter are given below:

- Random variable X with mean and covariance at time step k: $\hat{X}_{k|k}, P_{k|k}$. Statistics require: $\hat{X}_{k+1|k+1}, P_{k+1|k+1}$.
- Choose sigma point and corresponding weights:

$$\chi^0_{k|k} = \hat{X}_{k|k} \qquad W_0 = \lambda / (n+\lambda)$$

$$\chi^i_{k|k} = \hat{X}_{k|k} + \left(\sqrt{(n+\lambda)P_{k|k}}\right)_i \qquad W_i = 1/\{2(n+\lambda)\}$$

$$\chi^{i+n}_{k|k} = \hat{X}_{k|k} - \left(\sqrt{(n+\lambda)P_{k|k}}\right)_i \qquad W_{i+n} = 1/\{2(n+\lambda)\}$$

where $\left(\sqrt{(n+\lambda)P_{k|k}}\right)_i$ is the i-th row or column of the matrix square root of $\sqrt{(n+\lambda)P_{k|k}}$ and W_i is associated with the i-th point. The parameter λ is defined by $\lambda = \alpha^2(n+\kappa) - n$, where α is a scaling parameter that determines the spread of the sigma points and κ is a secondary scaling parameter.

- Apply nonlinear transition and observation functions:

$$\chi^i_{k+1|k} = f\left(\chi^i_{k|k}, U_k\right) + W_k$$

$$\varsigma^i_{k+1|k} = h\left(\chi^i_{k|k}\right) + V_k$$

- Predict mean and observation:

$$\hat{X}_{k+1|k} = \sum_{i=0}^{2n} W_i \chi^i_{k+1|k}$$

$$\hat{Z}_{k+1|k} = \sum_{i=0}^{2n} W_i \varsigma^i_{k+1|k}$$

- Compute predicted and innovation covariance matrices:

$$P_{k+1|k} = \sum_{i=0}^{2n} W_i \left(\chi^i_{k+1|k} - \hat{X}_{k+1|k}\right)\left(\chi^i_{k+1|k} - \hat{X}_{k+1|k}\right)^T$$

$$P^{ZZ}_{k+1|k} = \sum_{i=0}^{2n} W_i \left(\varsigma^i_{k|k-1} - \hat{Z}_{k+1|k}\right)\left(\varsigma^i_{k|k-1} - \hat{Z}_{k+1|k}\right)^T$$

- Compute the cross correlation matrix:

$$P^{XZ}_{k+1|k} = \sum_{i=0}^{2n} W_i \left(\chi^i_{k|k-1} - \hat{X}_{k+1|k}\right)\left(\varsigma^i_{k|k-1} - \hat{Z}_{k+1|k}\right)^T$$

- Update as a Kalman Filter:

$$\hat{X}_{k+1|k+1} = \hat{X}_{k+1|k} + W_{k+1} V_{k+1}$$

$$P_{k+1|k+1} = P_{k+1|k} + W_{k+1} P^{ZZ}_{k+1|k} W^T_{k+1}$$

$$V_{k+1} = Z_{k+1} - \hat{Z}_{k+1|k}$$

$$W_{k+1} = P^{XZ}_{k+1|k}\left(P^{ZZ}_{k+1|k}\right)^{-1}$$

In a UKF, samples are drawn deterministically, and not drawn at random as in the PF described in the following section.

8.3 PARTICLE FILTER (PF)

The particle filter (PF) is a simple and effective algorithm for estimating the state of a dynamic system over time where the state cannot be measured directly, but may be inferred from a set of observations at each time. The filter is also known as a sequential Monte Carlo (introduced in Handschin and Mayne, 1969), Condensation (CONditional DENSity propagATION) (Israd and Blake, 1998), Sampling Importance Resampling (SIR) (Doucet et al., 2001), the bootstrap filter (Gordon et al., 1993), the survival of the fittest (Kanazawa et al., 1995), etc. The PF is a nonparametric approach, and thus handles nonlinearities and multimodal distributions by approximating them via a finite weighted sum of N samples, named *particles*. With sufficiently many particles, an approximate conditional distribution can be obtained that is arbitrarily close to the true conditional distribution.

8.3.1 Basic Particle Filter

The particle filter (PF) is based on propagating a probability density function for the state $X_t \in \mathbb{R}^{n_x}$ of a dynamic system over time using the state equation

$$X_t = f(X_{t-1}, V_t)$$

and a set of observations $Y_t \in \mathbb{R}^{n_y}$ at each time step governed by the measurement equation

$$Z_t = h(X_t, W_t)$$

where $V_t \in \mathbb{R}^{n_v}$ and $W_t \in \mathbb{R}^{n_w}$ are independent white noise (f and h could be time-variant). The PF algorithm computes the conditional density $L_t = p(X_t \mid Z_{1:t})$ and also any functional of the state by the expectation $\mathrm{E}(g(X_t) \mid Z_{1:t})$. The prediction step is done according to the following equation:

$$\begin{aligned} p(X_t \mid Z_{1:t-1}) &= \int_{\mathbb{R}^{n_x}} p(X_t \mid X_{t-1} = x)\, p(X_{t-1} = x \mid Z_{1:t-1})\, dx \\ &= \int_{\mathbb{R}^{n_x}} p(X_t \mid X_{t-1} = x)\, L_{t-1}\, dx \end{aligned}$$

The term $p(X_t \mid X_{t-1})$ can be calculated using $X_t = f(X_{t-1}, V_t)$ as follows:

$$\begin{aligned} p(X_t \mid X_{t-1}) &= \int_{\mathbb{R}^{n_v}} p(X_t \mid X_{t-1}, V_t = v)\, p(V_t = v \mid X_{t-1} = x)\, dv \\ &= \int_{\mathbb{R}^{n_v}} \delta(X_t - f(X_{t-1}, v))\, p(V_t = v)\, dv \end{aligned}$$

where $\delta(x)$ denotes the Dirac delta function.[1] In the filtering step, the conditional density L_t is computed by applying the Bayes' rule as follows:

$$\begin{aligned} L_t(x_t) &= p(X_t \mid Z_{1:t}) \\ &= \frac{p(Z_t \mid X_t)\, p(X_t \mid Z_{1:t-1})}{p(Z_t \mid Z_{1:t-1})} \end{aligned}$$

The term $p(Z_t \mid X_t)$ can be calculated using $Z_t = h(X_t, W_t)$ as follows:

$$\begin{aligned} p(Z_t \mid X_t) &= \int_{\mathbb{R}^{n_w}} p(Z_t \mid X_t, W_t = w)\, p(W_t = w \mid X_{t-1})\, dw \\ &= \int_{\mathbb{R}^{n_w}} \delta(Z_t - h(X_t, w))\, p(W_t = w)\, dw \end{aligned}$$

---

[1] A function that has the value everywhere except at $x = 0$ where its value is infinitely large in such a way that $\int_{-\infty}^{\infty} \delta(x)\, dx = 1$.

The normalizing denominator $p(Z_t \mid Z_{1:t-1})$ is given by the following integral:

$$p(Z_t \mid Z_{0:t-1}) = \int_{\mathbb{R}^{n_x}} p(Z_t \mid X_t = x)\, p(X_t = x \mid Z_{0:t-1})\, dx$$

The above integrals cannot be solved analytically for many problems, and thus numerical methods must be used. The PF approach is to approximate these densities by a finite weighted sum of N samples, named *particles*. These particles are Dirac densities centered on elements of $\mathbb{R}^{n_x}$. Thus we have the following:

$$L_t = p(X_t \mid Z_{1:t}) \approx \sum_{i=1}^{N} w_t^i \delta\left(X_t - x_t^i\right)$$

where x_t^i is the i-th sample of X_t. The above discretization makes the PF adaptive, placing more particles in places where the probability density is higher. The target distribution $p(X_t \mid Z_{1:t})$ is hard to sample from; instead, we sample from a proposal or importance distribution $q(X_t \mid Z_{1:t})$, and weight the sample according to

$$\begin{aligned} q_t^i &\propto \frac{p(X_t^i \mid Z_{1:t})}{q(X_t^i \mid Z_{1:t})} \\ &\propto \frac{p(Z_t \mid X_t^i)\, p(X_t^i \mid X_{t-1}^i)\, p(X_{t-1}^i \mid Z_{1:t-1})}{q(X_t^i \mid X_{t-1}^i, Z_{1:t})\, q(X_{t-1}^i \mid Z_{1:t-1})} \\ &\propto \frac{p(Z_t \mid X_t^i)\, p(X_t^i \mid X_{t-1}^i)}{q(X_t^i \mid X_{t-1}^i, Z_{1:t})} q_{t-1}^i \end{aligned}$$

The weights need to be normalized using the relation $\sum_i q_t^i = 1$. The most common proposal is to sample from the transition prior:

$$q(X_t^i \mid X_{t-1}^i, Y_{1:t}) = p(X_t^i \mid X_{t-1}^i)$$

Thus, $q_t^i \propto p(Z_t \mid X_t^i) q_{t-1}^i$.

8.3.2 Particle Filter Algorithms

The application of the PF requires the following:

- Create the first particle set S_0 by drawing N independent samples from an initial prior marginal $p(X_0)$ and assign uniform weight $q_0^n = \frac{1}{N}$ to each sample x_0^n.
- Sample from $p(V_t)$ for all t.
- Compute $p(Z_t \mid X_t)$ via a known function $l(Z_t; X_t) \propto p(Z_t \mid X_t)$ for all t.

Based on the above, prediction, filtering, and estimation are recursively computed via the following algorithm:

ALGORITHM: Basic Particle Filter

Initialization

$$\begin{cases} x_0^n \sim p(X_0) \\ q_0^n = \dfrac{1}{N} \end{cases} \qquad n = 1,\dots,N$$

Prediction

$$\begin{cases} v_t^n \sim p(V_t) \\ x_{t|t-1}^n = f\left(x_{t-1}^n, v_t^n\right) \end{cases} \qquad n = 1,\dots,N$$

Filter

$$\begin{cases} q_t^n = \dfrac{l\left(z_t; x_{t|t-1}^n\right) q_{t-1}^n}{\sum_{i=1}^{N} l\left(z_t; x_{t|t-1}^i\right) q_{t-1}^i} \left\{ \quad n = 1,\dots,N \right. \\ x_t^n = x_{t|t-1}^n \end{cases}$$

Estimation

$$\text{Estimate } \mathrm{E}(X_t) \text{ by } \sum_{n-1}^{N} q_t^n x_{t|t-1}^n$$

The major drawback of the above algorithm is that after a few steps only a few particles keep high weights and the rest will have very small ones that do not carry any information. One solution to avoid this problem of degeneracy of the particle set, adopted in Gordon et al. (1993), is to apply the following resampling step at each time period:

ALGORITHM: Resampling of a Basic Particle Filter

Resampling

$$\begin{cases} x_t^n \sim \sum_{i=0}^{N} q_t^i \delta\left(X_t - x_{t|t-1}^i\right) \\ q_t^n = \dfrac{1}{N} \end{cases} \qquad n = 1,\dots,N$$

Instead of applying the resampling step at each time period, period resampling has been introduced (Kong et al., 1994) based on the effective sample size N_{eff}, which can be estimated as

$$\frac{1}{\sum_{n=1}^{N} \left(q_t^n\right)^2}$$

representing the number of meaningful particles. The modified resampling algorithm based on the degeneracy measure is given below:

ALGORITHM: Degeneracy Measure-Based Resampling

Resampling

Calculate $N_{eff} = \dfrac{1}{\sum_{i=1}^{N} \left(q_t^n\right)^2}$

If $N_{eff} < N_{threshold}$ then

$$\begin{cases} x_t^n \sim \sum_{i=0}^{N} q_t^i \delta\left(X_t - x_{t|t-1}^i\right) \\ q_t^n = \dfrac{1}{N} \end{cases} \quad n = 1, \ldots, N$$

Else

$$x_t^n = x_{t|t-1}^n \qquad n = 1, \ldots, N$$

8.3.3 Rao-Blackwellised Particle Filter (RBPF)

If the posterior distribution is close to Gaussian and the functions $f(.)$ and $h(.)$ are linear, the density propagation can be modeled analytically in an efficient manner, e.g., as a KF. The Rao-Blackwellised Particle Filter (RBPF) algorithm exploits this kind of analytical structure of the model. In particular, if part of a state vector is Gaussian then the computational complexity can be reduced using the RBPF technique. The algorithm combines a PF to compute the distribution of the non-Gaussian states X_t with a bank of KFs to compute the distribution of the continuous states Y_t. The principle is based on the following identity:

$$Var\left(\tau\left(X_t, Y_t\right)\right) = Var\left(\mathrm{E}\left(\tau\left(X_t, Y_t\right) \mid X_t\right)\right) + \mathrm{E}\left(Var\left(\tau\left(X_t, Y_t\right)\right) \mid X_t\right)$$

where $\tau\left(X_t, Y_t\right)$ is some estimator of the state vector composed of X_t and Y_t. Since $Var\left(\mathrm{E}\left(\tau\left(X_t, Y_t\right) \mid X_t\right)\right) \leq Var\left(\tau\left(X_t, Y_t\right)\right)$, $\mathrm{E}\left(\tau\left(X_t, Y_t\right) \mid X_t\right)$ is a lower variance estimator that requires only sampling for X_t and then compute the expectation of the whole state vector analytically. The reduced problem is of lower dimensionality and needs relatively fewer samples. In particular, X_t is sampled and then propagates the mean μ_t and covariance Σ_t of Y_t with a KF. First, we adopt the following linear Gaussian model:

$$\mu_{t|t-1} = A\left(X_t\right)\mu_{t-1|t-1} + F\left(X_t\right)U_t$$

$$\Sigma_{t|t-1} = A\left(X_t\right)\Sigma_{t-1|t-1}A\left(X_t\right)^T + B\left(X_t\right)B\left(X_t\right)^T$$

where

$$\mu_{t|t-1} = \mathrm{E}(Y_t \mid Z_{1:t-1})$$
$$\mu_{t|t} = \mathrm{E}(Y_t \mid Z_{1:t})$$
$$\Sigma_{t|t-1} = \mathrm{cov}(Y_t \mid Z_{1:t-1})$$
$$\Sigma_{t|t} = \mathrm{cov}(Y_t \mid Z_{1:t})$$

The measurement update and propagation of the mean and covariance with a KF are as follows:

$$S_t = C(X_t) S_{t-1} C(X_t)^T + D(X_t) D(X_t)^T$$
$$Z_{t|t-1} = C(X_t)\mu_t + G(X_t)U_t$$
$$\mu_{t|t} = \mu_{t|t-1} + \Sigma_{t|t-1} C(X_t)^T S_t^{-1} (Z_t - Z_{t|t-1})$$
$$\Sigma_{t|t} = \Sigma_{t|t-1} - \Sigma_{t|t-1} C(X_t)^T S_t^{-1} C(X_t) \Sigma_{t|t-1}$$

where $Z_{t|t-1} = \mathrm{E}(Z_t \mid Z_{1:t-1})$, and $S_t = \mathrm{cov}(Z_t \mid Z_{1:t-1})$. The new algorithm is shown below:

ALGORITHM: Rao-Blackwellised Particle Filter

Basic Particle Filter

$$\begin{cases} \bar{\mu}^n_{t|t-1} = \mu^n_{t|t-1} \\ \bar{\Sigma}^n_{t|t-1} = \Sigma^n_{t|t-1} \end{cases} \qquad n = 1,\ldots,N$$

$$\begin{cases} v^n_t \sim p(V_t) \\ x^n_{t|t-1} = f\left(x^n_{t-1}, v^n_t\right) \end{cases} \qquad n = 1,\ldots,N$$

$$q^n_t = \frac{l(z_t; x^n_{t|t-1}) q^n_{t-1}}{\sum_{i=1}^{N} l(z_t; x^i_{t|t-1}) q^i_{t-1}} \qquad n = 1,\ldots,N$$

Resampling

$$N_{eff} = \frac{1}{\sum_{n=1}^{N} (q^n_t)^2}$$

If $N_{eff} < N_{threshold}$

Multiple/discard particles $\left\{\bar{\mu}^n_{t|t-1}, \bar{\Sigma}^n_{t|t-1}, x^n_{t|t-1}\right\}_{n=1}^{N}$ wrt the weights q^n_t to obtain particles $\left\{\mu^n_{t|t-1}, \Sigma^n_{t|t-1}, x^n_t\right\}_{n-1}^{N}$

Kalman Filter

Use the one-step Kalman Filter to compute

$$\left\{\mu^n_{t+1|t}, \Sigma^n_{t+1|t}, z^n_{t+1|t}, S^n_{t+1}\right\}_{n=1}^N \text{ given } \left\{\mu^n_{t|t-1}, \Sigma^n_{t|t-1}, x^n_t\right\}_{n=1}^N$$

8.3.4 Multitarget Tracking and Particle Filters

The formalism and algorithm to be presented here for tracking multiple targets using the PF are based on Hue et al. (2002). Let the number of objects to be tracked be fixed and equal to *M*. The state vector is constructed by concatenating the state vector of each object, that is, $X_t = \left(X_t^1, ..., X_t^M\right)$, where each X_t^i is defined as follows:

$$X_t^i = f^i\left(X_{t-1}^i, V_t^i\right)$$

The noise vectors V_t^i are independent of each other. The observation vector Z_t is $\left(Z_t^1, ..., Z_t^{m_t}\right)$, where m_t is the number of observations at time *t*, which may differ from *M*. An object can produce zero or several measurements and one measurement can originate from at most one object or from the clutter. Consider the vector $K_t \in \{1, ..., M\}^{m_t}$, such that $K_t^j = i$ if Z_t^j is issued from the *i*-th object. The observation vector for multitarget tracking is the following:

$$Z_t^j = h^i\left(X_t^i, W_t^j\right) \text{ if } K_t^j = i$$

Again the noise vectors W_t^i are independent of each other. Let $\Pi_t \in [0,1]^{M+1}$ defined by $\Pi_t^i = p\left(K_t^j = i\right)$, for all $j = 1, ..., m_t$, where 0 represents false alarms. Thus $\sum_{i=0}^{M} \Pi_t^i = 1$. Under this formalism, the expanded state vector at each time point *t* takes the form $\left(X_t, K_t, \Pi_t\right)$.

The distribution $p\left(X_t, \Pi_t \mid Z_{0:t}\right)$ cannot be sampled directly. The Gibbs' sampling technique will be used at each time point to obtain samples alternatively from the conditional posterior distributions of K_t, Π_t, and X_t, which are, respectively, $p\left(K_t \mid X_t, \Pi_t\right)$, $p\left(\Pi_t \mid K_t\right)$, and $p\left(X_t \mid K_t, \Pi_t\right)$.

ALGORITHM: Multitarget Particle Filter

Initialization

$$\begin{cases} x_0^{n,i} \sim p\left(X_0^i\right) \\ q_0^n = \dfrac{1}{N} \end{cases} \qquad n = 1, ..., N;\ i = 1, ..., M$$

Prediction

$$\begin{cases} v_t^{n,i} \sim p\left(V_t^i\right) \\ x_{t|t-1}^{n,i} = f\left(x_{t-1}^{n,i}, v_t^{n,i}\right) \end{cases} \qquad n = 1,\ldots,N,\ i = 1,\ldots,M$$

Weighting via Gibbs' Sampling

$$\text{Initialization: } \begin{cases} \pi_{t,0}^i = \dfrac{1-\pi_t^0}{M} \\ x_{t,0}^i = \sum_{n=1}^{N} q_{t-1}^n x_{t|t-1}^{n,i} \end{cases} \qquad i = 1,\ldots,M$$

For $\tau = 0 : \tau_{end}$

$$K_{t,\tau+1}^j \sim p\left(K_{t,\tau+1}^j = i\right) \propto \begin{cases} \pi_{t,\tau}^i l_t^i\left(z_t^j ; x_{t,\tau}^i\right) & i = 1,\ldots,M \\ \dfrac{\pi_t^0}{V} & i = 0 \end{cases}$$

$$\pi_{t,\tau+1}^{1:M} \sim Dirichlet\left(\left(1 + n^i\left(K_{t,\tau+1}\right)\right)_{i=1,\ldots,M}\right)$$

$$\text{where } n^i\left(K\right) = \#\left\{j : K^j = i\right\}$$

For each i if there exist $j^1,\ldots,j^i$ for which $K_{t,\tau+1}^{j^i} = i$,

$\bar{z}$ is the centroid of $z_t^{j^1},\ldots,z_t^{j^i}$

$$x_{t,\tau+1}^i \sim p\left(X_t^i \mid \bar{z}\right)$$

For each i if there is no j for which $K_{t,\tau+1}^{j^i} = i$

$$x_{t,\tau+1}^i \sim \sum_{n=1}^{N} q_{t-1}^n \delta\left(X_t^i - x_{t|t-1}^{n,i}\right)$$

$$\text{Set } \hat{\pi}_t^i = \frac{1}{\tau_{beg} - \tau_{end}} \sum_{\tau=\tau_{beg}}^{\tau_{end}} \pi_{t,\tau}^i \qquad i = 1,\ldots,M$$

$$\text{Set } q_t^n \propto q_{t-1}^n \prod_{j=1}^{m_t} \left[\frac{\pi_t^0}{V} + \sum_{i=1}^{M} l_t^i\left(z_t^j ; x_{t|t-1}^{n,i}\right)\hat{\pi}_t^i\right],\ n = 1,\ldots,N$$

Estimation

$$\text{Estimate } \mathrm{E}\left(X_t\right) \text{ by } \sum_{n=1}^{N} q_t^n x_{t|t-1}^n$$

Resampling

Perform adaptive resampling.

8.3.5 Tracking a Variable Number of Targets via DBNs

Modeling an environment via a fixed-length state vector is not always appropriate. The environment will sometimes be densely populated with a mixture of unknown numbers and types of entities. Detection of new entities or disappearance of entities from the environment requires adjustments of the length of the representational vector. We must therefore allow the state space to grow and shrink dynamically. Here we propose a DBN-based approach that takes this variability into account, and deals dynamically with changes in the numbers and types of tracked entities.

The variables Π_t, K_t, X_t, Y_t are as defined above. Let M_t be a random variable with the following five possible values:

B = *birth*, D = *death*, S = *split*, M = *merge*, N = *no change*

The variable M_t specifies the ways in which length of the state vector X_t increases or decreases at each time step. The state of M_t at each time step is determined based on the current and the immediate past observation. For example, birth means the creation of a new entity and the state vector X_t is expanded based on X_{t+1}. As an example, the probability

$$P\left(M_t = (B, D, S, M, N) \mid K_t, K_{t-1}\right) = (0.9, 0, 0.1, 0, 0)$$

states that there is a new observation recorded in K_t and that it is most likely to be the birth of a new entity but could also be a split of an existing entity. In either case, the state vector X_t has to be expanded to accommodate the new entity. The state of the random variable M_t at time point t guides the formation of the state vector.

8.3.6 Particle Filter for DBN

The transition and observation models can also be represented by a complex DBN, instead of parametric functions f and h, exploiting the underlying causal structure among the random variables. The following routine can be used to apply PF to a DBN.

ALGORITHM: Particle Filter for DBN

Initialization

$$\begin{cases} x_0^n \sim p(X_0) \\ q_0^n = \dfrac{1}{n} \end{cases} \qquad n = 1, \ldots, N$$

Prediction & Filter

For $n = 1 : N$

Set x_t^n = empty vector of length equal to the

number of nodes in a time slice of the DBN

Set $q_t^n = 1$

For each node i in topological order in time slice t (ordering guarantees that parent node values are known)

Let u be the value of $pa(X_t^i)$ from x_{t-1}^n

If X_t^i not in Z_t

Sample $x_t^{n,i} \sim p(X_t^i \mid pa(X_t^i) = u)$

Else

Set $x_t^{n,i}$ = value X_t^i in Z_t

Set $q_t^n = q_t^n \times p(X_t^i = x_t^{n,i} \mid pa(X_t^i) = u)$

Estimation

Estimate (for continuous variables): $E(X_t)$ by $\sum_{n=1}^{N} q_t^n x_t^n$

8.3.7 Example DBN Inferencing by Particle Filtering

Consider the example Markovian DBN (Doucet et al., 2000a and 2000b) in Figure 8-1 containing three temporal Boolean variables A (*Traction*), B (*Rain*), and C (*Visibility*) representing, respectively, the road traction (*good*, *bad*), the rain status (*yes*, *no*), and the visibility (*clear*, *poor*). The Y-variables are observations. There are temporal dependencies among the variables; for example, the rain status at the current time step affects both the road traction and the visibility at the next time step.

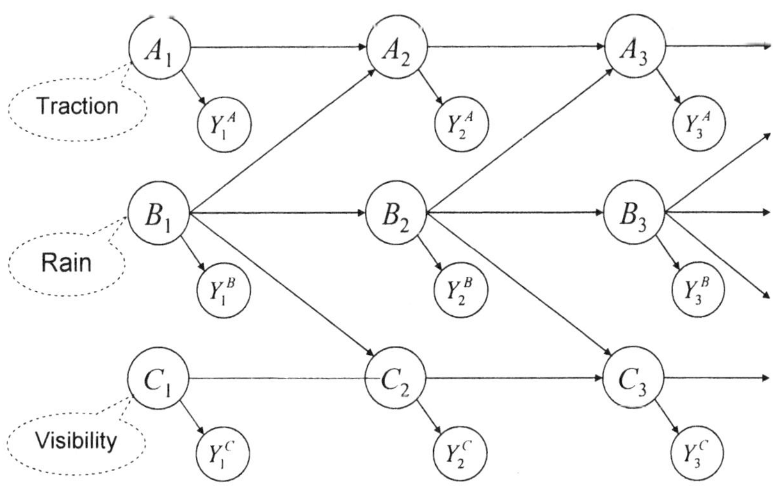

Figure 8-1: Example DBN

Bayesian Filter

Given the initial prior $p(A_0, B_0, C_0)$ for the joint distribution, the following pair of recursive prediction and filtering equations can be derived by applying a combination of Bayes' and marginalization rules:

$$
\begin{aligned}
&p(A_t, B_t, C_t \mid Y_{1:t-1}) \\
&\quad = \sum_{A_{t-1}} \sum_{B_{t-1}} \sum_{C_{t-1}} p(A_t, B_t, C_t \mid A_{t-1}, B_{t-1} C_{t-1}) \, p(A_{t-1}, B_{t-1}, C_{t-1} \mid Y_{1:t-1}) \\
&\quad = \sum_{A_{t-1}} \sum_{B_{t-1}} \sum_{C_{t-1}} p(A_t \mid A_{t-1}, B_{t-1}) p(B_t \mid B_{t-1}) \, p(C_t \mid B_{t-1} C_{t-1}) \, p(A_{t-1}, B_{t-1}, C_{t-1} \mid Y_{1:t-1}) \\
&\quad = \sum_{B_{t-1}} p(B_t \mid B_{t-1}) \sum_{A_{t-1}} p(A_t \mid A_{t-1}, B_{t-1}) \sum_{C_{t-1}} p(C_t \mid B_{t-1} C_{t-1}) p(A_{t-1}, B_{t-1}, C_{t-1} \mid Y_{1:t-1})
\end{aligned}
$$

$$
\begin{aligned}
&p(A_t, B_t, C_t \mid Y_{1:t}) \\
&\quad = \frac{p(Y_t \mid A_t, B_t, C_t) \, p(A_t, B_t, C_t \mid Y_{1:t-1})}{\sum_{A_t} \sum_{B_t} \sum_{C_t} p(Y_t \mid A_t, B_t, C_t) p(A_t, B_t, C_t \mid Y_{1:t-1})} \\
&\quad = \frac{p(Y_t^A \mid A_t) \, p(Y_t^B \mid B_t) \, p(Y_t^C \mid C_t) \, p(A_t, B_t, C_t \mid Y_{1:t-1})}{\sum_{A_t} \sum_{B_t} \sum_{C_t} p(Y_t \mid A_t, B_t, C_t) p(A_t, B_t, C_t \mid Y_{1:t-1})}
\end{aligned}
$$

Particle Filter

The following transition priors are used to sample particles at time t:

$$p(B_t \mid B_{t-1}), p(A_t \mid A_{t-1}, B_{t-1}), p(C_t \mid C_{t-1}, B_{t-1})$$

Then evaluate the importance weight as

$$w_t \, \alpha \, p(Y_t \mid A_t, B_t, C_t)$$

Rao-Blackwellised Filter

Only sample from $p(B_t \mid B_{t-1})$ and evaluate the importance weight as

$$w_t \, \alpha \, p(Y_t \mid Y_{1:t-1}, B_{1:t})$$

The equations for predictions and filtering are then derived as follows:

$$
\begin{aligned}
p(A_t \mid Y_{1:t-1}, B_{1:t}) &= \sum_{A_{t-1}} p(A_t \mid Y_{1:t-1}, B_{1:t}, A_{t-1}) p(A_{t-1} \mid Y_{1:t-1}, B_{1:t}) \\
&= \sum_{A_{t-1}} p(A_t \mid Y_{1:t-1}, A_{t-1}) p(A_{t-1} \mid Y_{1:t-1}, B_{1:t-1}) \\
p(C_t \mid Y_{1:t-1}, B_{1:t}) &= \sum_{A_{t-1}} p(C_t \mid Y_{1:t-1}, C_{t-1}) p(C_{t-1} \mid Y_{1:t-1}, B_{1:t-1})
\end{aligned}
$$

$$p(A_t \mid Y_{1:t}, B_{1:t}) \propto p(Y_t \mid Y_{1:t-1}, B_{1:t}, A_t)\, p(A_t \mid Y_{1:t-1}, B_{1:t})$$
$$\propto p(Y_t^A \mid A_t)\, p(A_t \mid Y_{1:t-1}, B_{1:t})$$
$$p(C_t \mid Y_{1:t}, B_{1:t}) \propto p(Y_t^C \mid C_t)\, p(C_t \mid Y_{1:t-1}, B_{1:t})$$

A Matlab implementation of this example can be obtained from Doucet et al. (2000b). A Java implementation by the author of this book can be obtained by request.

8.3.8 Particle Filter Issues

One problem of the PF is its state vector-based formalism is usually very high in dimension. The higher the dimension is in an application domain, the larger the number of particles required to cover the area in the high-dimensional space (which increases exponentially). Consequently, there is a need for a large number of particles to approximate conditional distributions from high-dimensional spaces.

The RBPF exploits some of the analytical structure of the model by dividing the set of random variables into two: one subset is handled by the PF and the other set by a bank of KFs. Sometimes a partitioned or factored sampling approach is adopted when the underlying system dynamics can be separated into several independent components (Das et al., 2005; Pfeffer et al., 2007).

If the conditional distribution $p(Z_t \mid X_t)$ has severe outliers then many samples will be allocated to parts of the distribution with high predicted probability but low observational probability, and relatively few samples will be allocated to parts of the distribution with low predicted probability but high observational probability. Consequently, a large number of samples are needed for approximation accuracy, which creates a problem as dimensionality increases.

8.4 FURTHER READING

For an overview of the state of the art in PF, see Doucet et al. (2001), Arulampalam et al. (2002), and Djuric et al. (2003). More details on UKF can be found in Julier and Uhlmann (1997).

Chapter 9

Decision Support

This chapter describes two approaches to decision support: influence diagram (ID) and Symbolic Argumentation. By incorporating the concepts of action and utility, Bayesian belief networks (BNs) are converted to IDs, subsuming the functionality of the normative theory of decision making under expected utility theory and decision trees. For inferencing in IDs, we extend the junction tree algorithm for BNs presented in Chapter 6. The extended algorithm presented in this chapter compiles an ID into a strong junction tree in which the computation of maximum expected utility can be done by local message passing in the tree.

Arguments for and against decision options in symbolic argumentation (as opposed to natural argumentation) are expressed in high-level logical if-then rules incorporating subjective probabilities from decision makers representing the argument strengths. To aggregate a set of such probabilistic arguments for and against the decision options, we apply the Dempster-Shafer theory to compute degrees of belief for decision options. Incoming evidence is used to compute degrees of belief, and hence rankings, among the decision options. Decision making based on such degrees of belief is therefore based on a strong mathematical foundation.

Both approaches to decision support are illustrated in the context of a Theater Missile Defense (TMD) scenario.

9.1 EXPECTED UTILITY THEORY AND DECISION TREES

A *decision* is a choice between several alternative courses of risky or uncertain action. Expected Utility Theory (EUT) states that the decision maker chooses between alternative courses of action by comparing their *expected utility* values, which is the weighted sum obtained by adding the utility values of possible outcomes multiplied by their respective probabilities. Formally, if $A = \{a_1, ..., a_n\}$ is the set of all possible actions as decision alternatives and $W = \{w_1, ..., w_n\}$ is the corresponding set of possible outcomes or world states, then the *expected utility* (EU) for action a_k is the following:

$$EU(a_k) = \sum_i U(w_i) p(w_i \mid a_k)$$

where $U(w_i)$ is the utility of the outcome w_i and $p(w_i \mid a_k)$ is the probability that the outcome is w_i if the action a_k is taken. The *maximum expected utility* (MEU) operation is used to choose the best alternative course of action:

$$MEU(A) = \max_k EU(a_k)$$

In cases where an outcome w_i does not depend on the decision maker's action but on other context-dependent factors that the decision maker has no control over, then the expected utility is simply defined without actions, that is,

$$EU = \sum_i U(w_i) p(w_i)$$

The computation of MEU is not relevant in this case.

A *decision tree* poses an ordered list of systematic questions that leads the decision maker through a series of decisions to a logical endpoint that results in a solution. These likely outcomes of possible solutions to the problem are projected as utilities. The systematic exploration of likely outcomes is organized in a tree in which each branch node represents a choice between a number of alternatives, and each leaf node represents a solution. A branch node is either a *chance node* or an *action node*. MEUs are computed at action nodes and EUs are computed at chance nodes.

Example

Figure 9-1 shows an example decision tree in the context of our vehicle-journey example in which ovals, rectangles, and diamonds represent chance nodes, decision or action nodes, and terminal nodes, respectively. The root node is a chance node representing the status of rain, and branches are its uncertain outcomes. It is 70% likely that rain will be present and 30% likely that rain will be absent. These two possible outcomes are represented as two branches coming out of the chance node for rain. Given either of these two alternatives of rain status, the next step is to decide whether to start the journey or cancel it. This decision requires exploring four possible combinations of options between the rain status and the journey decision. For each of these combinations, the likely mobility of the road, affecting the duration, fuel consumption, and safety of the journey, needs to be considered to compute earned credits for the journey that are projected as utilities.

For example, two combinations corresponding to the top two branches are that the rain is present, the action taken is to start the journey, and there is a 10% chance that the mobility will be a go and 90% chance that the mobility will be a no ("no" is a combination of "slow go" and "no go," for the sake of simplicity). Each of these two possibilities yields different credit as utility and is attached to the terminal node. So the top branch of the tree in the figure encodes a possible outcome, which states that if the rain is present, a decision to start the journey is taken, and the mobility is a go, then the utility is 900. Eight possible outcomes are explored in this manner in the context of the example.

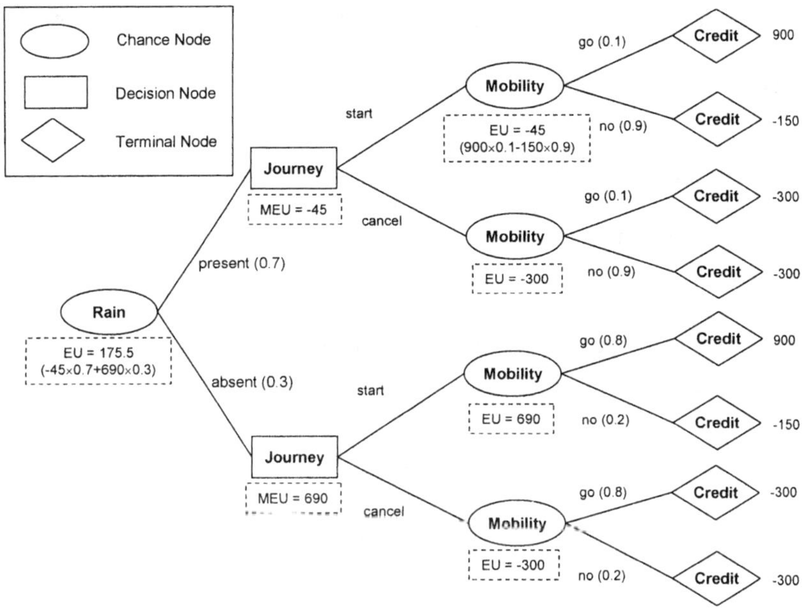

Figure 9-1: An example decision tree

EUs and MEUs at branch nodes are computed bottom-up along the tree (right to left in the figure). For example, the EU at the *Mobility* node of the topmost branch is computed using the formula $\sum_i U(w_i)p(w_i)$, yielding $900 \times 0.1 + (-150) \times 0.9$, that is, –45. To compute the MEU at the node *Journey* in the upper branch of the tree, EUs of the two decisions *start* and *cancel*, *EU*(*start*) and *EU*(*cancel*), need to be computed using the formula $\sum_i U(w_i)p(w_i \mid a_k)$. Now, the EUs of the two *Mobility* nodes of the upper branch are –45 and –300. Therefore, the MEU of the node *Journey* is the maximum of –45 and –300, that is, –45. Finally, the EU of the node *Rain* is computed using the formula $\sum_i U(w_i)p(w_i)$, yielding 175.5.

The procedure for computing utilities using decision trees, as explained above, is a simple statistical procedure. Decision trees also provide easy-to-understand graphic representations. But laying out all possible options is not feasible when there are several factors, each with multiple outcomes. Moreover, the procedure provides no mechanism for incorporating subjective knowledge. IDs address some of these issues.

9.2 INFLUENCE DIAGRAMS FOR DECISION SUPPORT

Influence diagrams (IDs) are BNs augmented with decision variables and utility functions, and are used to solve decision problems. There are three types of nodes in an ID:

- *Chance nodes* (i.e., BN nodes): represented by ovals
- *Decision nodes*: represented by rectangles
- *Value* or *utility nodes*: represented by diamonds

As opposed to chance nodes representing probabilistic variables, decision nodes represent actions that are under the full control of decision makers and hence no CPT is attached to a decision node.

Example

Figure 9-2 shows an example ID from our vehicle-journey example.

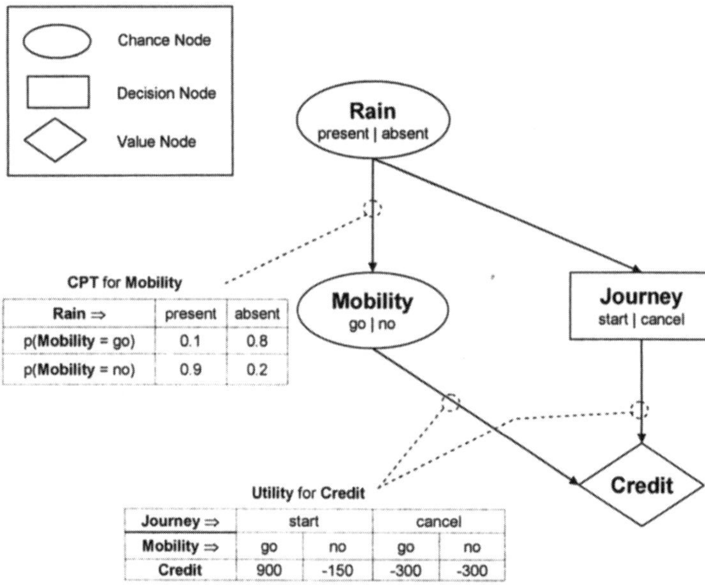

Rain ⇒	present	absent
p(Mobility = go)	0.1	0.8
p(Mobility = no)	0.9	0.2

Journey ⇒	start		cancel	
Mobility ⇒	go	no	go	no
Credit	900	-150	-300	-300

Figure 9-2: An example influence diagram

There are two chance nodes (*Rain* and *Mobility*), one decision node (*Journey*), and one value node (*Credit*). The arrow from the node *Rain* to the node *Mobility* represents the causal influence of the status of rain on the road mobility. The CPT quantifies this causality as in usual BNs. As we mentioned before, there is no CPT attached to a decision node, but the arrow from the chance node *Rain* to the decision node *Journey* represents the fact that the knowledge of the status of rain should be known before making the decision (whether to *start* or *cancel*) about the journey. The value node *Credit* has two parents, representing the causal influence of the road mobility and the journey status on the credit score. The table represents a *utility function* whose definition quantifies the utilities for all possible combinations of the parent values. For example, if the decision maker starts the vehicle journey and the mobility is a no then the credit is negative 150.

Inferences in IDs involve computing the EU for each of the action alternatives. This kind of computation is sensitive to the temporal order in which the decisions are made. Therefore, IDs require a directed path connecting all the decision nodes sequentially. EUs are evaluated on the condition that an optimal choice has been made in all previous steps.

Suppose, $A_1,...,A_m$ is the ordered set of actions to be decided (A_m is the last action), where each A_i is a set of mutually exclusive action options; $V_0,...,V_m$ is a partition of the chance variables $X_1,...,X_n$ such that V_{i-1} is the set of variables instantiated before deciding on the action A_i. Therefore we have the following ordering:

$$V_0 \prec A_1 \prec V_1 \prec ... \prec A_m \prec V_m$$

Then the MEU for a chosen set of alternatives for $A_1,..., A_m$ is given below:

$$MEU(A_1,..., A_m) =$$
$$\sum_{V_0} \max_{A_1} \sum_{V_1} ... \sum_{V_{m-1}} \max_{A_m} \sum_{V_m} \sum_{U} U \times p(V_0, V_1,..., V_m \mid A_1,..., A_m)$$

The MEU for action A_k is given as

$$MEU(A_k) =$$
$$\max_{A_k} \sum_{V_k} ... \sum_{V_{m-1}} \max_{A_m} \sum_{V_m} \sum_{U} U \times p(V_k,..., V_m \mid V_0,..., V_{k-1}, A_1,..., A_m)$$

Thus the computation of the MEU in an ID involves a series of alternating sum-marginalization and max-marginalization to eliminate the variables.

9.2.1 Inferencing in Influence Diagrams

In this section, we illustrate inferencing in IDs considering the following two special cases of interactions between decisions and other types of variables:

- *Nonintervening interactions*: Actions that have no impact on variables (or probability distribution) in the network.
- *Intervening interactions*: Actions that have an impact on variables in the network that then affect the beliefs of other variables, or affect the utilities of value nodes.

Next, we present a junction tree-based algorithm for inferencing in IDs containing arbitrary sequences of interactions.

Example

Figure 9-3 shows an ID with one node of each type: chance, decision, and value. The decision node represents a nonintervening decision *Journey*, which has no impact on $p(Rain)$, that is, $p(Rain \mid Journey) = p(Rain)$. Given the likelihood *e* as shown in the figure, the posterior probability for *Rain* and expected utilities for each of the actions are computed as follows:

$$p(Rain \mid e) = \begin{bmatrix} p(Rain = present \mid e) \\ p(Rain = absent \mid e) \end{bmatrix} = \alpha \begin{bmatrix} 0.9 \\ 0.1 \end{bmatrix} \begin{bmatrix} 0.2 \\ 0.8 \end{bmatrix} = \begin{bmatrix} 0.692 \\ 0.308 \end{bmatrix}$$

EUs for each of the decisions based on the posterior probabilities are computed as follows:

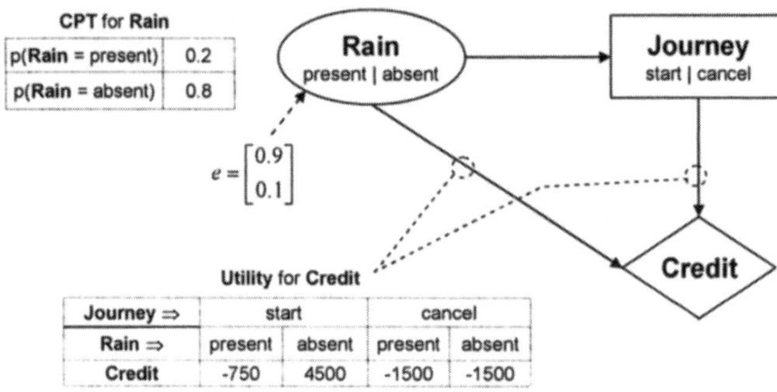

Figure 9-3: ID with nonintervening action

$$\begin{aligned} &EU(Journey = start \mid e) \\ &\quad = U(Journey = start, Rain = present) \times p(Rain = present \mid e) + \\ &\quad\quad U(Journey = start, Rain = absent) \times p(Rain = absent \mid e) \\ &\quad = (-750) \times 0.692 + 4500 \times 0.308 \approx 865 \end{aligned}$$

$$\begin{aligned} &EU(Journey = cancel \mid e) \\ &\quad = (-1500) \times 0.692 + (-1500) \times 0.308 = -1500 \end{aligned}$$

The MEU operation to choose the best alternative course of decision is computed below:

$$\begin{aligned} &MEU(Journey \mid e) \\ &\quad = \max\{EU(Journey = start \mid e), EU(Journey = cancel \mid e)\} \\ &\quad = \max\{865, -1500\} = 865 \end{aligned}$$

This MEU computation can be executed in a decision tree equivalent to the ID in Figure 9-3. One such tree is shown in Figure 9-4, where the decision and chance variables are systematically laid out (parent variables before their children) to explore all possible combinations of states. Therefore, there will be as many leaf nodes in the tree as the number of states in the joint distribution of the set of all decision and chance variables. Each branch yields some utility in terms of positive or negative credit obtained from the utility table in Figure 9-3. The computation of EU in the tree is performed in the usual manner starting at the leaf nodes and moving backwards.

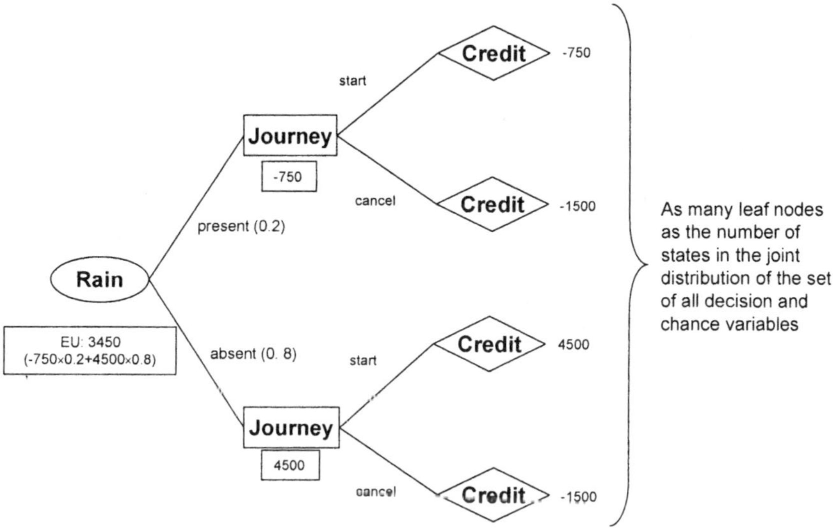

Figure 9-4: Decision tree equivalent to the ID in Figure 9-3

When evidence on the variable *Rain* is obtained, its posterior probability is computed by component-wise multiplication followed by normalization. The EU is then computed based on the posterior probability of *Rain* as shown in Figure 9-5.

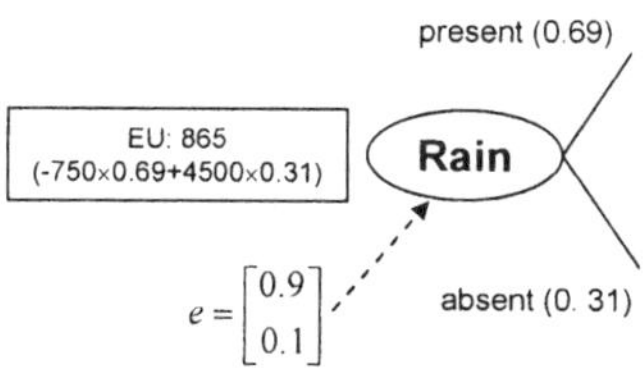

Figure 9-5: Decision tree of Figure 9-4 with evidence entered

Example

Figure 9-6 shows an ID with a set of nonintervening actions. The EU is computed by summing up the MEU of each of the constituent actions. In other words,

$$MEU\left(ServiceVehicle, Deliver\ Logistics \mid e\right) = MEU(ServiceVehicle \mid e) + MEU(Deliver\ Logistics \mid e)$$

The MEU for the variable *Deliver Logistics* can be computed as follows:

$$EU\left(Deliver\ Logistics = proceed \mid e\right) = \left(-750\right) \times 0.692 + 4500 \times 0.308 \approx 865$$

$$EU\left(Deliver\ Logistics = abandon \mid e\right) = \left(-1500\right) \times 0.692 + \left(-1500\right) \times 0.308 = -1500$$

$$MEU\left(Deliver\ Logistics \mid e\right) \approx 865$$

Similarly, the MEU of *ServiceVehicle* can be computed, giving $MEU(ServiceVehicle \mid e) = -200$.

Therefore, $MEU\left(ServiceVehicle, Deliver\ Logistics \mid e\right) \approx 665$.

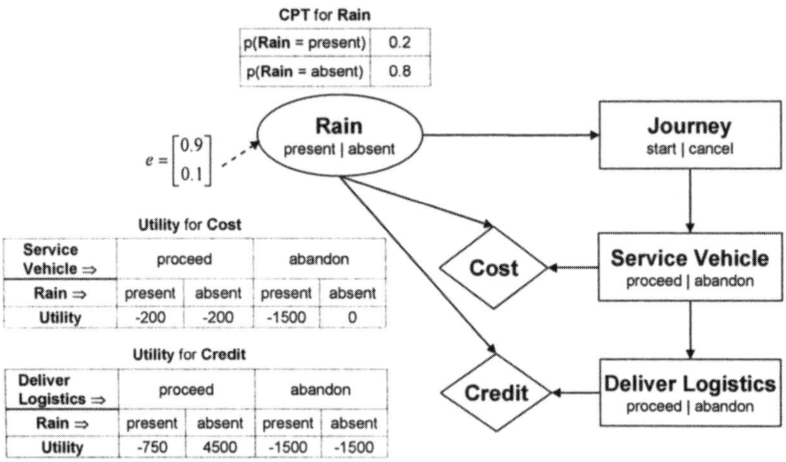

Figure 9-6: ID with a set of nonintervening actions

A decision tree equivalent to the ID of Figure 9-6 is shown in Figure 9-7. The utility along a branch of the tree is computed by summing up the cost and credit corresponding to the actions specified along the branch. For example, the topmost branch states that if rain is present and the option *start* is chosen for the decision variable *Journey*, and the option *proceed* is chosen for each of the two decision variables *ServiceVehicle* and *Deliver Logistics*, then the overall utility is –950. This utility is obtained by adding the cost of proceeding with the *ServiceVehicle* action in the presence of rain (–200) to the credit for proceeding with *Deliver Logistics* action in the presence of rain (–750).

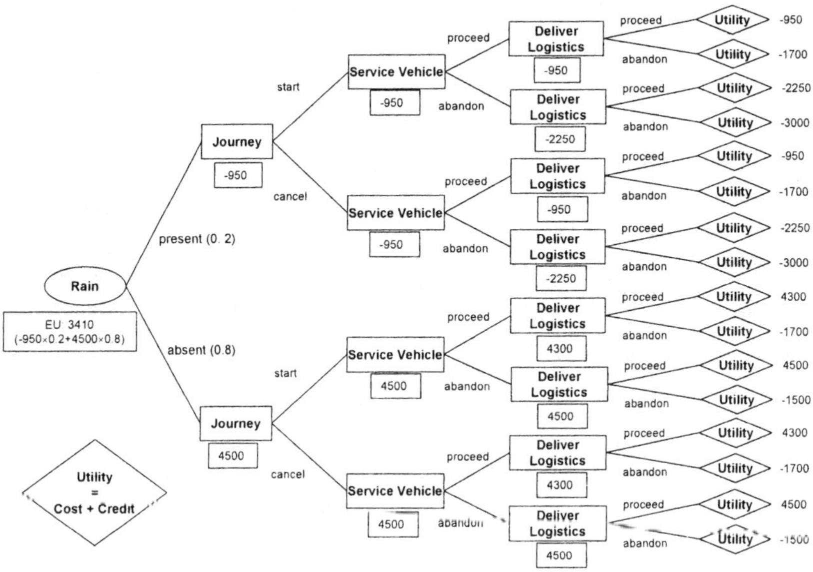

Figure 9-7: Decision tree equivalence to the ID in Figure 9-6

Example

Figure 9-8 shows an ID with one decision node representing the intervening decision *Journey* that directly influences the variable *Mobility*. The posterior probability and EU for the variable *Mobility* when the option *start* is chosen is computed as follows:

$$p(Mobility \mid e, Journey = start) = \begin{bmatrix} 0.42 \\ 0.51 \\ 0.07 \end{bmatrix}$$

$$EU(Journey = start \mid e)$$
$$= 4500 \times 0.42 + (-500) \times 0.51 + (-1500) \times 0.07 \approx 1504$$

Similarly, EU for the option *cancel* is computed similarly, yielding the following:

$$EU(Journey = cancel \mid e) = -1500$$

Therefore, $MEU(Journey \mid e) = 1504$

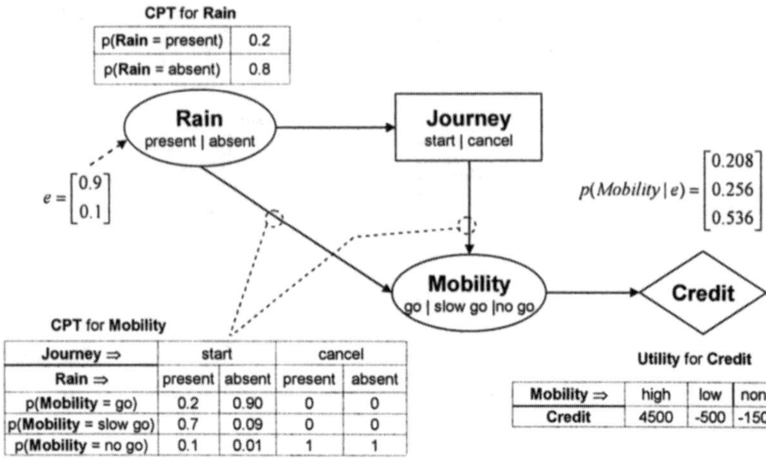

Figure 9-8: ID with intervening action

A decision tree equivalent to the ID in Figure 9-8 is shown in Figure 9-9.

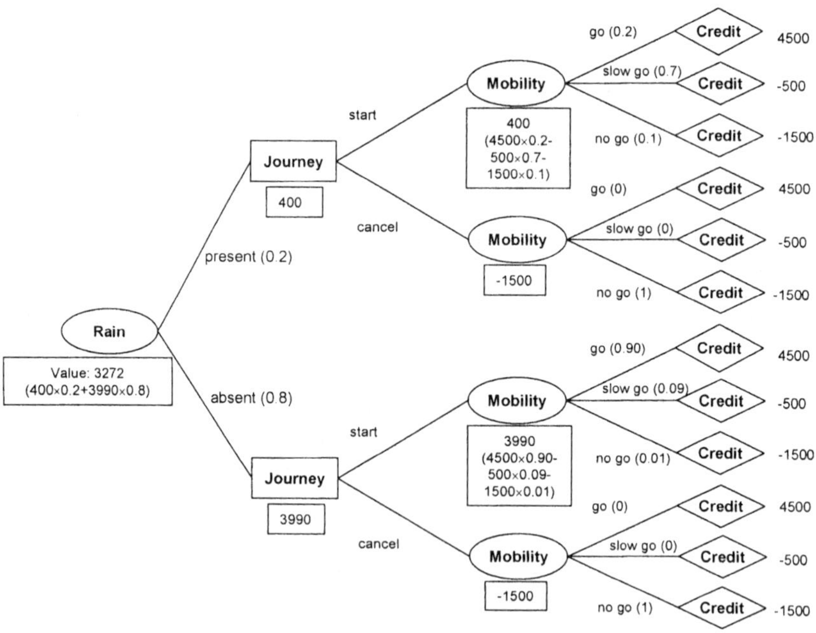

Figure 9-9: Decision tree equivalent to the ID in Figure 9-8

9.2.2 Compilation of Influence Diagrams

Compiling an ID involves transforming it to a *strong junction tree* (or rooted junction tree), that maintains a special ordering relative to the root clique. Just to recall, a *junction tree* is a tree of cliques if for each pair $\mathbf{C}_1$ and $\mathbf{C}_2$ of cliques, $\mathbf{C}_1 \cap \mathbf{C}_2$ is contained in every clique on the path connecting $\mathbf{C}_1$ and $\mathbf{C}_2$. A *separator* of the two adjacent cliques $\mathbf{C}_1$ and $\mathbf{C}_2$ is the intersection $\mathbf{C}_1 \cap \mathbf{C}_2$.

The compilation steps for IDs are shown on the right side of Figure 9-10. Steps to compile BNs into junction trees are shown on the left. Details of each step are explained below.

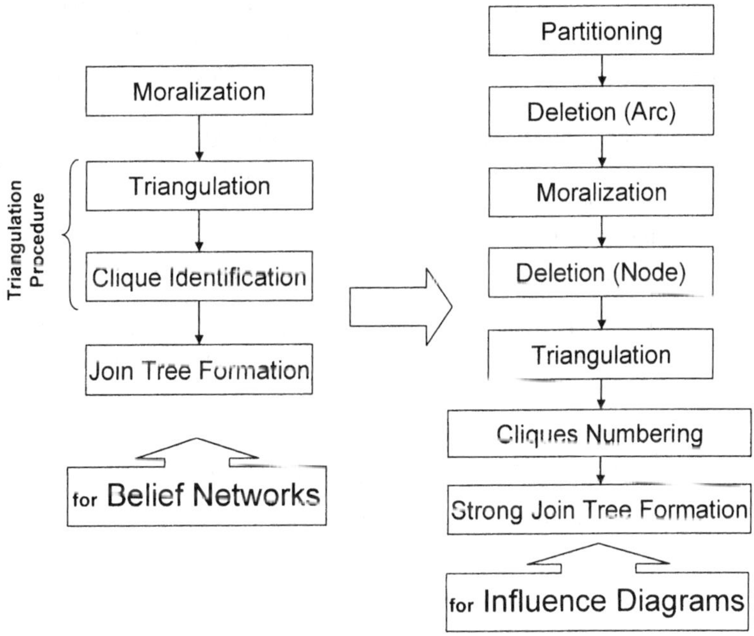

Figure 9-10: Steps in the compilation process of IDs

- *Partitioning*: Partition the chance variables $X_1,...,X_n$ into $V_0,...,V_p$ such that V_{i-1} is the set of variables instantiated before deciding on the action A_i, yielding the ordering $V_0 \prec A_1 \prec V_1 \prec ... \prec A_m \prec V_m$.
- *Deletion (Arc)*: Delete edges pointing into decision nodes.
- *Moralization*: Marry parents with common children including parents of utility nodes and drop the directions of the original links.
- *Deletion (Node)*: Delete utility nodes along with their edges.
- *Triangulation*: Triangulate in such a way that it facilitates the computation of MEU. Do this by adopting a special elimination order based on the

ordering found during partitioning. Start eliminating variables using the triangulation procedure in the following order: eliminate variables from the set V_m, then the decision variable A_m, then variables from the set V_{m-1}, then the decision variable A_{m-1}, and so on. The elimination sequence constitutes a numbering where the first eliminated variable is assigned the highest number.

- *Clique Numbering*: A clique is numbered after its variable with the highest number k (otherwise 1) such that the rest of the variables in the clique have a common neighbor outside the clique with a number lower than k.
- *Strong Join Tree Formation*: A *strong root* of a junction tree is a distinguished clique **R** such that for each pair $\mathbf{C}_1$ and $\mathbf{C}_2$ of adjacent cliques with $\mathbf{C}_1$ closer to **R** than $\mathbf{C}_2$, there is an ordering in $\mathbf{C}_2$ that respects $\prec$ and with the vertices of the separator $\mathbf{C}_1 \cap \mathbf{C}_2$ precedes the vertices of $\mathbf{C}_2 - \mathbf{C}_1$. A junction tree is *strong* if it has a strong root. The ordering within a junction tree ensures that the computation of the MEU can be done by local message passing in the tree. The following steps are followed to form a strong junction tree from the sequence $\mathbf{C}_1,...,\mathbf{C}_n$ of cliques arranged in increasing order according to their indices:
  - Start with $\mathbf{C}_1$ as the root clique with number 1.
  - Connect the cliques as follows in the increasing order of their numbers: first, compute the following for each clique C_k $(k > 1)$:

$$S_k = C_k \cap \bigcup_{i=1}^{k-1} C_i$$

Connect the clique C_k with another clique that contains S_k

  - Variables that were eliminated first are placed farthest from the root in the strong junction tree.

The following example illustrates these steps to construct a strong junction tree.

Example

Figure 9-11 is an example ID modified from our original BN for the vehicle journey example presented in Chapter 6. The chance variable *Visibility* is the only parent of the decision variable *Journey*, indicating that information about the visibility condition is required before the decision to start the journey or cancel the journey. A fixed cost will be incurred if the decision to cancel the journey is taken. If the journey decision is to start, there will be a cost due to fuel consumption depending on the road mobility condition, but credits can also be earned depending on the amount of delay in arrival.

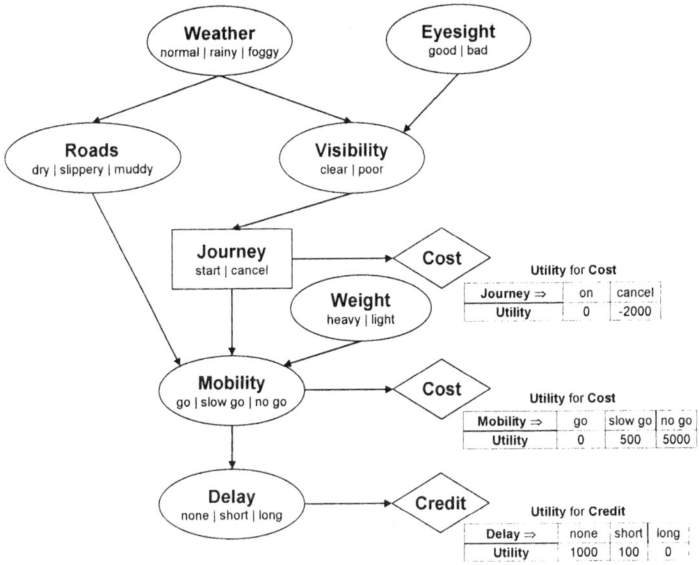

Figure 9-11: An example influence diagram

Consider the following ordering of the chance and decision variables of the BN:

$$\{Eyesight, Weather, Visibility, Roads\} \prec \{\mathbf{Journey}\} \prec \{Weight, Mobility, Delay\}$$

The sequence in which the variables are eliminated during the triangulation step is shown in Figure 9-12, along with the assigned numbers.

Variable	Number
Delay	8
Mobility	7
Weight	6
Journey	5
Roads	4
Visibility	3
Weather	2
Eyesight	1

Figure 9-12: Numbering of variables during triangulation

The five cliques identified during the triangulation step and their numbers are given below:

$\mathbf{C}_1$: {*Weather, Eyesight, Visibility*} → 3
$\mathbf{C}_2$: {*Weather, Roads, Visibility*} → 4
$\mathbf{C}_3$: {*Roads, Visibility, Journey*} → 5
$\mathbf{C}_4$: {*Roads, Journey, Weight, Mobility*} → 7
$\mathbf{C}_5$: {*Mobility, Delay*} → 8

To construct a strong junction tree, we start with $\mathbf{C}_1$ as the root. Figure 9-13 shows the strong junction tree based on these cliques. The variables that were eliminated first (such as *Delay* and *Mobility*) are placed farthest from the root (clique number 1) in the strong junction tree.

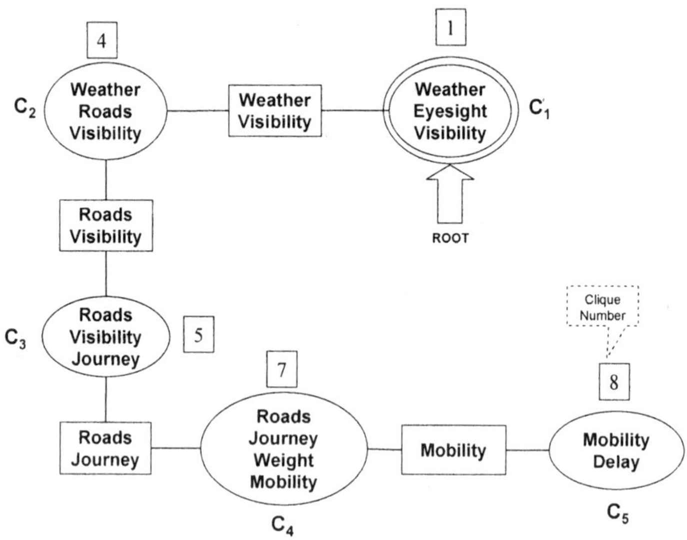

Figure 9-13: Strong junction tree

9.2.3 Inferencing in Strong Junction Trees

Inferencing in a strong junction tree employs a special collect operation from the leaves to the root of the tree. This operation is similar to the one for computing EU at the root node of a decision tree as shown in Figure 9-1. The difference is that a node here takes the form of a clique rather than a single variable. In addition to associating a probability potential with each clique of the tree, a utility potential is also associated with each clique. The utility potential for a clique is the sum of the utility functions assigned to it. The utility potential for a clique is a null function if no utility functions are assigned to it.

Example

Continuing with our earlier example, Figure 9-14 shows an assignment of probability distributions and utility functions. The cost utility function associated to the node Mobility and the credit utility function are assigned to the clique $\mathbf{C}_5$ since the parents of the corresponding two utility nodes belong to $\mathbf{C}_5$. Similarly, the cost utility function associated to the node Journey could have been assigned to any of the two cliques $\mathbf{C}_3$ and $\mathbf{C}_4$ containing the variable *Journey*, and $\mathbf{C}_3$ was chosen arbitrarily.

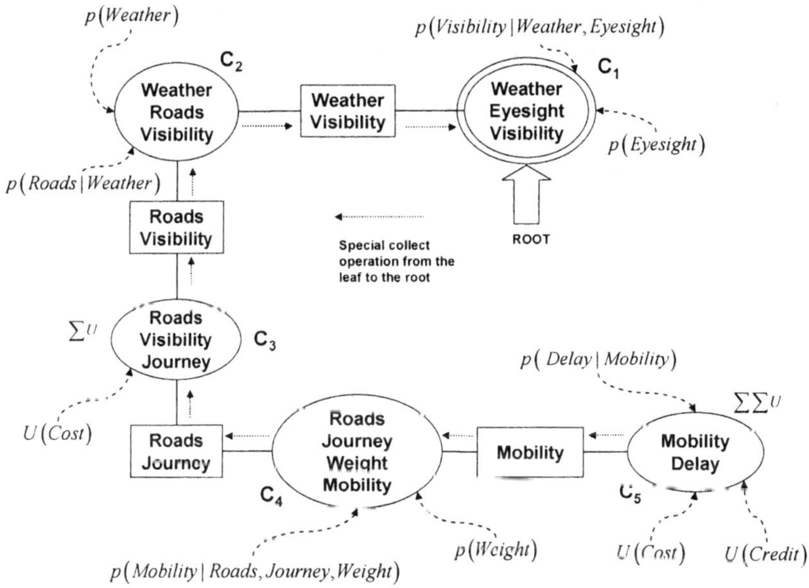

Figure 9-14: Assignment of probability distributions and utility functions to cliques

The computation of the utility potential for the node $\mathbf{C}_5$ is given below in Figure 9-15 based on the two utility functions for the two nodes for cost and credit.

Consider the following ordering of the chance and decision variables as before:

$$\{Eyesight, Weather, Visibility, Roads\} \prec \{\boldsymbol{Journey}\} \prec \{Weight, Mobility, Delay\}$$

Then the MEU for *Journey* based on this ordering is (only the initial letter has been used from each variable name, except *Wt* stands for *Weight*):

$$MEU(Journey) = \sum_{E,W,V,R} \max_{J} \sum_{\{Wt,M,D\}} \sum_{U} U \times p(Wt, M, D, E, W, V, R \mid J)$$

Utility for Cost

Mobility ⇒	go	slow go	no go
Utility	5000	500	0

Utility for Credit

Delay ⇒	none	short	long
Utility	1000	100	0

Mobility	Delay	Utility
go	none	6000
go	short	5100
go	long	5000
slow go	none	1500
slow go	short	600
slow go	long	500
no go	none	1000
no go	short	100
no go	long	0

Figure 9-15: Utility potential

But the special collect operation on the strong junction tree in Figure 9-14 allows the MEU to be computed at node 5 (clique $\mathbf{C}_3$) as follows:

$$MEU(Journey) =$$

$$\max_J \sum_{\{R,V\}} U(Cost) \times p(R,V,J) \sum_{\{Wt,M\}} p(R,J,M,Wt) \sum_D \sum_U U(P) \times p(M,D)$$

where $U(P)$ is the sum utility as shown in Figure 9-15. In this example, node 5 could also have been considered as the root node.

9.2.4 An Example Influence Diagram for Theater Missile Defense

We consider the ID in Figure 9-16 to illustrate the recommendation of interceptor type and number in case of a missile launch. The set of three hypotheses or possible recommendations are due to the two types of interceptors and no interception at all. The PAC-3 (Patriot Advanced Capability 3)-type interceptor is more expensive than the PAC-2/GEM (Patriot Advanced Capability 2/Guidance Enhanced Missile) type, but has higher probability of kill. The total loss of lives as a result of a missile launch depends on the probability of kill of an interceptor and on the type of the missile's warhead. A missile with an N-B-C warhead is likely to kill more people than with a conventional warhead. Whether or not a missile has an N-B-C warhead can be analyzed from previous intelligence on the country from which the missile has been launched.

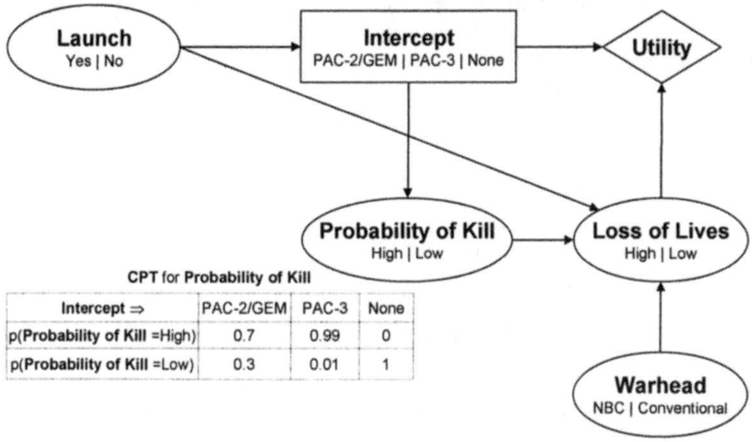

Utility

Intercept ⇒	PAC-2/GEM		PAC-3		None	
Loss of Lives ⇒	High	Low	High	Low	High	Low
Utility	-11	-12	-10	9	8	10

CPT for Probability of Kill

Intercept ⇒	PAC-2/GEM	PAC-3	None
p(Probability of Kill =High)	0.7	0.99	0
p(Probability of Kill =Low)	0.3	0.01	1

Figure 9-16: An example ID in the Theater Missile Defense (TMD) domain

An alternative argumentation approach can also be employed to aid in selecting an interceptor. Table 9-1 shows example arguments for selecting an interceptor. The following section describes the symbolic argumentation-based approach to decision support.

IF	country possesses biological weapons
AND	country does NOT possess submunition warheads
THEN	select 2 PAC-2/GEM
WITH	certainty 0.9
IF	country possesses nuclear weapons
THEN	select 2 PAC-3
WITH	certainty 0.9

Table 9-1: Argumentation approach to the selection of interception

9.3 SYMBOLIC ARGUMENTATION FOR DECISION SUPPORT

Table 9-2 below illustrates the symbolic argumentation in the context of our TMD scenario.

- Situation
    - A missile has been launched from a TEL located in the far northwest of country X
    - It is moving on an east to west track
- Essential Element of Information
    - Launch time: 03.45 GMT
    - Launch location: 30N40E
    - Vector (direction): 281 degrees
    - Predicted point of impact: City Y
    - Maximum ordinate (apogee): 9200ft
    - Missile type: Z
    - Type of warhead: HE
- Possible Responses (hypotheses)
    - Ground-to-Air interceptor
    - Air-to-Air interceptor
    - None

Table 9-2: Summary of symbolic argumentation for a TMD scenario

As shown in Table 9-2, a situation occurs and one or more sensor assets and intel sources report the essential element of information. Possible responses are then listed and the argumentation process recommends the one with the highest support from the accumulated evidence and existing data. Doctrinal knowledge used to construct and evaluate arguments is represented by rules. Example arguments for the current scenario are shown in Table 9-3. The values of the *strengths* of arguments for ($s1$, $s2$, $s3$, $s4$) and against ($n1$, $n2$) can be obtained through a knowledge elicitation process with expert tactical decision makers.

Justifications for and against candidate responses are based on the arguments in Table 9-3 using relevant information such as the missile flight stage, probability of kill, target value, etc. The evaluation of these arguments is performed through a process of *aggregation*. This process includes not only an assessment of the number and strength of each independent argument, and the collective force of argument for each hypothesis, but also the plausibility of the evidence and justification for each line of argument. For example, two arguments for the response "allocate Ground-to-Air interceptor" will be aggregated to produce a combined strength that is expected to be higher than both $s1$ and $s3$. If we use the probabilistic dictionary $[0,1]$ to represent source reliability, then the two arguments can be aggregated by using Dempster's rule of combination as presented in Chapter 3 and illustrated in Chapter 5. On the other hand, if we use the dictionary $[-1,+1]$, then the arguments can be combined by using formulae from the certainty factor formalism as presented in Chapter 3.

Arguments for Candidates	
IF	the attacking missile is at an early flight stage
AND	an optimal tracking acquisition point is available
AND	probability of kill (PK) is high
THEN	allocate Ground-to-Air (s1)
IF	the attacking missile is at a high altitude
AND	the missile in the air defense is in close range
THEN	allocate Air-to-Air (s2)
IF	target value is high
AND	it is being defended by X (Ground-to-Air or Air-to-Air) interceptor as a point target
THEN	allocate X (s3)
IF	target value is low
THEN	allocate None (s4)
Arguments against Candidates	
IF	the flight is at a late stage of detection
AND	PK is low
THEN	do NOT allocate Ground-to-Air or Air-to-Air (n1)
IF	the attacking missile is carrying an N-B-C warhead
THEN	do NOT allocate None (n2)

Table 9-3: Example arguments for and against

All arguments for and against each response are combined in the above manner to accumulate total evidence for and against each response. Evidence against each response discounts evidence for the response. For example, total evidence for and against the response "allocate Ground-to-Air interceptor" is, respectively, $s3$ and $n1$, and therefore the aggregated evidence is $s3(1-n1)/(1-s3*n1)$. If this aggregation process associates more weights to the response "allocate Ground-to-Air interceptor" than any other response then it will be considered as a best option in the current situation.

9.3.1 Measuring Consensus

The need to measure consensus in a decision-making context with the Dempster-Shafer theory as the underlying theoretical foundation is first explained in this section with the help of a concrete example.

Example

Suppose some source of evidence heavily suggests against the use of an *Air-to-Air* interceptor (that is, either the *Ground-to-Air* interceptor or *None*) and the values 0.9 and 0 together constitute a belief function. The focal element is {*Ground-to-Air*, *None*} and the mass distribution is m_1({*Ground-to-Air*, *None*}) = 0.9. We know nothing about the remaining probability 0.1; so it is allocated to the whole frame of discernment Ω = {*Ground-to-Air*, *Air-to-Air*, *None*}. Another source of evidence heavily suggests against the use of *Ground-to-Air*. The mass distribution in this case is m_2({*Air-to-Air*, *None*}) = 0.9. The remaining probability, as before, is allocated to the whole frame of discernment. Dempster's rule can then be used to combine the masses as follows:

Ground-to-Air (*GtA*) *Air-to-Air* (*AtA*)	$m_2(\{AtA, None\}) = 0.9$	$m_2(\Omega) = 0.1$
$m_1(\{GtA, None\}) = 0.9$	$m_{1,2}(\{None\}) = 0.81$	$m_{1,2}(\{GtA, None\})$ $= 0.09$
$m_1(\Omega) = 0.1$	$m_{1,2}(\{AtA, None\}) = 0.09$	$m_{1,2}(\Omega) = 0.01$

Therefore, the combined belief and plausibility are computed in the following table:

Focal Element (A)	$Bel(A)$	$Pl(A)$
$\{None\}$	0.81	1.0
$\{GtA, None\}$	0.09	0.91
$\{AtA, None\}$	0.09	0.91
Ω	1.0	1.0

The above result is counterintuitive, given that none of the two sources of evidence explicitly supported no interception. Moreover, the table above does not reflect the underlying high degree of disagreement. The consensus metric provided below will be able to highlight this case. The "entropy"-based algorithm presented here generates a very low (2.3 out of 3) degree of consensus (zero or minimum entropy is the highest consensus). The consensus-measuring criterion is based on the generalization of the concept of entropy from point function to set function (Stephanou and Lu, 1988). This generalization is composed of three measurements: belief entropy, core entropy, and partial ignorance.

Suppose Ω is a frame of discernment, m is a basic probability assignment, and $\mathbf{F} = \{A_1, ..., A_n\}$ is the set of focal elements.

The *belief entropy* of the pair $\langle m, \mathbf{F} \rangle$ is defined as follows (log is base 2):

$$E_b(m, \mathbf{F}) = -\sum_{i=1}^{n} m(A_i) \log m(A_i)$$

The belief entropy is a measure of the degree of confusion in the decision maker's knowledge about the exact fraction of belief that should be committed to each focal element in **F**. Thus, the belief entropy is naturally equal to zero if the entire belief is committed to a single focal element, that is, $m(A_k)=1$, for some k. The maximum belief entropy occurs when belief is divided in equal fractions among the focal elements.

The *core entropy* of the pair $\langle m, \mathbf{F}\rangle$ is defined as follows:

$$E_c(m, \mathbf{F}) = -\sum_{i=1}^{n} r_i \log r_i$$

where

$$r_i = \frac{\| A_i \| - 1}{\sum_{j=1}^{n} (\| A_i \| - 1)}$$

and $\| A_i \|$ is the cardinality of A_i ($0 \log 0$ is considered 0). The core entropy is a measure of the degree of confusion in the decision maker's knowledge of which focal elements the true value might be in. If belief is committed to a single focal element A_k, then r_k is one, and therefore the core entropy is equal to zero. The core entropy is maximum when belief is divided among a number of focal elements with the same cardinality.

The *partial ignorance* of the pair $\langle m, \mathbf{F}\rangle$ is defined as follows:

$$I(m, \mathbf{F}) - \sum_{i=1}^{n} m(A_i) s_i(A_i)$$

where $s_i(A_i) = \dfrac{\| A_i \| - 1}{\| \Omega \| - 1}$ and $\| A_i \|$ is the cardinality of A_i, and the assumption is that the frame of discernment has more than one element. Partial ignorance is a measure of the inability to confine the true value within a small-sized focal element. Consequently, partial ignorance is large when a large belief is committed to large focal elements. It is zero when the entire belief is committed to a singleton.

The *generalized entropy* of the pair $\langle m, \mathbf{F}\rangle$ is then defined as follows:

$$E(m, \mathbf{F}) = E_b(m, \mathbf{F}) + E_c(m, \mathbf{F}) + I(m, \mathbf{F})$$

Now suppose that we have two basic probability assignments m_1 and m_2, corresponding to two different expert knowledge sources, defining focal elements $\{A, \Omega\}$ and $\{B, \Omega\}$, respectively, where $A \subset \Omega$ and $B \subset \Omega$. Let us consider four different cases as shown in Figure 9-17.

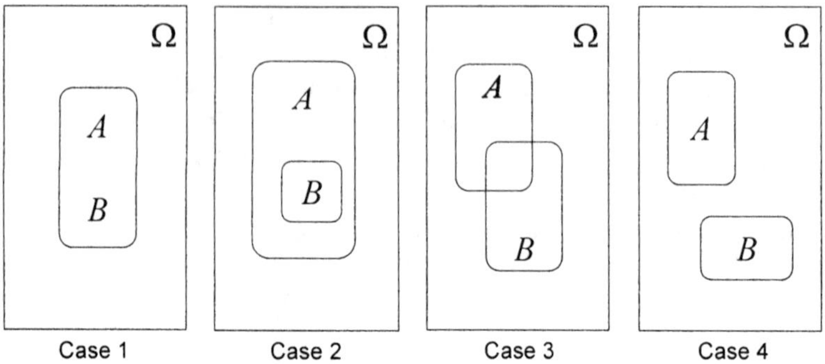

Figure 9-17: Combining expert knowledge sources (Stephanou and Lu, 1988)

It can be shown that the generalized entropy of the knowledge sources via Dempster's rule of combination for each of the first three cases is smaller than the sum of the individual entropies of the knowledge sources. In other words, the following relation holds for the first three cases:

$$E(m_{12}, \mathbf{F}) \leq E(m_1, \mathbf{F}) + E(m_2, \mathbf{F})$$

The above implies that the pooling of two concordant bodies of evidence reduces entropy. The disjointedness between A and B in the last case indicates that there are two bodies of evidence supporting two conflicting propositions, and hence the combined generalized entropy does not necessarily get reduced.

Example

To illustrate reduction of entropy, consider the frame of discernment $\Omega = \{GtA, AtA, None\}$, the two focal elements $\{GtA, None\}$ and $\{AtA, None\}$, and their corresponding basic probability assignments:

$$A = \{GtA, None\} \qquad B = \{AtA, None\}$$

$$\mathbf{F}_1 = \{A, \Omega\} \qquad \mathbf{F}_2 = \{B, \Omega\}$$

$$m_1(A) = 0.7, m_1(\Omega) = 0.3 \qquad m_2(B) = 0.6, m_2(\Omega) = 0.4$$

The entropies and ignorance are computed as follows:

Belief Entropy:

$$\begin{aligned} E_b(m_1, \mathbf{F}_1) &= -m_1(A)\log m_1(A) - m_1(\Omega)\log m_1(\Omega) \\ &= -0.7\log 0.7 - 0.3\log 0.3 = 0.88 \end{aligned}$$

$$\begin{aligned} E_b(m_2, \mathbf{F}_2) &= -m_2(B)\log m_2(B) - m_2(\Omega)\log m_2(\Omega) \\ &= -0.6\log 0.6 - 0.4\log 0.4 = 0.97 \end{aligned}$$

Core Entropy:

$$E_c(m_1, \mathbf{F}_1) = -r_1(A)\log r_1(A) - r_1(\Omega)\log r_1(\Omega)$$
$$= -\frac{1}{3}\log\frac{1}{3} - \frac{2}{3}\log\frac{2}{3} = 0.92$$
$$E_c(m_2, \mathbf{F}_2) = -r_2(B)\log r_2(B) - r_2(\Omega)\log r_2(\Omega)$$
$$= -\frac{1}{3}\log\frac{1}{3} - \frac{2}{3}\log\frac{2}{3} = 0.92$$

Partial Ignorance:

$$I(m_1, \mathbf{F}_1) = m_1(A)s_1(A) + m_1(\Omega)s_2(\Omega)$$
$$= 0.7 \cdot \frac{1}{2} + 0.3 \cdot \frac{2}{2} = 0.65$$
$$I(m_2, \mathbf{F}_2) = m_2(B)s_1(B) + m_2(\Omega)s_2(\Omega)$$
$$= 0.6 \cdot \frac{1}{2} + 0.4 \cdot \frac{2}{2} = 0.7$$

Generalized Entropy:

$$E(m_1, \mathbf{F}_1) = 0.88 + 0.92 + 0.65 = 2.45$$
$$E(m_2, \mathbf{F}_2) = 0.97 + 0.92 + 0.70 = 2.59$$

Now, consider Case 3 in Figure 9-17. The Dempster combination rule yields a new set of focal elements and basic probability assignments as follows:

$$A \cap B = \{None\}$$
$$\mathbf{F} = \{A, B, A \cap B, \Omega\}$$
$$m_{1,2}(A) = 0.18, m_{1,2}(B) = 0.28, m_{1,2}(A \cap B) = 0.42, m_{1,2}(\Omega) = 0.12$$

The entropies and ignorance for the combined set of focal elements are computed as follows:

Belief Entropy:

$$E_b(m_{1,2}, \mathbf{F}) = -0.18\log 0.18 - 0.28\log 0.28$$
$$-0.42\log 0.42 - 0.12\log 0.12 = 1.85$$

Core Entropy:

$$E_c(m_{1,2}, \mathbf{F}) = -\frac{1}{4}\log\frac{1}{4} - \frac{1}{4}\log\frac{1}{4} - \frac{0}{4}\log\frac{0}{4} - \frac{2}{4}\log\frac{2}{4} = 1.5$$

Partial Ignorance:

$$I(m_{1,2}, \mathbf{F}) = 0.18 \cdot \frac{1}{2} + 0.28 \cdot \frac{1}{2} + 0.42 \cdot \frac{0}{2} + 0.12 \cdot \frac{2}{2} = 0.35$$

Generalized Entropy:

$$E(m_{1,2},\mathbf{F}) = 1.85 + 1.50 + 0.35 = 3.70$$

Thus we observe that the relation $E(m_{12},\mathbf{F}) \le E(m_1,\mathbf{F}) + E(m_2,\mathbf{F})$ holds.

9.3.2 Combining Sources of Varying Confidence

Here we present a credibility transformation function (Yager, 2004) for combining sources with various confidences. This approach discounts the evidence with a credibility factor α and distributes remaining evidence $1-\alpha$ equally among the n elements of the frame of discernment. The transformed credibility function takes the following form:

$$m_\alpha(A) = m(A)\cdot\alpha + \frac{1-\alpha}{n}$$

In the Bayesian formalism where each focal element A is a singleton set, we distribute the remaining evidence $1-\alpha$ as per their prior probabilities. In other words, we modify the credibility function as the following:

$$p_\alpha(A) = p(A)\cdot\alpha + p_0(A)\cdot\frac{1-\alpha}{n}$$

When prior probabilities are uniformly distributed among elements of the frame of discernment, this case becomes a special case of the credibility function for the Dempster-Shafer case. Moreover, if the prior probability of A is zero, then it remains zero after the transformation via the credibility function.

Example

$$\Omega = \{GtA, AtA, None\}$$

$$A = \{GtA, AtA\}$$

$$m_1(A) = 0.7$$

If the reliability of the source of m_1 is 0.8, then

$$m_{0.8}(A) = 0.7\cdot 0.8 + \frac{1-0.8}{2} = 0.66$$

$$m_{0.8}(\Omega) = 0.3\cdot 0.8 + \frac{1-0.8}{2} = 0.34$$

9.4 FURTHER READING

More details on the strong junction tree algorithm presented in this chapter can be found in Jensen et al. (1994) and Shenoy (1992). From the historical perspective, Howard and Matheson (1981) first introduced IDs, and their ability to model decision problems with uncertainty by transforming them into decision trees, for a single decision maker. Shachter (1986) described a reduction-based approach for evaluating IDs. More details on symbolic argumentation, and the underlying theoretical foundation, can be found in Fox and Das (2000) and Das (2006, 2008).

Chapter 10

Learning of Fusion Models

This chapter presents technologies for learning and adapting some of the high-level fusion models presented in the previous chapters. We start with learning Naïve Bayesian Classifiers (NBCs) from unstructured text documents, and then present algorithms for learning decision tree (DT) classifiers. Once a DT has been constructed from training data, propositional probabilistic rules can be extracted from the tree. We then move to learning both parameters and structures for Bayesian Belief Networks (BNs) considering cases for complete and incomplete data sets. Finally, we present the Baum-Welch algorithm for learning Hidden Markov Models (HMMs). One needs to have understanding in NBCs (Chapter 5), rules (Chapter 5), BNs (Chapter 6), and HMMs (Chapter 7) before reading their respective learning techniques.

10.1 LEARNING NAÏVE BAYESIAN CLASSIFIERS

This section describes how to learn Naïve Bayesian Classifiers (NBCs) from text documents to be used for, for example, classifying or indexing unstructured HUMINT reports. NBC is also useful for classifying targets as shown in Chapter 5. Learning NBCs from structured relational databases is just a special case of learning from unstructured documents. Each child node in an NBC corresponds to an attribute or feature. For text-based document indexing, these attributes usually indicate the presence or absence of a word in a text document. The parent node determines the output of the discriminant function, which is the class to which the document belongs.

To index items within a document, the child nodes in an NBC correspond to "tokens," or words, in a set of items. The selection of tokens is generally accomplished in three steps:

- *Stop words* are removed from the set of documents. Stop words are those words that carry little semantic content (e.g., a, and, the, that, it).

- Documents are *stemmed*. Stemming refers to a process by which words are replaced with their root; thus raining, rains, and rained would all be replaced with the common root "rain."
- *Tokens* are selected to omit both low-frequency and high-frequency words. This is generally done by computing the TF/IDF (term frequency/inverse document frequency) value for each word, as described in Quinlan (1986). Cutoff values are used to select the set of tokens to be used.

Let D be the set of all training documents that are labeled with members from the set $C = \{C_1, ..., C_n\}$ of classes. Following the above steps, collect all distinct words and other tokens that occur in D and call them $V = \{v_1, ..., v_m\}$. Consider an NBC as shown in Figure 10-1 with root node C whose states are $C_1, ..., C_n$. For each member v_j of V, add a binary child node of C.

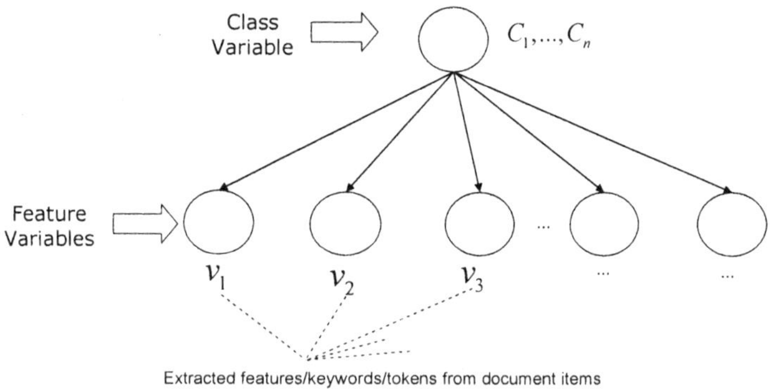

Figure 10-1: Generic structure of an NBC

Compute the following probabilities applying the Maximum Likelihood Estimate (MLE), which is simply the relative frequency and corresponds to the most likely value of each parameter given the training data:

$$p(C_i) = \frac{|\text{Subset of } D \text{ for which the label is } C_i|}{|D|}$$

$$p(v_j \mid C_i) = \frac{\text{Number of times } v_j \text{ occurs in all the training documents labeled } C_i}{\text{Total number of words that occur in all the training documents labeled } C_i}$$

$$= \frac{N(v_j, C_i)}{\sum_j N(v_j, C_i)}$$

where $N(v_j, C_i)$ is the number of times v_j occurs in all the training documents labeled C_i.

The problem with the MLE estimate is that $p(v_j \mid C_i)$ is zero if v_j did not occur in any document labeled C_i in the training data. So a document with occurrences of v_j, and perhaps genuinely belongs to the class C_i, will get a conditional probability of zero for C_i since we're multiplying the conditional probabilities for all words. To avoid this, Laplace smoothing, which simply adds one to each count, is applied as follows:

$$p(v_j \mid C_i) = \frac{1 + N(v_j, C_i)}{\sum_j (1 + N(v_j, C_i))} = \frac{1 + N(v_j, C_i)}{|V| + \sum_j N(v_j, C_i)}$$

Let *d* be a document that we want to classify and the given set of classes is $C = \{C_1, \ldots, C_n\}$. We want to compute $p(C_i \mid d)$, for every *i*.

$$p(C_i \mid d) = \frac{p(C_i)\, p(d \mid C_i)}{p(d)} = \frac{p(C_i) \prod_j p(v_j \mid C_i)}{\sum_{k=1}^{n} p(C_k) \prod_j p(v_j \mid C_k)}$$

where v_js are tokens in document *d*. The product of conditional probabilities comes from the assumption that tokens in a document are independent given the document class. Note that the conditional independence assumption of features does not hold in most cases. For example, word cooccurrence is a commonly used feature for text classification.

We don't need estimated posterior $p(C_i \mid d)$ to be correct. Instead, we only need

$$\arg\max_{C_i} p(C_i \mid d) = \arg\max_{C_i} p(C_i) \prod_j p(v_j \mid C_i)$$

The score for each class can be expressed in the following tractable form for analytical purposes:

$$\log(p(C_i)) + \sum_j \log(p(v_j \mid C_i))$$

The score is not a probability value, but sufficient for the purpose of determining the most probable class. It reduces the round-off errors due to a product of small fractions caused by a large number of tokens.

NBC ignores dependencies among different features. There are variations of NBC that consider the dependence among features. For example, a general framework for limited dependence BNs, called *k*-dependence Bayesian classifiers, has been presented in Sahami (1996). Open source tools exist (e.g., the jNBC toolkit, http://jnbc.sourceforge.com) that allow NBCs to be generated from any labeled training set.

10.2 RULE LEARNING FROM DECISION TREE ALGORITHMS

Learning of decision trees (DTs) is probably the most widely used learning technique. It works by building a DT (a similar structure is shown in Chapter 9 on influence diagrams) from a set of instances where the input decision variables and the correct decision (or classification) is known. Once built, a DT can be reinterpreted as a set of "if-then" rules (syntax as shown in Chapter 5 on target classification) if required. One rule is generated for each leaf node in the tree.

Each node in a DT specifies a test of some attribute and each branch descending from the node corresponds to one possible value of the node. Each leaf node corresponds to a classification. A newly arrived instance is classified by tracking it down the tree from the root to some leaf.

We apply the technique to an example related to the vehicle-journey example adopted in Chapter 6 on BNs. In fact, the DT that we learn here resembles one of the DTs in Chapter 9. We construct a DT to classify circumstances under which a journey took place or canceled based on past journey records in a structure containing visibility, mobility, weather, and ground truth. Table 10-1 presents sample classification performance from the past.

Weather	Visibility	Journey	Mobility
normal	clear	start	go
rainy	poor	start	go
normal	clear	start	slow go
foggy	poor	cancel	slow go
normal	clear	start	no go
foggy	clear	cancel	no go
rainy	clear	start	slow go
foggy	poor	start	slow go
normal	clear	cancel	no go
rainy	poor	cancel	no go
normal	clear	start	go
rainy	poor	cancel	go
foggy	poor	cancel	no go
normal	clear	start	go
normal	clear	start	slow go
foggy	poor	cancel	no go

Table 10-1: Records for classifying actionable hypotheses

Each row in the table represents a classification to a journey whose type is either *start* or *cancel*. Thus the first row is interpreted as a journey that took place on a day when the weather was normal, the visibility was clear, and the mobility was a go. Given these past results, one problem is to predict the journey (i.e., the *Journey* column as the output) given the weather, visibility, and mobility conditions. In other words, the learning algorithm has to discover patterns in the form of rules by learning from the past records as shown in the table. Another problem could be to predict the mobility (i.e., the *Mobility* column as the output) based on weather and visibility information.

10.2.1 Algorithms for Constructing Decision Trees

ID3 (Quinlan, 1986) is a basic DT learning algorithm, and C4.5 (Quinlan, 1993) is an extension of ID3 to account for unavailable values, continuous attribute value ranges, pruning of DTs, rule derivation, and so on. ID3 is a top-down, greedy search algorithm through the space of possible DTs. The algorithm is given below.

ALGORITHM ID3

Inputs

- *Training Set*: The training examples (e.g., the records in Table 10-1).
- *Target Attribute*: The attribute whose value is to be predicted by the learned DT (e.g., *Journey*).
- *Input Attributes*: A list of other attributes that may be tested by the learned DT (e.g., *Visibility*, *Mobility*, and *Weather*).

Output

- A learned DT that classifies the given training set

```
Create a new node Root for the tree;
If all members of Training Set are in the same class C
    Set Root = single node tree with label C;
Else If Input Attributes is empty
    Set Root = single node tree with label as the most common
            value of Target Attribute in Training Set;
Else
    Set A = member of Input Attributes that maximizes information gain
            Gain(Training Set, A) (explained below);
    A is the decision attribute for Root;
    For each possible value v of A
            Add a new branch below Root, testing for A = v;
            Set Training Set(v) = subset of Training Set with A = v;
            If Training Set(v) is empty
                    Add a leaf below the new branch with label that is most
                    common value of Target Attribute in Training Set;
            Else
                    Below the new branch add sub-tree
                            obtained by the recursively call
            ID3(Training Set(v), Target Attribute, Input Attributes - {A});
Return Root;
```

Within the above algorithm, a measure of effectiveness, called *information gain*, of an attribute A is computed via the procedure $Gain(Training\,Set, A)$. The attribute that provides the maximum information gain is placed at the root of the DT. The information gain metric is an information theoretic measure of how much "entropy" is revealed by a specific attribute. Given a collection S of c class labels, the entropy is defined as

$$Entropy(S) = -\sum_{i=1}^{c} p_i \log_2(p_i)$$

where p_i is the proportion of S belonging to class i. The formula for computing information gain for an attribute A with respect to a set of instances S is

$$Gain(S, A) = Entropy(S) - \sum_{v \in Values(A)} \frac{|S_v|}{|S|} Entropy(S_v)$$

where the sum is taken over all possible values of the attribute A, and S_v is the subset of A for which the attribute A has the value v.

Example

For the problem at hand, we have four attributes that are candidates for the root node. The attribute with the highest information gain will be selected. We first consider A as *Visibility*, S as the instances in the table, and compute the information gain for S.

$$Values(Visibility) = \{clear, poor\}$$

$$S = \{start(9), cancel(7)\}$$

$$Entropy(S) = -\frac{9}{16}\log_2\frac{9}{16} - \frac{7}{16}\log_2\frac{7}{16} = 0.99$$

where $start(9)$ means 9 instances of *start* and $cancel(7)$ means 7 instances of *cancel* in the set S. Similarly,

$$S_{clear} = \{start(7), cancel(2)\}$$

$$Entropy(S_{clear}) = -\frac{7}{9}\log_2\frac{7}{9} - \frac{2}{9}\log_2\frac{2}{9} = 0.76$$

$$S_{poor} = \{start(2), cancel(5)\}$$

$$Entropy(S_{poor}) = -\frac{2}{7}\log_2\frac{2}{7} - \frac{5}{7}\log_2\frac{5}{7} = 0.86$$

$$Gain(S,Visibility) = Entropy(S) - \frac{9}{16} Entropy(S_{clear}) - \frac{7}{16} Entropy(S_{poor})$$
$$= 0.99 - \frac{9}{16} \times 0.76 - \frac{7}{16} \times 0.86$$
$$= 0.18$$

Similarly,

$$Gain(S,Weather)$$
$$= Entropy(S) - \frac{7}{16} Entropy(S_{sunny}) - \frac{4}{16} Entropy(S_{rainy}) - \frac{5}{16} Entropy(S_{foggy})$$
$$= Entropy(S) - \frac{7}{16} Entropy\left(-\frac{6}{7}\log_2\frac{6}{7} - \frac{1}{7}\log_2\frac{1}{7}\right) - \frac{4}{16} Entropy\left(-\frac{2}{4}\log_2\frac{2}{4} - \frac{2}{4}\log_2\frac{2}{4}\right) - \frac{5}{16} Entropy\left(-\frac{1}{5}\log_2\frac{1}{5} - \frac{4}{5}\log_2\frac{4}{5}\right)$$
$$= 0.23$$

$$Gain(S,Mobility)$$
$$= Entropy(S) - \frac{5}{16} Entropy(S_{go}) - \frac{5}{16} Entropy(S_{slow\ go}) - \frac{6}{16} Entropy(S_{no\ go})$$
$$= Entropy(S) - \frac{5}{16} Entropy\left(-\frac{4}{5}\log_2\frac{4}{5} - \frac{1}{5}\log_2\frac{1}{5}\right) - \frac{5}{16} Entropy\left(-\frac{4}{5}\log_2\frac{4}{5} - \frac{1}{5}\log_2\frac{1}{5}\right) - \frac{6}{16} Entropy\left(-\frac{1}{6}\log_2\frac{1}{6} - \frac{5}{6}\log_2\frac{5}{6}\right)$$
$$= 0.29$$

The attribute *Mobility* achieves the highest information gain and should be used as the root for the DT. In fact, inspecting the partial trees in Figure 10-2, it makes most sense to include *Mobility* at the root since it discriminates most of the samples split along its branches.

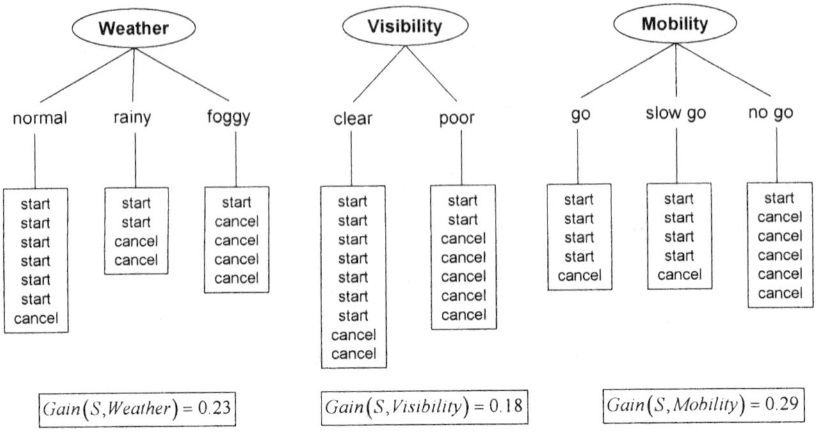

Figure 10-2: Alternative splits from the root node and information gains

Having found the best split, the whole procedure is repeated with each subset of the example training set. As shown in Figure 10-3, the information gain is the same for each of the two attributes *Weather* and *Visibility* and either of them can be used to split the DT further for $Mobility = slow\,go$.

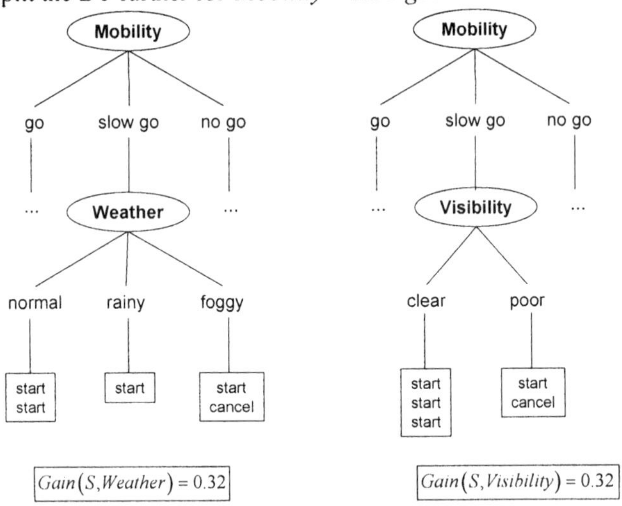

Figure 10-3: Alternative splits and information gains

The splitting process continues for those branches containing samples from different classes (i.e., containing both *start* and *cancel*). One can now easily extract rules from the tree constructed so far. The leftmost branch yields the following rule (assuming that *Visibility* is chosen for splitting the branch for *slow go*):

IF *Mobility* = *slow go*
AND *Visibility* = *clear*
THEN *Journey* = *start*

The support value in this case is 1.0 as no counterexample has been considered for the specific conditions of the rule above, but one can easily extract support from the probability scale [0, 1] by taking the ratio between the number of correct classification cases to the total number of cases. For example, the support value is 0.5 for each of the following two rules:

IF *Mobility* = *slow go*
AND *Visibility* = *poor*
THEN *Journey* = *start* (0.5)

IF *Mobility* = *slow go*
AND *Visibility* = *poor*
THEN *Journey* = *cancel* (0.5)

On the other hand, if the tree is pruned (to be introduced later) after the *Mobility* node as in the rightmost tree in Figure 10-2 then the following two rules are generated:

IF *Mobility* = *slow go*
THEN *Journey* = *start* (0.8)

IF *Mobility* = *slow go*
THEN *Journey* = *cancel* (0.2)

Information gain has a natural bias for attributes with a large domain (i.e., the "date" attribute). This can result in selecting an attribute that works well with training data but will not generalize. One alternative approach is to use the following *GainRatio* function (Quinlan, 1986).

$$SplitInformation(S, A) = -\sum_{i=1}^{c} \frac{|S_i|}{|S|} \log_2 \frac{|S_i|}{|S|}$$

$$GainRatio(S, A) = \frac{Gain(S, A)}{SplitInformation(S, A)}$$

where S_i is the subset of S for which A has value v_i.

10.2.2 Overfitting in Decision Trees

During the construction of a DT from a training data set, there is always a danger that the process is learning only the characteristics of the specific training set, as opposed to the general concepts involved. If that is indeed the case then the tree "overfits" to the training set, that is, the tree has been specialized to the training set as opposed to what will occur generally. Formally, suppose H is the hypotheses set of all possible DTs. Then a hypothesis h in H *overfits* training data if there is an alternative hypothesis h_1 in H such that h has a smaller error than h_1 over the training data set, but h_1 has a smaller error than h over the entire distribution of instances.

There are two ways to prevent overfitting:

- Stop growing when splitting data is no more statistically significant.
- Grow a full tree and then post-prune to remove the parts that overfit.

In the first case, all available data for training are used and a statistical test is applied to estimate whether pruning or expanding is likely to produce improvement. In the second case, a separate test data set is used to evaluate the utility of post-pruning. There are two types of pruning:

- *Reduced Error Pruning*: Decision nodes are pruned from the full tree. Pruning a node removes the sub-tree rooted at the node and makes it a leaf node. The most common classification of the training examples associated with the node is assigned to the node. The resulting pruned tree has to perform no worse than the original tree over the test set.
- *Rule Postpruning*: Convert the full tree into equivalent set of rules and then prune each rule by removing preconditions that result in improving its estimated accuracy over the test set.

Note that an advantage of converting a DT to rules before pruning is that the distinction between attribute tests that occur near the root of the tree and those that occur near the leaves is removed.

10.2.3 Handling Continuous Attributes

For a classification problem with only discrete attributes, each nonleaf node of a constructed DT is a test and its number of descendants is equal to the domain size of the attribute (i.e., number of possible values for the discrete attribute). But for a continuous attribute A, its range is partitioned into a discrete set of intervals. A new Boolean attribute A_c is created to test if the value of A is less that a threshold c or not ($A < c$). To choose c, instances of A are sorted first and then the boundaries where the output or target classification changes are identified. Candidate thresholds are then generated between two boundaries. The information gains for different choices of thresholds are computed.

Example

Consider the records in Table 10-2 created with the discrete attributed *Visibility* and *Journey* columns from Table 10-1 and adding the continuous counterpart of *Visibility*. Table 10-3 is the sorted version of the table based on the continuous attribute *Visibility* (*C*).

Visibility	Visibility (C)	Journey
clear	5000	start
poor	400	start
clear	3000	start
poor	50	cancel
clear	1000	start
clear	750	cancel
clear	800	start
poor	100	start
clear	600	cancel
poor	40	cancel
clear	4500	start
poor	200	cancel
poor	25	cancel
clear	4500	start
clear	2500	start
poor	30	cancel

Table 10-2: Continuous attribute *Visibility* (*C*) of its discrete counterpart in Table 10-1

Visibility (C)	25	30	40	50	100	200	400	600	750	800	1000	2500	3000	4500	4500	5000
Visibility	p	p	p	p	p	p	p	c	c	c	c	c	c	c	c	c
Journey	c	c	c	c	s	c	s	c	c	s	s	s	s	s	s	s

Table 10-3: Sorting by the continuous attribute in Table 10-2

The discrete values of *Visibility* have been obtained from its continuous counterpart based on the threshold 500 ft (i.e., visibility is less than 500 ft is poor, otherwise clear). Now if the continuous attribute *Visibility* (*C*) is given instead of the discrete attribute *Visibility* then, for example, a list of thresholds between boundaries where the target classification value of the attribute *Journey* changes is 50, 100, 200, 400, and 750. Therefore, the test intervals for the new discretized attribute are:

$$Visibility_c = \begin{cases} Visibility < 50 \\ 50 \le Visibility < 100 \\ 100 \le Visibility < 200 \\ 200 \le Visibility < 400 \\ 400 \le Visibility < 750 \\ Visibility > 750 \end{cases}$$

The information gains for these choices of thresholds are then computed.

10.2.4 Advantages and Disadvantages of Decision Tree Techniques

The advantages of DT techniques are the following:

- Employ a fast learning process;
- Generate intuitive rules in natural language-like syntax;
- Handle both continuous and discrete variables;
- Indicate clearly which fields are most important for prediction.

The disadvantages of DT techniques are the following:

- Not suitable for prediction tasks where the output attribute is continuous;
- Linear rectangular classifier and does not handle nonlinear data sets well;
- No backtracking in core ID3 during the construction of a tree,

10.3 BAYESIAN BELIEF NETWORK LEARNING

Modeling of a high-level fusion problem based on Bayesian Belief Networks (BNs) requires a specification of the network structure and a set of probabilities (or parameters) for each node (or variable), called a Conditional Probability Table (CPT). Therefore, the task of defining a BN model can be divided into two subtasks: 1) defining or learning the structure of the network, and 2) defining or learning the set of probabilities. Usually, learning structure is much harder than learning probabilities. However, a structure is much easier to elicit from experts than probabilities. Eliciting probabilities is especially difficult for a node with a large number of parents (in which case its CPT becomes very complex). Also, since the world is rarely fully observable, we must consider cases where we have to derive a BN structure when some of the nodes are hidden, or some data are missing. Deriving the structure is much harder when everything we wish to model is only partially observed.

In general, the framework for learning structure and/or probabilities for BNs given some data set gives rise to the following four cases:

- *Known Structure and Fully Observable Variables*: The network structure is given and assumed to be complete, and the given data set is complete in the sense that it includes observations of all the variables in the network. The goal here is to learn the probabilities of the variables from the data set.
- *Known Structure and Partially Observable Variables*: The network structure is given and assumed to be complete, and the data set includes observations of some (but not all) of the variables in the network. The variables whose observations are not specified in the data set are called *hidden variables*. The goal here is to learn the probabilities of the network from the data set, in spite of the missing data.
- *Unknown Structure and Fully Observable Variables*: Neither the network structure nor the probabilities of the network are known. But the network that is to be learned will contain only those variables that are in the data set; there are no hidden variables. The goal here is to learn both the structure and the probabilities of the network.

- *Unknown Structure and Partially Observable Variables*: Neither the network structure nor the probabilities of the network are known, and the network may contain hidden variables that are not in the data set. The goal here is to learn both the structure and the probabilities of the network.

In the rest of this section, we summarize the most widely used techniques for deriving BNs with only discrete variables (a discrete variable can take on one of a limited number of values, which are often referred to as the *states* of that variable). These techniques are summarized in Table 10-4:

Structure ⇒ Observability ⇓	Known Network Structure	Unknown Network Structure
Fully Observable Variables	**Parameter Estimation** Maximum Likelihood (ML) Estimation Bayesian Dirichlet (BD)	**Discrete Search** Entropy, Bayesian, or MDL Score-Based (e.g., K2) Dependency Constraint-Based
Partially Observable Variables	**Parametric Optimization** Expectation Maximization (EM) Gibbs' Sampling Gradient Descent-Based (e.g., APN)	**Search and Optimizaton** Expectation Maximization (EM) (e.g., Structural EM)

Table 10-4: Cases for learning BNs

10.3.1 Learning Probabilities: Brief Survey

Learning probabilities from a complete data set is discussed in Spiegelhalter and Lauritzen (1990) and Buntine (1991). Lauritzen (1995) describes how to apply the Expectation Maximization (EM) algorithm (Dempster et al., 1977) to BNs. Learning using Gibbs' sampling (Geman and Geman, 1984) is discussed in Thomas et al. (1992) and Gilks et al. (1996). Learning probabilities from incomplete data using gradient methods is discussed by Thiesson (1995), Russel et al. (1995), and Binder et al. (1997).

10.3.2 Learning Probabilities from Fully Observable Variables

A common approach here is to use the Maximum Likelihood (ML) estimates for the probabilities, which reduces to simple counting of relative frequencies of the values of the variable (Spiegelhalter and Lauritzen, 1990). To illustrate the technique, let $\mathbf{X} = \{X_1,...,X_n\}$ (or $(X_1,...,X_n)$ based on an ordering) be the set of all variables in the given network structure S. The vector $\theta_S = (\theta_1,...,\theta_n)$ denotes the probability distribution in the structure and its corresponding vector of variables is $\Theta_S = (\Theta_1, ..., \Theta_n)$. Each X_i is discrete with states $x_i^1, ..., x_i^{r_i}$, and the variable θ_{ijk} represents a specific CPT entry, the probability that variable X_i takes on its k-th possible value assignment given that its parents pa_i take on their j-th possible value assignment. Therefore,

$$\sum_{k=1}^{r_i} \theta_{ijk} = 1$$

where r_i is the number of states in the variable X_i. To learn the probabilities given the structure S and data set $\mathbf{D} = \{D_1, ..., D_N\}$, we need to compute the posterior distribution $p(\theta_S \mid \mathbf{D}, S)$. Two assumptions are made to compute this distribution. The first assumption is that the data set $\mathbf{D}$ is complete and the second assumption is that the parameter vectors $\theta_{ij} = (\theta_{ij1}, ..., \theta_{ijr_i})$ are mutually independent as follows:

$$p(\theta_S \mid S) = \prod_{i=1}^{n} \prod_{j=1}^{q_i} p(\theta_{ij} \mid S)$$

where q_i is the number of possible parent states for the node X_i. θ_{ij} are the parameters for the local likelihood function. These parameters remain independent given the data set $\mathbf{D}$ as follows:

$$p(\theta_S \mid \mathbf{D}, S) = \prod_{i=1}^{n} \prod_{j=1}^{q_i} p(\theta_{ij} \mid \mathbf{D}, S)$$

Therefore, the parameters θ_{ij} can be updated independently. Often, a Dirichlet distribution is used to estimate these parameters. In general, the probability density of the Dirichlet distribution for variables $\mathbf{p} = (p_1, ..., p_n)$ with parameters $\mathbf{u} = (u_1, ..., u_n)$ is defined by

$$p(\mathbf{p}) = Dirichlet(\mathbf{p}; \mathbf{u}) = \frac{\Gamma\left(\sum_{i=1}^{n} u_i\right)}{\prod_{i=1}^{n} \Gamma(u_i)} \prod_{i=1}^{n} p_i^{u_i - 1}$$

where each $p_i \geq 0$, $\sum_{i=1}^{n} p_i = 1$, and each $u_i > 0$ is interpreted as the prior observation counts for events governed by p_i. Γ represents the Gamma function, which satisfies $\Gamma(n+1) = n\Gamma(n)$ and $\Gamma(1) = 1$. In our case, assume that each vector θ_{ij} has a Dirichlet distribution with the prior observation counts α_{ijk}, that is,

$$p(\theta_{ij} \mid S) = c \cdot \prod_{k=1}^{r_i} \theta_{ijk}^{\alpha_{ijk} - 1}$$

where c is a normalization constant. If N_{ijk} is the number of times in **D** the variable X_i takes on its k-th possible value assignment $X_i = v_{ik}$ given that its parents pa_i take on their j-th possible value assignment, then

$$p(\theta_{ij} \mid \mathbf{D}, S) = c \cdot \prod_{k=1}^{r_i} \theta_{ijk}^{\alpha_{ijk} + N_{ijk} - 1}$$

Therefore, learning probabilities from a complete data set is based on simple counting of relative frequencies of occurrences of the values of the variable.

10.3.3 Learning Probabilities from Partially Observable Variables

Various approaches exist to maximize the posterior distribution $p(\mathbf{D} \mid \theta_S, S)$, the probability assigned by the network to the observed data set when the CPT parameters are set to θ_S. The Adaptive Probabilistic Network (APN) (Russel et al., 1995) is an approach that uses a gradient descent algorithm for maximization. A widely studied stochastic method is Gibbs' sampling, which can be used to approximate any function of an initial joint distribution $p(\mathbf{X})$, provided certain conditions are met. An alternative approximation algorithm is the EM algorithm, which can be viewed as a deterministic version of Gibbs' sampling.

The gradient-based algorithms view the probability $p(\mathbf{D} \mid \theta_S, S)$ as a function of the CPT entries θ_S, thereby reducing the probability learning problem to one of finding the maximum of a multivariate nonlinear function. Equivalently, the log-likelihood function of $p(\mathbf{D} \mid \theta_S, S)$ is maximized via the gradient ascent approach where a small step is taken to the direction of the gradient to the point $\theta_S + \alpha \Delta \theta_S$, for some step-size parameter α.

The gradient-ascent approach is modified to take into account the constraint that θ_S consists of conditional probability values $\theta_{ijk} \in [0,1]$ and $\sum_k \theta_{ijk} = 1$. The contribution of each data case D_l to the gradient can be computed separately, and the results are summed as illustrated below:

$$\frac{\partial \ln p(\mathbf{D} \mid \theta_S, S)}{\partial \theta_{ijk}}$$

$$= \frac{\partial \ln \prod_{l=1}^{N} p(D_l \mid \theta_S, S)}{\partial \theta_{ijk}} \qquad \text{(data points are independent)}$$

$$= \sum_{l=1}^{N} \frac{\partial \ln p(D_l \mid \theta_S, S)}{\partial \theta_{ijk}} = \sum_{l=1}^{N} \frac{\dfrac{\partial p(D_l \mid \theta_S, S)}{\partial \theta_{ijk}}}{p(D_l \mid \theta_S, S)}$$

$$= \sum_{l=1}^{N} \frac{\frac{\partial}{\partial \theta_{ijk}} \left(\sum_{j'k'} p(D_l \mid x_{ik'}, pa_{ij'}, \theta_S, S) p(x_{ik'} \mid pa_{ij'}, \theta_S, S) p(pa_{ij'}, \theta_S, S) \right)}{p(D_l \mid \theta_S, S)}$$

$$= \sum_{l=1}^{N} \frac{p(D_l \mid x_{ik}, pa_{ij}, \theta_S, S) p(pa_{ij}, \theta_S, S)}{p(D_l \mid \theta_S, S)}$$

$$\left[\theta_{ijk} \text{ appears once and } \theta_{ijk} = p(x_{ik'} \mid pa_{ij'}, \theta_S, S) \right]$$

$$= \sum_{l=1}^{N} \frac{p(x_{ik}, pa_{ij} \mid D_l, \theta_S, S) p(D_l \mid \theta_S, S) p(pa_{ij}, \theta_S, S)}{p(x_{ik}, pa_{ij}, \theta_S, S) p(D_l \mid \theta_S, S)}$$

$$= \sum_{l=1}^{N} \frac{p(x_{ik}, pa_{ij} \mid D_l, \theta_S, S)}{\theta_{ijk}}$$

Each term $\frac{p(x_{ik}, pa_{ij} \mid D_l, \theta_S, S)}{\theta_{ijk}}$ in the summation gets computed in a network when the evidence D_l is posted. In a junction tree implementation of evidence propagation (discussed in Chapter 6), a node and its parents always appear together in at least one clique (*cliques* are groupings of nodes formed when performing graph transformations on a BN, as called for by the junction tree algorithm), and the required probability can be found by marginalizing the other variables in the clique.

To approximate $p(\theta_S \mid \mathbf{D}, S)$ via Gibbs' sampling, first the values of θ_S and the states of unobserved variables in $\mathbf{D}$ are initialized, producing a complete data set $\mathbf{D}'$. Then, for each variable X_i in D_l that is not observed in the original data set $\mathbf{D}$, that variable's state is reassigned according to the distribution $p(x_{il} \mid \mathbf{D}' \setminus x_{il} \mid \theta_S, S)$. This produces another complete dataset $\mathbf{D}''$. Next, the posterior distribution $p(\theta_S \mid \mathbf{D}'', S)$ is computed as in the case of learning probabilities using a complete data set, and finally the values of θ_S are reassigned. These two steps are iterated to compute the average of $p(\theta_S \mid \mathbf{D}'', S)$ as an approximation.

10.3.4 Online Adjustment of Parameters

Unlike the EM technique, there is no iterative process for convergence in online learning and adaptation of BN parameters. Here, for evidence at time t, the algorithm is run over the network nodes once and the CPTs are updated with the new values.

$$\theta_{ijk}^{t+1} = \eta \frac{p_{\theta_t}(x_{ik} pa_{ij} \mid z_t)}{p_{\theta_t}(pa_{ij} \mid z_t)} + (1 - \eta) \theta_{ijk}^{t}$$

where

- θ_{ijk}^{t} represents the state of a specific CPT entry at the current time t;
- z_t is evidence at time t;
- η is the learning rate (e.g., values of 0.05 and lower); and
- $p_{\theta_t}(.)$ represents the probability value at time t.

10.3.5 Brief Survey of Structure Learning

Approaches to learning the structure of a BN can broadly be classified into two categories:

- A *dependency analysis*-based approach captures dependencies among nodes, and more importantly, leaves unconnected nodes that are independent of each other. Some of the approaches in this category are in Srinivas et al. (1990), Fung and Crawford (1990), Spirtes et al. (1990), and Verma and Pearl (1992).
- A *searching and scoring*-based approach defines a "score" that describes the fitness of each possible searched structure to the observed data. Commonly used scores include Entropy (Chow and Liu, 1968; Rebane and Pearl, 1987; Herskovits and Cooper, 1990), Bayesian (Cooper and Herskovits, 1992; Heckerman et al., 1995; Singh and Valtorta, 1995), and MDL (Lam and Bacchus, 1994; Suzuki, 1996; Friedman and Goldszmidt, 1996).

In cases when the number of variables is large, the dependency analysis-based methods are usually more efficient. However, when the sample size is small and the data are noisy, the searching- and scoring-based algorithms can often give more accurate results via an exhaustive search of the whole model space to find the optimal model. There are also algorithms that can handle data sets with missing values (Ramoni and Sebastiani, 1997; Singh, 1997; Friedman, 1998).

The notion of equivalence of network structures is important and should be kept in mind when pruning the search for structures. Consider the networks in Figure 10-4 constructed from the domain of three variables X, Y, and Z. All are equivalent in that they can capture the same set of data, given appropriate conditional probability tables.

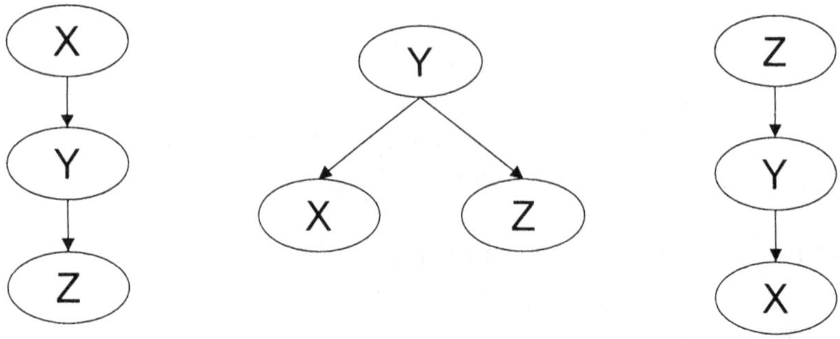

Figure 10-4: Equivalent network structures

The joint probability distributions decompose respectively as the following:

$$p(X,Y,Z) = p(X)\,p(Y \mid X)\,p(Z \mid Y)$$
$$p(X,Y,Z) = p(Y)\,p(X \mid Y)\,p(Z \mid Y)$$
$$p(X,Y,Z) = p(Z)\,p(Y \mid Z)\,p(X \mid Y)$$

Applying Bayes' rule repeatedly, we can derive the following:

$$\begin{aligned} p(X)\,p(Y \mid X)\,p(Z \mid Y) &= p(Y,X)\,p(Z \mid Y) = p(Y)\,p(X \mid Y)\,p(Z \mid Y) \\ &= p(Z,Y)\,p(X \mid Y) = p(Z)\,p(Y \mid Z)\,p(X \mid Y) \end{aligned}$$

The above network structures are shown to be distribution equivalent, as the variable probabilities must be equal at any state of information. The methods for learning equivalent classes of network structures can be found in Chickering (1995).

It is also important here to contrast between statistical correlation and causal semantics in BNs to illustrate ambiguities in BN learning. The BN fragment in Figure 10-5 illustrates that the presence of certain assets will cause the generation of SIGINT and IMINT reports. The two children nodes SIGINT and IMINT are correlated but one does not cause the other.

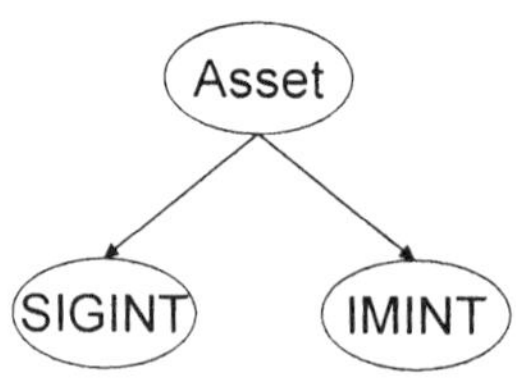

Figure 10-5: Correlated SIGINT and IMINT nodes

Figure 10-6 shows two data sets and their associated BN structures. The data set on the left seems to support both the structures (rain causes poor visibility, and vice versa), but only the first relationship is a causal one that is more apparent in the second data set.

Rain	Visibility
yes	clear
yes	poor
yes	poor
yes	poor
yes	poor
no	poor
no	clear

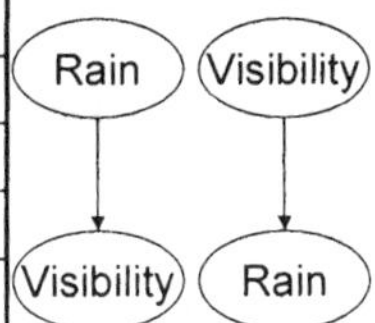

Rain	Visibility
no	clear
yes	poor
yes	poor
yes	poor
yes	poor
no	poor
no	clear

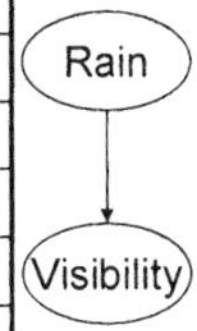

Figure 10-6: BNs representing data sets

These ambiguities can be resolved effectively in a mixed-initiative learning system with human-in-the-loop.

10.3.6 Learning Structure from Fully Observable Variables

The technique proposed in Cooper and Herskovits (1992) to learn structure from a complete data set uses a greedy search algorithm through a space of possible BN structures. Aliferis and Cooper (1994) evaluate the accuracy of K2, a specific instantiation of a greedy search algorithm. To select the most probable structure S of conditional dependencies among the variables in a complete data set $\mathbf{D}$ of m independent cases $D_1, ..., D_m$, the selection of the structure is based on the evaluation of

$$p(S, \mathbf{D}) = c \cdot p(S) p(\mathbf{D} \mid S)$$

where $p(\mathbf{D} \mid S)$ is the marginal likelihood of $\mathbf{D}$ given S, derived in Cooper and Herskovits (1992), as follows:

$$p(\mathbf{D} \mid S) = \prod_{i=1}^{n} \prod_{j=1}^{q_i} \frac{\Gamma(\alpha_{ij})}{\Gamma(\alpha_{ij} + N_{ij})} \cdot \prod_{k=1}^{r_i} \frac{\Gamma(\alpha_{ijk} + N_{ijk})}{\Gamma(\alpha_{ijk})}$$

The above marginal likelihood is also known as the Bayesian-Dirichlet scoring function. Suppose that the possible models are equally likely a priori, and that the user can formulate a partial ordering of the variables so that if $X_i \prec X_j$ then X_i cannot be a parent of X_j. Let pa_i be the current set of parents of the node X_i. Then the local contribution of a node X_i and its parents to the joint probability $p(S, \mathbf{D})$ is measured by the local marginal likelihood as follows:

$$g(X_i, pa_i) = \prod_{j=1}^{q_i} \frac{\Gamma(\alpha_{ij})}{\Gamma(\alpha_{ij} + N_{ij})} \cdot \prod_{k=1}^{r_i} \frac{\Gamma(\alpha_{ijk} + N_{ijk})}{\Gamma(\alpha_{ijk})}$$

where $N_{ij} = \sum_{k=1}^{r_i} N_{ijk}$. If we assume a uniform distribution of the prior observation counts (i.e., $\alpha_{ijk} = c$, for some c) then the above expression can be simplified to

$$g(X_i, pa_i) = \prod_{j=1}^{q_i} \frac{(r_i - 1)!}{(N_{ij} + r_i - 1)!} \cdot \prod_{k=1}^{r_i} N_{ijk}!$$

Here we describe how this expression is derived. If P is the vector whose values denotes the conditional probability assignments associated with the BN structure S, and f is the conditional density function over P given S then

$$\begin{aligned} p(S,\mathbf{D}) &= \int_P p(\mathbf{D} \mid P, S) f(P \mid S) p(S) dP \\ &= p(S) \int_P p(\mathbf{D} \mid P, S) f(P \mid S) dP \\ &= p(S) \int_P \left[\prod_{l=1}^{N} p(D_l \mid P, S) \right] f(P \mid S) dP \\ &= p(S) \int_P \left[\prod_{l=1}^{N} \prod_{i=1}^{n} p\left(X_i = D_{il} \mid P, pa_{i\sigma(i,l)}\right) \right] f(P \mid S) dP \end{aligned}$$

where pa_{ij} is the j-th possible value assignment of parents pa_i, and $\sigma(i,l)$ is an index function such that the instantiation of parents in case l is the $\sigma(i,l)$-th possible value assignment of pa_i. Suppose the possible values of variable X_i are $v_{i1}, \ldots, v_{ir_i}$, then we can rewrite the above equation by grouping terms as

$$p(S,\mathbf{D}) = p(S) \int_P \left[\prod_{i=1}^{n} \prod_{j=1}^{q_i} \prod_{k=1}^{r_i} p\left(X_i = v_{ik} \mid P, pa_{ij}\right)^{N_{ijk}} \right] f(P \mid S) dP$$

Recall that N_{ijk} is the number of times in $\mathbf{D}$ the variable X_i takes on its k-th possible value assignment $X_i = v_{ik}$ given that its parents pa_i take on their j-th possible value assignment. Since, $f\left(p\left(X_i \mid pa_{ij}\right)\right)$ is independent of $f\left(p\left(X_{i'} \mid pa_{i'j'}\right)\right)$, we have

$$f(P \mid S) = \prod_{i=1}^{n} \prod_{j=1}^{q_i} f\left(\theta_{ij1}, \ldots, \theta_{ijr_i}\right)$$

Therefore,

$$\begin{aligned} p(S,\mathbf{D}) &= p(S) \int_{\theta_{111}} \cdots \int_{\theta_{ijk}} \cdots \int_{\theta_{nq_n r_n}} \left[\prod_{i=1}^{n} \prod_{j=1}^{q_i} \prod_{k=1}^{r_i} \theta_{ijk}^{N_{ijk}} \right] \\ &\qquad \left[\prod_{i=1}^{n} \prod_{j=1}^{q_i} f\left(\theta_{ij1}, \ldots, \theta_{ijr_i}\right) \right] d\theta_{111} \ldots \theta_{ijk} \ldots \theta_{nq_n r_n} \\ &= p(S) \prod_{i=1}^{n} \prod_{j=1}^{q_i} \int_{\theta_{ij1}} \cdots \int_{\theta_{ijr_i}} \left[\prod_{k=1}^{r_i} \theta_{ijk}^{N_{ijk}} \right] f\left(\theta_{ij1}, \ldots, \theta_{ijr_i}\right) d\theta_{ij1} \ldots \theta_{ijr_i} \\ &= p(S) \prod_{i=1}^{n} \prod_{j=1}^{q_i} C_{ij} \int_{\theta_{ij1}} \cdots \int_{\theta_{ijr_i}} \prod_{k=1}^{r_i} \theta_{ijk}^{N_{ijk}} d\theta_{ij1} \ldots \theta_{ijr_i} \end{aligned}$$

where the distribution $f\left(\theta_{ij1},...,\theta_{ijr_i}\right)$ is assumed to be uniform and is equal to some constant C_{ij}. Since, $f\left(\theta_{ij1},...,\theta_{ijr_i}\right)$ is a probability density function

$$\int_{\theta_{ij1}}...\int_{\theta_{ijr_i}} C_{ij} d\theta_{ij1}...\theta_{ijr_i} = 1$$

$$\text{i.e., } C_{ij} = \frac{1}{\int_{\theta_{ij1}}...\int_{\theta_{ijr_i}} d\theta_{ij1}...\theta_{ijr_i}}$$

But we have the following Dirichlet's integral:

$$\int_{\theta_{ij1}}...\int_{\theta_{ijr_i}} \prod_{k=1}^{r_i} \theta_{ijk}^{\ N_{ijk}} d\theta_{ij1}...\theta_{ijr_i} = \frac{\prod_{k=1}^{r_i} N_{ijk}!}{\left(N_{ij}+r_i-1\right)!}$$

Substituting, $N_{ijk}=0$, we obtain

$$\int_{\theta_{ij1}}...\int_{\theta_{ijr_i}} d\theta_{ij1}...\theta_{ijr_i} = \frac{1}{\left(r_i-1\right)!}, \text{ that is, } C_{ij} = \left(r_i-1\right)!$$

Therefore,

$$p\left(S,\mathbf{D}\right) = p\left(S\right)\prod_{i=1}^{n}\prod_{j=1}^{q_i}\frac{\left(r_i-1\right)!}{\left(N_{ij}+r_i-1\right)!}\prod_{k=1}^{r_i} N_{ijk}!$$

Example

Consider the data set **D** in Table 10-5 to illustrate the computation of $p\left(S,\mathbf{D}\right)$.

Weather	Visibility	Journey
sunny	clear	start
rainy	poor	start
sunny	clear	start
sunny	clear	start
rainy	clear	start
sunny	clear	cancel
rainy	poor	cancel
sunny	clear	start
rainy	poor	cancel
sunny	clear	start
sunny	clear	start

Table 10-5: Records for deriving BN structures

We consider the two structures S_1 and S_2 as shown in Figure 10-7.

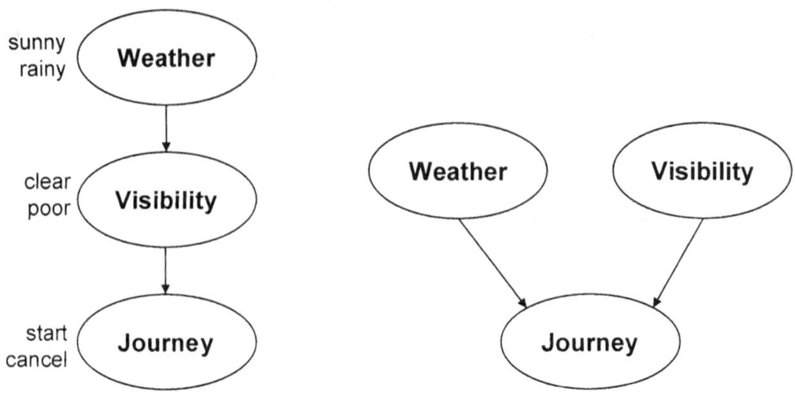

Figure 10-7: Possible BN structures for records in Table 10-4

$$n = 3$$
$$X_1 = Weather, X_2 = Visibility, X_3 = Journey$$

There are 25 possible structures and thus $p(S_1) = p(S_2) = \frac{1}{25}$. The probabilities of the two structures $p(S_1, \mathbf{D})$ and $p(S_2, \mathbf{D})$ are computed as follows:

$$q_1 = 0, q_2 = 2, q_3 = 2$$
$$r_1 = 2, r_2 = 2, r_3 = 2$$

$$N_{211} = 7; N_{212} = 0; N_{221} = 1; N_{222} = 3; N_{311} = 7; N_{312} = 1; N_{321} = 1; N_{322} = 2$$
$$N_{21} = 7; N_{22} = 4; N_{31} = 8; N_{32} = 3$$

$$\begin{aligned} p(S_1, \mathbf{D}) &= p(S_1) \prod_{i=1}^{n} \prod_{j=1}^{q_i} \frac{(r_i - 1)!}{(N_{ij} + r_i - 1)!} \prod_{k=1}^{r_i} N_{ijk}! \\ &= \frac{1}{25} \frac{(2-1)!\,7!\,0!}{(7+2-1)!} \frac{(2-1)!\,1!\,3!}{(4+2-1)!} \frac{(2-1)!\,7!\,1!}{(8+2-1)!} \frac{(2-1)!\,1!\,2!}{(3+2-1)!} \\ &= 0.3 \times 10^{-8} \end{aligned}$$

$$q_1 = 0, q_2 = 0, q_3 = 4$$
$$r_1 = 2, r_2 = 2, r_3 = 2$$

$$N_{311} = 6; N_{312} = 1; N_{321} = 0; N_{322} = 0; N_{331} = 1; N_{332} = 0; N_{341} = 1; N_{342} = 2$$
$$N_{31} = 7; N_{32} = 0; N_{33} = 1; N_{34} = 3$$

$$p(S_2, \mathbf{D}) = p(S_2) \prod_{i=1}^{n} \prod_{j=1}^{q_i} \frac{(r_i - 1)!}{(N_{ij} + r_i - 1)!} \prod_{k=1}^{r_i} N_{ijk}!$$
$$= \frac{1}{25} \frac{(2-1)!\,6!\,1!}{(7+2-1)!} \frac{(2-1)!\,0!\,0!}{(0+2-1)!} \frac{(2-1)!\,1!\,0!}{(1+2-1)!} \frac{(2-1)!\,1!\,2!}{(3+2-1)!}$$
$$= 0.3 \times 10^{-6}$$

Therefore, the second structure is 100 times more likely than the first structure.

K2 Algorithm

The K2 algorithm proceeds by adding one parent at a time and computing $g(X_i, pa_i)$. The set pa_i is expanded to include the parent nodes that give the largest contribution to $g(X_i, pa_i)$ and stops inclusion when the probability settles.

Input

- An ordered set of nodes $X_1, ..., X_n$, such that if $i < j$ $(1 \le i, j \le n)$ then a structure with X_j as an ancestor of X_i is not allowed.
- An upper bound u on the number of parents a node can have.
- A data set **D** of m cases.

Output

- A BN structure specifying the parents, if any, of each node

For each node $X_i, i = 1, ..., n$, perform the following steps:

  Set $pa_i = \varnothing$, where pa_i is the set of parents of the node X_i.

  Set $g_{old} = g(X_i, pa_i)$.

  While $|pa_i| < u$, perform the following steps:

    Set node X from $\{X_1, ..., X_{i-1}\} - pa_i$ that maximizes $g(X_i, pa_i \cup \{X\})$.

    If $g(X_i, pa_i \cup \{X\}) > g_{old}$ then

      $g_{old} = g(X_i, pa_i \cup \{X\})$

      $pa_i = pa_i \cup \{X\}$

    Else

      Break out of the while loop

  End

End

10.3.7 Learning Structure from Partially Observable Variables

The most difficult case occurs when the structure is unknown and there are hidden variables and/or missing data. Most of the above techniques could be adapted to discover variables, but with a high search cost. The hidden nodes have to be marginalized to compute the Bayesian score. Since this is usually intractable, it is common to use an asymptotic approximation to the posterior called BIC (Bayesian Information Criterion), which is easy to use and does not require the evaluation of prior distribution. Friedman (1998) developed a Structural EM algorithm that combines the standard EM algorithm to optimize parameters with a structure search for model selection. Structural EM learns networks based on penalized likelihood scores, which include the BIC/MDL score.

10.3.8 Use of Prior Knowledge from Experts

There are approaches to include prior knowledge from experts in BN structure learning to reduce the search space of all possible models. The simplest choice is a uniform prior over structures (Heckerman, 1996), but an expert can provide structural knowledge of the following kind (Srinivas et al., 1990) that can constrain the search space:

- Declaring that a variable must be a root node;
- Declaring that a variable must be a leaf node;
- Declaring that one variable must be a parent or ancestor of another node;
- Declaring conditional independence among variables.

A common constraint is a total ordering on variables (Buntine, 1991; Cooper and Herskovits, 1992). As an alternative to total ordering on variables, Hojsgaard and Thiesson (1995) use a partition of the available variables into "blocks" provided by an expert. Another common constraint is to limit the number of parents any variable may have (Cooper and Herskovits, 1992; Friedman and Koller, 2003).

10.4 BAUM-WELCH ALGORITHM FOR LEARNING HMM

Given an HMM $M = \langle S, O, A, B, \Pi \rangle$ and an observation sequence $Y_1 ... Y_T$, we want to re-estimate M as $\bar{M} = \langle S, O, \bar{A}, \bar{B}, \bar{\Pi} \rangle$ so that $\bar{M}$ is more likely than M in the sense that $p(Y_1 ... Y_T \mid M) > p(Y_1 ... Y_T \mid \bar{M})$. In other words, the new model is more likely to produce the given observation sequence. This is done via an iterative Baum-Welch or forward-backward algorithm.

The forward algorithm considered a forward variable $\alpha_t(s_i)$ as defined below:

$$\alpha_t(s_i) = p(Y_1 ... Y_t, X_t = s_i)$$

The forward variable represents the joint probability of the partial observation sequence $Y_1 ... Y_t$ and state s_i at time t. The forward algorithm computes the forward variable iteratively. In a similar manner, we define a backward variable $\beta_t(s_i)$ as follows:

$$\beta_t(s_i) = p(Y_{t+1}...Y_T \mid X_t = s_i)$$

The backward variable represents the probability of the partial observation sequence from $t+1$ to the end given the state at time point t is s_i. The variable is inductively computed as follows:

Step 1: Initialization

$$\beta_T(s_i) = 1 \qquad 1 \le i \le n$$

Step 2: Induction

$$\beta_t(s_i) = \sum_{j=1}^{n} a_{ij} b_j(Y_{t+1}) \beta_{t+1}(s_j) \qquad t = T-1, T-2, ..., 1;\ 1 \le i \le n$$

Figure 10-8 illustrates the computation of a backward variable.

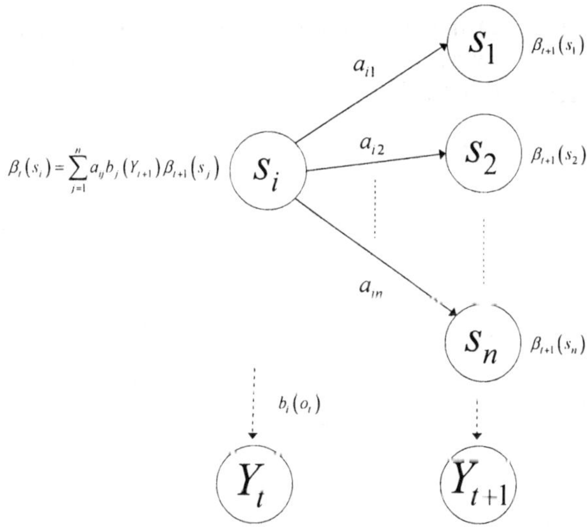

Figure 10-8: Illustration of the computation of a backward variable

Given the sequence $Y_1...Y_T$ of observations and any two states s_i and s_j, the re-estimation $\bar{a}_{ij}$ of the parameter is simply the ratio of the expected number of transitions from state s_i to state s_j and the expected number of transitions from state s_i as a whole. Therefore, we define the following variable representing the posterior probability of transition from state s_i to state s_j given the observation sequence:

$$\xi_t(s_i, s_j) = p(X_t = s_i, X_{t+1} = s_j \mid Y_1...Y_T)$$

Then the expected number of transitions from state s_i has to be computed for every time point between 1 and $T-1$ as follows:

$$\text{Expected number of transitions from state } s_i \text{ to state } s_j = \sum_{t=1}^{T-1} \xi_t(s_i, s_j)$$

In a similar argument,

$$\text{Expected number of transitions from state } s_i = \sum_{t=1}^{T-1}\sum_{j=1}^{n}\xi_t(s_i,s_j)$$

Application of Bayes' rule converts $\xi_t(s_i,s_j)$ as follows:

$$\xi_t(s_i,s_j) = \frac{p(X_t = s_i, X_{t+1} = s_j, Y_1...Y_T)}{p(Y_1...Y_T)}$$

$$= \frac{p(X_t = s_i, X_{t+1} = s_j, Y_1...Y_T)}{\sum_{i=1}^{n}\sum_{j=1}^{n} p(X_t = s_i, X_{t+1} = s_j, Y_1...Y_T)}$$

Now the computation of the numerator of the above expression, as illustrated in Figure 10-9, is the following:

$$p(X_t = s_i, X_{t+1} = s_j, Y_1...Y_T)$$

$$= p(Y_1...Y_t, X_t = s_i)a_{ij}b_j(Y_{t+1})p(Y_{t+2}...Y_T \mid X_{t+1} = s_j)$$

$$= \alpha_t(s_i)a_{ij}b_j(Y_{t+1})\beta_{t+1}(s_j)$$

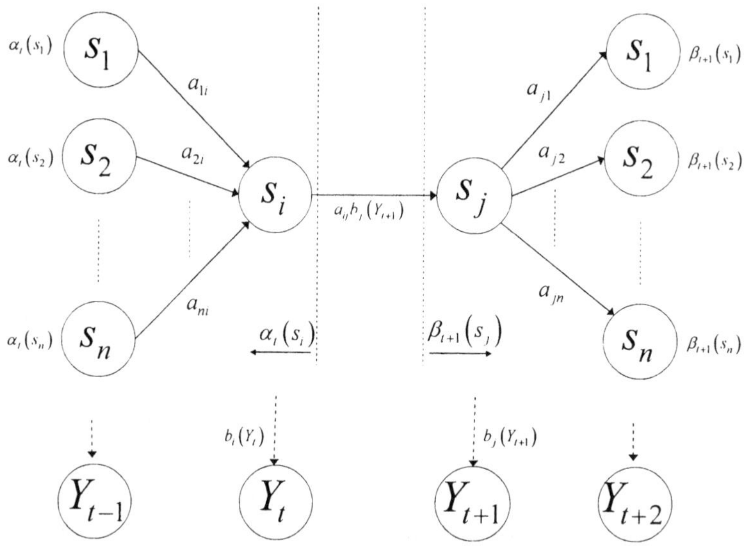

Figure 10-9: Computation of the joint probability of being in state s_i at time t and state s_j at time $t+1$

Therefore,

$$\xi_t(s_i, s_j) = \frac{\alpha_t(s_i) a_{ij} b_j(Y_{t+1}) \beta_{t+1}(s_j)}{\sum_{i=1}^{n}\sum_{j=1}^{n} \alpha_t(s_i) a_{ij} b_j(Y_{t+1}) \beta_{t+1}(s_j)}$$

Define the following variable $\gamma_t(s_i)$ as the expected number of transitions from the state s_i at time t:

$$\gamma_t(s_i) = \sum_{j=1}^{n} \xi_t(s_i, s_j)$$

The re-estimated model parameters are the following:

$\overline{\pi}_i$ = expected number of times in state s_i at the state (i.e., t = 1)

$= \gamma_1(s_i) \qquad 1 \le i \le n$

$$\overline{a}_{ij} = \frac{\text{expected number of transitions from state } s_i \text{ to state } s_j}{\text{expected number of transitions from state } s_i}$$

$$= \frac{\sum_{t=1}^{T-1} \xi_t(s_i, s_j)}{\sum_{t=1}^{T-1}\sum_{j=1}^{n} \xi_t(s_i, s_j)} = \frac{\sum_{t=1}^{T-1} \xi_t(s_i, s_j)}{\sum_{t=1}^{T-1} \gamma_t(s_i)}$$

$$\overline{b}_{s_i}(o_k) = \frac{\text{expected number of times in state } s_i \text{ and observing symbol } o_k}{\text{expected number of times in state } s_i}$$

$$= \frac{\sum_{t=1}^{T-1} \gamma_t(s_i, o_k)}{\sum_{t=1}^{T-1} \gamma_t(s_i)}$$

where $\gamma_t(s_i, o_k)$ is $\gamma_t(s_i)$ for the observation sequence $Y_1 \ldots Y_{t-1} o_k Y_t \ldots Y_T$.

The probability of the observation sequence can be improved by recursively re-estimating the parameters $\overline{A}, \overline{B}, \overline{\Pi}$ until some limiting point is reached.

10.4.1 Generalized Baum-Welch Algorithm for HHMM

Similar to the definition of the forward variable α, we consider the backward variable β as defined below:

$\beta(t, t+k, s_k^{d-1}, s_i^d)$ = the probability that the partial observation sequence $Y_t \ldots Y_{t+k}$ was generated by state s_k^{d-1}, given the sequence $Y_1 \ldots Y_T$ of observations, and that s_i^d started at time t and s_k^{d-1} finished at $t+k$ during the generation of the sequence. By definition,

$$\beta\left(t,t+k,s_k^{d-1},s_i^d\right)=p\left(Y_t...Y_{t+k} \mid X_t^d=s_i^d,X_{t+k}^{d-1}=s_k^{d-1}\right)$$

Now $\beta(.)$ can be recursively computed as below:

$$\beta\left(t,t,s_k^{D-1},s_i^D\right)=b^{Di}\left(Y_t\right)a_{ie}^{(D-1)k}$$

$$\beta\left(t,t+\tau,s_k^{D-1},s_i^D\right)=b^{(D-1)i}\left(Y_t\right)\left(\sum_{j\neq e}^{\left|s_k^{D-1}\right|}a_{ij}^{(D-1)k}\beta\left(t+1,t+\tau,s_k^{D-1},s_j^D\right)\right)$$

$$\beta\left(t,t,s_k^{d-1},s_i^d\right)=\left(\sum_{j=1}^{\left|s_i^d\right|}\pi_j^{di}\beta\left(t,t,s_i^d,s_j^{d+1}\right)\right)a_{ie}^{(d-1)k}$$

$$\beta\left(t,t+\tau,s_k^{d-1},s_i^d\right)=\sum_{l=0}^{\tau-1}\left(\sum_{j=1}^{\left|s_i^d\right|}\pi_j^{di}\beta\left(t,t+l,s_i^d,s_j^{d+1}\right)\right)$$
$$\left(\sum_{j=1}^{\left|s_k^{d-1}\right|}a_{je}^{di}\beta\left(t+l+1,t+\tau,s_k^{d-1},s_j^d\right)\right)$$
$$+\left(\sum_{j=1}^{\left|s_i^d\right|}\pi_j^{di}\beta\left(t,t+\tau,s_i^d,s_j^{d+1}\right)\right)a_{ie}^{(d-1)k}$$

To take into account stochastic vertical transitions that recursively generate observations, we need to consider, in addition to generalized backward and forward variables, variables χ and ξ to correspond, respectively, to horizontal and vertical transitions at different levels.

$\xi\left(t,s_k^{d-1},s_i^d,s_j^d\right)$ represents, given the sequence $Y_1...Y_T$ of observations, the probability of performing a horizontal transition from s_i^d to s_j^d at time t after the production of Y_t and before the production of Y_{t+1}, where s_i^d and s_j^d are substates s_k^{d-1}. Therefore, the expected number of horizontal transitions from s_i^d to s_j^d is $\sum_{t=1}^{T}\xi\left(t,s_k^{d-1},s_i^d,s_j^d\right)$.

$\chi\left(t,s_k^{d-1},s_i^d\right)$ represents, given the sequence $Y_1...Y_T$ of observations, the probability that the state s_k^{d-1} was entered at time t before Y_t was generated, and activated state s_i^d. Therefore, the expected number of vertical transitions from s_k^{d-1} to s_i^d is $\sum_{t=1}^{T}\chi\left(t,s_k^{d-1},s_i^d\right)$.

Therefore, the HHMM parameters are re-estimated as the following:

$\hat{\pi}_i^{(d-1)k}$ = probability that s_k^{d-1} will initally activate s_i^d

$$= \frac{\text{expected number of vertical transitions from } s_k^{d-1} \text{ to } s_i^d}{\text{expected number of vertical transitions from } s_k^{d+} \text{ to any of its sub-states in Level } d}$$

$$= \frac{\sum_{t=1}^{T} \chi\left(t, s_k^{d-1}, s_i^d\right)}{\sum_{l=1}^{\left|s_k^{d-1}\right|} \sum_{t=1}^{T} \chi\left(t, s_k^{d-1}, s_l^d\right)} \qquad (1 < d < D)$$

Note that $\hat{\pi}_i^{11}$ is $\chi\left(1, s_1^1, s_i^2\right)$, since there is only one state at Level 1 and nothing can be generated before entering into Level 2. So, $\sum_{t=1}^{T} \chi\left(t, s_k^{d-1}, s_i^d\right)$ is $\chi\left(1, s_1^1, s_i^2\right)$ and, consequently, the denominator of $\hat{\pi}_i^{(d-1)k}$ is $\sum_{i=1}^{\left|s_k^{d-1}\right|} \chi\left(1, s_1^1, s_i^2\right)$, that is, 1.

$\hat{a}_{ij}^{(d-1)k}$ = probability of making a horizontal transition at depth d from state s_i^d to state s_j^d, both are sub-states of s_k^{d-1}

$$= \frac{\text{expected number of horizontal transitions from } s_i^d \text{ to } s_j^d}{\text{expected number of horizontal transitions from } s_i^d \text{ to any of its neighboring states in Level } d}$$

$$= \frac{\sum_{t=1}^{T} \xi\left(t, s_k^{d-1}, s_i^d, s_j^d\right)}{\sum_{l=1}^{\left|s_k^{d-1}\right|} \sum_{t=1}^{T} \xi\left(t, s_k^{d-1}, s_i^d, s_l^d\right)} \qquad (1 < d < D)$$

$\hat{b}^{Di}\left(o_j\right)$ = Probability of observing o_j at state s_i^D, for some $1 \le j \le T$

$$= \frac{\begin{array}{l}\text{expected number of vertical transitions at any time } t \text{ from} \\ \text{any state } s_k^{D-1} \text{ to } s_i^D \text{ with } o_j = Y_t + \text{expected number of} \\ \text{horizontal transitions from any state } s_k^D \text{ to } s_i^D \text{ with } o_j = Y_t\end{array}}{\begin{array}{l}\text{expected number of vertical transitions at any time } t \text{ from} \\ \text{any state } s_k^{D-1} \text{ to } s_i^D + \text{expected number of horizontal} \\ \text{transitions at any time } t \text{ from any state } s_k^D \text{ to } s_i^D\end{array}}$$

$$= \frac{\sum_{t=1}^{T} \left(\chi\left(t, s_k^{d-1}, s_i^d\right)\right)_{o_j = Y_t} + \sum_{l=1}^{\left|s_k^{d-1}\right|} \sum_{t=2}^{T} \left(\xi\left(t-1, s_k^{D-1}, s_l^D, s_i^D\right)\right)_{o_j = Y_t}}{\sum_{t=1}^{T} \chi\left(t, s_k^{d-1}, s_i^d\right) + \sum_{l=1}^{\left|s_k^{d-1}\right|} \sum_{t=2}^{T} \xi\left(t-1, s_k^{D-1}, s_l^D, s_i^D\right)} \qquad (1 < d < D)$$

Note that a horizontal transition from state s_l^D to state s_i^D cannot happen before $t=2$ as a vertical transition to s_l^D is required at $t=1$. We need the following two η variables to compute the above ξ and χ variables.

$\eta_{in}\left(t, s_k^{d-1}, s_i^d\right)$ = Probability that $Y_1 ... Y_{t-1}$ was generated by s_k^{d-1} and then s_i^d was activated at t

$= p\left(Y_1 ... Y_{t-1}, s_i^d \text{ started at } t\right)$

$\eta_{out}\left(t, s_k^{d-a}, s_i^d\right)$ = Probability that s_i^d was the last state activated by s_k^{d-1} before generating $Y_{t+1} ... Y_T$

$= p\left(s_i^d \text{ finished at } t, Y_{t+1} ... Y_T\right)$

Now $\eta_{in}(.)$ can be recursively computed as follows:

$$\eta_{in}\left(1, s_1^1, s_i^2\right) = \pi_{2i}^{11}$$

$$\eta_{in}\left(t, s_1^1, s_i^2\right) = \sum_{j=1}^{\left|s_1^1\right|} \alpha\left(1, t-1, s_1^1, s_j^2\right) a_{ji}^{11} \qquad (1 < t \le T)$$

$$\eta_{in}\left(1, s_k^{d-1}, s_i^d\right) = \eta_{in}\left(1, s_l^{d-2}, s_k^{d-1}\right) \pi_{di}^{(d-1)k}$$

$$\eta_{in}\left(t, s_k^{d-1}, s_i^d\right) = \sum_{t'=1}^{t-1} \eta_{in}\left(t', s_l^{d-2}, s_k^{d-1}\right) \left(\sum_{j=1}^{\left|s_k^{d-1}\right|} \alpha\left(t', t-1, s_k^{d-1}, s_j^d\right) a_{ji}^{(d-1)k} \right)$$

$$+ \eta_{in}\left(t, s_l^{d-2}, s_k^{d-1}\right) \pi_{di}^{(d-1)k} \qquad (1 < t \le T)$$

Now $\eta_{out}(.)$ can be recursively computed as follows:

$$\eta_{out}\left(t, s_1^1, s_i^2\right) = \sum_{j=1}^{\left|s_1^1\right|} a_{ij}^{11} \beta\left(t+1, T, s_1^1, s_j^2\right) \qquad (1 < t \le T)$$

$$\eta_{out}\left(t, s_k^{d-1}, s_i^d\right) = \sum_{t'=t+1}^{T} \left(\sum_{j=1}^{\left|s_k^{d-1}\right|} a_{ji}^{(d-1)k} \beta\left(t+1, t', s_k^{d-1}, s_j^d\right) \right) \eta_{out}\left(t', s_l^{d-2}, s_k^{d-1}\right)$$

$$+ a_{ie}^{(d-1)k} \eta_{out}\left(t, s_l^{d-2}, s_k^{d-1}\right) \qquad (1 < t \le T)$$

$$\eta_{out}\left(T, s_k^{d-1}, s_i^d\right) = a_{ie}^{(d-1)k} \eta_{out}\left(T, s_l^{d-2}, s_k^{d-1}\right)$$

Finally, $\xi(.)$ and $\chi(.)$ are recursively computed as below in terms of $\eta_{in}(.)$ and $\eta_{out}(.)$:

$$\xi\left(t,s_1^1,s_i^2,s_j^2\right)=\frac{\alpha\left(1,t,s_1^1,s_i^2\right)a_{ij}^{11}\beta\left(t+1,T,s_1^1,s_j^2\right)}{p\left(Y_1...Y_t\right)}$$

$$\xi\left(T,s_1^1,s_i^2,s_j^2\right)=\frac{\alpha\left(1,T,s_1^1,s_i^2\right)a_{ij}^{11}}{p\left(Y_1...Y_t\right)}$$

$$\xi\left(t,s_k^{d-1},s_i^d,s_j^d\right)=\frac{1}{p\left(Y_1...Y_t\right)}\left(\sum_{t'=1}^{t}\eta_{in}\left(t',s_{k'}^{d-2},s_k^{d-1}\right)\alpha\left(t',t,s_k^{d-1},s_i^d\right)\right)$$
$$a_{ij}^{(d-1)k}\left(\sum_{t'=t+1}^{T}\beta\left(t+1,t',s_k^{d-1},s_j^d\right)\eta_{out}\left(t',s_{k'}^{d-2},s_k^{d-1}\right)\right)$$

$$\xi\left(t,s_k^{d-1},s_i^d,s_e^d\right)=\frac{1}{p\left(Y_1...Y_t\right)}\left(\sum_{t'=1}^{t}\eta_{in}\left(t',s_{k'}^{d-2},s_k^{d-1}\right)\alpha\left(t',t,s_k^{d-1},s_i^d\right)\right)$$
$$a_{ie}^{(d-1)k}\eta_{out}\left(t,s_{k'}^{d-2},s_k^{d-1}\right)$$

$$\chi\left(1,s_1^1,s_i^2\right)=\frac{\pi_i^{11}\beta\left(1,T,s_1^1,s_i^2\right)}{p\left(Y_1...Y_t\right)}$$

$$\chi\left(t,s_k^{d-1},s_i^d\right)=\frac{1}{p\left(Y_1...Y_t\right)}\left(\sum_{t'=1}^{t}\eta_{in}\left(t',s_{k'}^{d-2},s_k^{d-1}\right)\alpha\left(t',t,s_k^{d-1},s_i^d\right)\right)a_{ij}^{(d-1)k}$$
$$\left(\sum_{t'=t+1}^{T}\beta\left(t+1,t',s_k^{d-1},s_j^d\right)\eta_{out}\left(t',s_{k'}^{d-2},s_k^{d-1}\right)\right)$$

10.5 FURTHER READING

Mitchell's book (1997) is a good introduction to DT learning, and, more generally to machine learning. The book discusses inductive logic programming, another learning technique that can extract rules in first-order logic syntax. For surveys on BN learning, please consult Buntine (1996), Heckerman (1996), and Krause (1998). Neapolitan (2003) discusses the Bayesian method for learning the values of discrete and continuous parameters, and both the Bayesian and constraint-based methods for learning structure. For a discussion on HMM and HHMM learning, see Rabiner (1989) and Fine et al. (1998), respectively.

Chapter 11

Towards Cognitive Agents for Data Fusion

Agent technology is increasing in importance as it naturally fits within modern Network Centric Warfare (NCW) environments for autonomous local processing at the sensor level to save bandwidth, information filtering for human operators to avoid the information overloading problem, and collaborative situation and threat assessments for shared situational awareness. This chapter presents agent technology from the architectural perspective. We describe a cognitive agent architecture with human behavioral elements such as perception, goal generation, situation assessment, hypotheses generation, and action planning and scheduling.

We describe *Envelope*, an adaptive, symbolic, and hybrid cognitive architecture. Envelope integrates reactive and skill-based human behavior, which relies primarily on working memory, with conscious and knowledge-based problem-solving behavior, which relies heavily on long-term memory. We compare Envelope's foundational assumptions with those of well-known architectures in the field of cognitive science. Envelope relaxes various lower-level psychological constraints and becomes logically omniscient in its inference-based implementation. On the other hand, Envelope imposes somewhat high-level constraints on some aspects of intelligence in its use of hierarchical task and goal decompositions. This difference makes Envelope especially suitable for producing intelligent agents for a variety of goal-oriented information processing tasks involving large volumes of data.

The specific implementation framework described here includes a high-level knowledge representation language for specifying tasks, goals, and knowledge, and various logical and probabilistic techniques discussed in this book for realizing the desired behavior of the architectural components. The generic nature of Envelope ensures its applicability across a variety of fusion-related application domains. Envelope's generality is tested via its use in various battlefield information processing tasks, including situation assessment, decision making, and planning for generating suitable Courses of Action (COAs).

11.1 MOTIVATION AND SCOPE

An *architecture* of a system specifies its key modules and their interactions. *Cognitive architectures* are developed for systems to model human intelligence where the key modules are those supporting cognition, including memory, visual, perceptual, and motor systems.

In an endeavor to develop a general architecture for supporting "practical" systems to be deployed within battlefield information processing environments, we did not seek to provide an accurate quantitative model of human perceptual and cognitive capabilities. Such an environment contains thousands of sensors and generates hundreds of thousands of records every hour, but fast reaction, goal-oriented problem solving, and planning are some of the key behavioral elements required of the analysts working in the environment to perform various tasks. Such tasks include filtering out relevant information from large streams to cope with information overload, and the generation of suitable COAs that require exploration and evaluation of a large number of options. So a system based on Envelope would probably filter out thousands of irrelevant records each second, a speed that is out of reach of human beings, but the stages the system deploys for filtering out an individual record will be the same as those taken by a human analyst.

Therefore, what we need is an architecture that supports fatigueless, emotionless processing of information that is orders of magnitude faster than human analysts in terms of its sensing and cognitive processing, but which also creates a computational account of its human-like behavior in terms of the stages it invokes for cognitive processing. Thus, we purposely ignored some of the low-level constraints posed by human cognition (e.g., visual, auditory, motor, memory, cognition) and other factors (e.g., emotion, fatigue, attention) affecting cognition, but adopted some high-level representational constraints (e.g., hierarchical goal and task decomposition) and parallelism in processing wherever needed. A better term for the developed architecture might be "agent architecture," which would include both those systems that make an explicit attempt to model human intelligence and those that simply explore some aspects of the general intelligent behavior of an integrated system capable of supporting intelligence. Consequently, Envelope is less relevant to various human computer interaction experiments that require simulation of behavior as close to human as possible.

Envelope is essentially classical or *symbolic* in nature (e.g., Newell, 1990; Lewis, 1999) in that understanding cognition involves operations in symbols, as opposed to the *connectionist* approach (e.g., Rumelhart, 1984) where such a system exhibits intelligent behavior without storing, retrieving, or otherwise operating on structured symbolic expressions (see Fodor and Pylyshyn (1988) for a detailed comparison). Nor does Envelope relate to *macrocognition*, which studies cognition in naturalistic contexts (Klein et al., 2003). Rather, we assign semantic content to expressions, and assume not only causal relations among the semantically evaluable objects, but also a range of other structural relations. The choice of one over the other is not about representational issues related to the faithfulness of cognitive behavior, but more to do with some practical issues in terms of transparency in the human behavior modeling demanded by the specific application domains considered.

To put the development in perspective, we first present the architecture in detail in the following section, using examples drawn from the data fusion domain. Then we carry out a comprehensive analysis of the architecture with respect to other well-established work in the field, including: 1) GOMS (Card et al., 1983), Soar (Laird et al., 1987; Lehman et al., 2005), ACT-R (Anderson, 1996; Anderson et al., 2004), and EPIC (Kieras and Meyer, 1997) from classical cognitive architectures to simulate human behavior; 2) horizontally and vertically layered architectures (Ferguson, 1992), PRODIGY (Minton, 1990), BDI (Rao and Georgeff, 1991), Subsumption (Brooks, 1986), and ICARUS (Langley et al., 1991) that have been developed to produce intelligent agents; and 3) various C4I architectures and frameworks, including OODA and the JDL model presented in Chapter 1. The study reveals the distinctness of Envelope in terms of representation, handling uncertainty, and practicality. We then present an implementation framework for Envelope that combines multiple AI technologies, including production rules and Bayesian Belief Networks (BNs) (Chapter 6) for implementing skill-based reactive behavior, modal logical inferencing for reasoning with beliefs and goals (Das, 2008), Influence Diagrams and Symbolic Argumentation (Chapter 9) for resolving options, and planning.

11.2 ENVELOPE MODEL OF HUMAN COGNITION

Envelope is along the line of a *unified theory of cognition* (Newell, 1990), which means trying to find a set of computationally realizable mechanisms and structures that can answer all the questions we might want to ask about cognitive behavior, such as problem solving, decision making, memorization, and learning. The symbolic approach of Envelope ensures that the underlying content (note that, behavior = architecture + content) can be elicited by a set of symbols and their abstractions. We view the architecture as a theory of what is common across cognitive behaviors:

- It is goal-oriented; specifically the commander's overall goal is to achieve his/her high-level mission objectives.
- It is placed in a rich, complex, detailed environment, such as a battlefield environment, where there is a huge number of actors (e.g., warfighters) and objects (e.g., friendly and enemy sensor assets and weapon systems, the characteristics of these objects, and their actions), any of which may be key to understanding how to achieve mission objectives.
- It requires a large amount of knowledge, for example, the commander draws on a vast amount of knowledge to describe all the actors and objects we know and their interrelationships.

- It requires the use of symbols and abstractions to facilitate a computationally realizable mechanism that can answer all relevant questions. Since every battlefield environment presents new challenges, the commander can draw from his/her previous experience only by abstracting away the current situation.
- It is flexible, and a function of the environment, for example, the commander responds to his/her own perception of the current battlefield environment to incorporate changes in situations and threats as events occur.
- It requires learning from the environment and experience.

A somewhat different and more elaborated set of essential desiderata for developing cognitive architectures has been described in Sun (2004), which additionally includes reactivity and modularity. Sun also views that "future cognitive architecture development should take an integrative approach, incorporating as much as possible various prior perspectives, ..."

The architecture Envelope (dubbed metaphorically as Envelope due to its shape as an envelope, which contains the main components of cognitive activities) is shown in Figure 11-1. Two different general behaviors are captured: the top half of the architecture represents reactive, skill-based behavior; the bottom half represents deliberative, knowledge-based problem-solving behavior. The behavioral components are connected by directed edges along the flow of information. Envelope is an extension of the generic *domino model* (Fox and Das, 2000) with the reactive capability based on its perception of the environment.

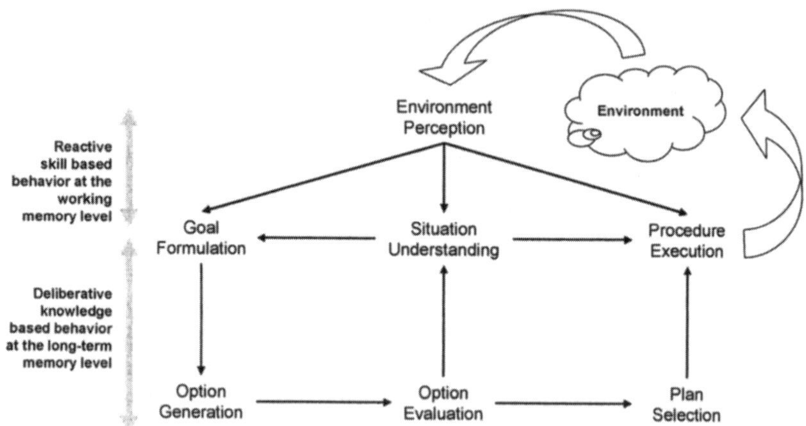

Figure 11-1: The Envelope architecture

The perception of the environment takes place at the *Environment Perception* component, and as a result symbolic abstractions of perceived objects are generated. A battlefield environment is perceived via its distributed sensor assets, and the results of perception are sensor reports about various friendly, hostile, and neutral units, and about terrain, weather, and so on. Only the environment can influence the behavior of this perception component. Consequently, there is no incoming arrow to this component in the architecture.

An implementation of any of the components makes use of the underlying knowledge stored in both long- and short-term memories. Therefore, if one wants to implement specific goal-driven perception and attention then the necessary logic should be encoded into the Environment Perception component. Envelope's long-term memory contains knowledge about ways to argue, plan, and act to achieve goals. But an agent architecture also requires short-term memory containing dynamic beliefs and intentions. Like declarative memory in ACT-R, Envelope's short-term memory is simply the "active" portion of its long-term memory.

Perceived objects are simultaneously fed into three components: Goal Formulation, Situation Understanding, and Procedure Execution. The *Goal Formulation* component generates one or more goals to be achieved based on the situation. For example, within a time-sensitive targeting domain a report corresponding to a perceived moving unit generates a goal for the commander of identifying the unit immediately. This type of goal generation is essentially reactive in nature based on the presence of, for example, a production rule in the working memory:

IF Suspected Object Reported
THEN Goal is to Identify the Object

Note that we are not limiting ourselves only to "deductive" production rules to generate reactive, skill-based, unconscious behavior. Their use here is just for the purpose of illustration. Case-based reasoning (Leake, 1996) or analogical reasoning (Sowa and Majumdar, 2003) is probably a better framework to emulate this kind of behavior, as it takes into account situation context better than simple Boolean matching of rule conditions. More complex reactive behavior can be emulated via BNs for both deductive and diagnostic reasoning, or even neural networks at the sub-symbolic level. We will discuss this issue in detail later.

Perceived objects may also cause execution of a procedure in the *Procedure Execution* component. Such a procedure is either an atomic action or a series of atomic actions that comprise a well-practiced procedure. For example, within a MOUT (Military Operation in Urban Terrain) environment the direct view of a moving unit by a soldier triggers a firing action or a series of actions involving hiding, positioning, and then firing. This process is also reactive in nature based on the presence of, for example, a production rule in the working memory:

IF Moving Object in View
THEN Fire at Object

Perceived objects can also generate further abstractions in the *Situation Understanding* component. For example, while looking at the display, the commander might infer the presence of a platoon-sized enemy armor unit based on the presence of certain other observables, such as tanks and soldiers that are in a certain spatial distribution. Here the abstraction representing inferred "platoon-sized enemy armor unit" is not directly observable. Each abstraction itself causes further action (e.g., zooming in on the screen) or generates further goals (e.g., destroy the unit) following one of the two outward arrows from the Situation Understanding component of the architecture.

Once a goal is generated by the Goal Formulation component it is then fed to the *Option Generation* component. This component generates various options or hypotheses through each of which the goal can be achieved. Generated options are then evaluated by the *Option Evaluation* component to analyze their merits. When a goal is to identify an enemy unit then the options are, for example, SAM, TEL, Tank, etc. These options are so-called "properties" of the unit (Allen, 1984). The Option Evaluation component selects the most suitable option to pass it to the Situation Understanding component, causing an update to the working memory. On the other hand, if the goal is to destroy the enemy then options are various plans such as ground attack and air attack, each of which helps to achieve the goal. In this case the Option Evaluation component sends the best plan to the Plan Generation component.

The *Plan Generation* component hierarchically breaks down a selected plan into sub-plans and actions, where each action is either atomic or a series of actions. In general, we call them *procedures* when they are grounded with concrete temporal intervals. The generated procedures are executed in the environment by the Procedure Execution component, thus changing the state of the environment. For example, a ground attack plan will involve surveillance, massing the troops, and then firing, along with continuous assessment of the situation. The surveillance sub-plan might involve multiple requests for national and theater assets, assets deployment and management, and intelligence collection.

11.3 COMPARATIVE STUDY

11.3.1 Classical Cognitive Architectures and Envelope

The earliest work that had a significant impact on the subsequent direction of cognitive psychology in general and cognitive architectures in particular is the *General Problem Solver* (GPS) (Ernst and Newell, 1969; Newell and Simon, 1972). GPS is a theory of human problem solving stated in the form of a simulation program. GPS analyzes differences between the current state and the goal state. It chooses operators to reduce the differences between these states until there were no differences remaining. The fundamental premise of GPS is that all cognitive activities are interpreted in terms of searching a problem space. Like GPS and many of its variants that were developed subsequently, Envelope's problem solving involves search and is goal-oriented.

The cognitive structure of GOMS (Card et al., 1983), which is based on GPS, consists of four components: 1) a set of Goals, 2) a set of Operators, 3) a set of Methods for achieving the goals, and 4) a set of Selection rules for choosing among competing methods. In the current logical implementation of Envelope, the rules of inference take the form of operators, and methods are just the underlying knowledge by which the problem-solving activities are carried out. Moreover, Envelope, like GOMS, explicitly encodes hierarchical task decompositions (Dean et al., 1988), starting with a top-level task goal, plus a number of methods, or plans, for achieving various types of goals and subgoals. Each goal's plan specifies a series of actions (called operators) invoking subgoals or primitive actions to complete the goal. Envelope, like GOMS, imposes high-level constraints on aspects of intelligence in its use of hierarchical goal decomposition. On the other hand, ACT-R and Soar impose somewhat lower-level constraints (detailed below). If there were to be such constraints on Envelope, they would be integrated in the upper half of the Envelope in the figure.

Three of the most prominent cognitive architectures, namely Soar, EPIC (Executive-Process/Interactive Control), and ACT-R (Adaptive Control of Thought – Rational), are production systems (see Byrne (2001) for a detailed comparison).

In Soar's decision cyclic operation, all productions that match the current contents of declarative memory fire, causing changes in declarative memory, and this continues until no new productions fire. Soar models human performance via low-level productions constituting a decision cycle of 50 ms. An impasse occurs if Soar's decision process is unable to select an action unambiguously among the set of actions the decision cycle proposes. Soar sets up a goal of resolving the impasse. Soar creates a new production rule whenever an impasse is resolved. Soar maintains an explicit goal stack. The actions that are possible at each step are attached to the state corresponding to each goal. Soar creates subgoals to select an operator when more than one is possible.

EPIC (Kieras and Meyer, 1997) was explicitly designed to pair high-fidelity models of perception and motor mechanisms with a production system. All the processors (visual, cognitive, motor) in EPIC run in parallel with one another. EPIC's parallel Cognitive Processor fires multiple rules on a production cycle. Unlike Envelope, EPIC has no planning and goals for choosing between actions and therefore explicit control is necessary to avoid such conflicts. EPIC has no in-built learning mechanism.

The ACT-R cognitive architecture contains two separate but interacting knowledge stores. The declarative knowledge is made up of chunks of symbolic information, and encodes simple facts (e.g., the hill is in the south of the town), current goals (e.g., secure the hill), and even ephemeral situational information (e.g., an enemy unit is present on the hill). The procedural knowledge is made up of production rules that manipulate declarative knowledge as well as the environment via actions. When all the conditions of a production rule (which specifies the current goal and a chunk from declarative memory) match and the rule fires, the result is an alteration to declarative memory, setting of a new current goal, and/or issuing of perceptual or motor commands. One of the most important constraints for the ACT-R model is that it can only "think" one thing at a time. Since ACT-R will only fire one production per cycle, when multiple productions match existing conditions, a conflict resolution process (which is a subsymbolic process) is needed to select the one that will fire. This means that the cognitive processor is serial but the perceptual and motor processes can run in parallel with cognition. ACT-R also forbids action parallelism, and uses a Bayesian mechanism to select one preferred action.

Imposing ACT-R-type constraints on Envelope amounts to forbidding parallel running of the processes implementing the arrows. On the other hand, an agent version of Envelope can, for example, generate goals, situations, and actions from a single perception following the three arrows in the reactive part of Envelope. Also, the deliberative component will try to achieve several goals in parallel, provided updates are synchronized with those goals that have dependencies on the updates. Envelope does not impose any constraint on action parallelism since the Plan Generation component will produce effective synergistic procedures (a procedure is a sequence of actions) for the Procedure Execution component to be executed in the environment in parallel (i.e., two independent actions to be executed in parallel) if needed.

Other notable and relevant mentions of classical cognitive architectures are 3CAPS, Construction-Integration, DEM, COGNET, and CLARION. For example, 3CAPS (Capacity-Constrained Collaborative Activation-based Production System) (Just and Carpenter, 1992) is also a production system where capacity-constrained pools of activation cause slower processing, thus producing the effect of individual difference.

Rasmussen's three-tier model of human information processing (Rasmussen, 1983, 1986) along with its overall similarity to Envelope is shown Figure 11-2. See Section 1.2 of Chapter 1 for more details on the model.

The deliberative portion of Envelope clearly corresponds to Rasmussen's knowledge-based processing, but Envelope also provides a systematic set of steps to achieve goals that are raised in the context. Moreover, the Situation Understanding and Procedure Execution components of Envelope, together with their interactions, produce the effect of rule-based processing behavior in Rasmussen's model. In the Situation Understanding component, Envelope does not confine itself only to the use of production rules but can also use more general well-practiced inference procedures implemented in BNs (examples provided later). Generated higher-level situations are then passed onto the Procedure Execution component to trigger procedures of action to be executed. In Envelope, skill-based processing not only refers to physical actions (occurring within the Procedure Execution component) but also cognitive processing for generating implicit properties (occurring within the Situation Understanding component) of the underlying knowledge base via rules based on the perceived environment.

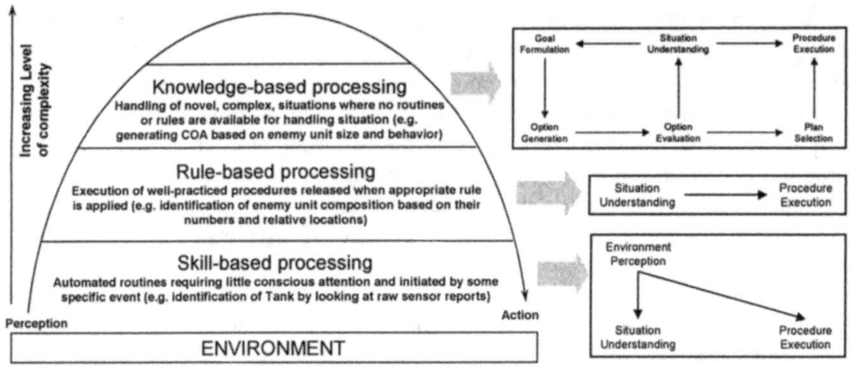

Figure 11-2: Mapping Rasmussen's human information processing hierarchy to Envelope

11.3.2 Agent Architectures and Envelope

Often an agent, be it a software agent or a robot, is created to perform specific tasks implementing aspects of human-like behavior. Thus the complexity of the agents varies depending on their tasks. Purely reactive agents only perform a simple mapping from sensor data representing the perceived environment to actions in the environment. The behavior of a purely reactive agent can be specified in Envelope by a collection of perception-action rules along the edge from the Environment Perception component to the Procedure Execution component.

In contrast to reactive agents, deliberative agent architectures, like the lower half of Envelope and like most other "classical" agent architectures, are more sophisticated, usually contain an explicitly represented model of the world that is used for decision making, and modularize cognitive abilities. Several different deliberative agent architectures have been proposed in the literature, and two of them are most prominent: horizontally layered architecture (Ferguson, 1992) and vertically layered architecture (Muller and Pischel, 1994). Each of the two-layered approaches models an agent architecture as consisting of several hierarchical functional modules representing different requirements of an agent, possibly incorporating communication, reaction, inference for planning or scheduling, perception, knowledge maintenance, etc. Each layer in a horizontally layered architecture has access to both the perception and the action components, whereas in a vertical approach, only one layer has direct interface to the perception and action.

These agent architectures divide the task of interacting with an environment into a number of vertically layered major components, including perception, world modeling, planning, task execution, and motor control. The perceptual component handles the sensing devices connected to the agent. The modeling component converts the sensor input into a symbolic description. The planning component attempts to work out how to achieve its goals given the current world state. Task execution breaks down the plan into detailed motion commands and finally the motor control causes these commands to be executed. In Envelope, the perceptual and modeling components reside within the Environment Perception component, producing a symbolic description of the world perceived from raw signals. The planning and the task execution components correspond, respectively, to the Plan Generation and Procedure Execution components of Envelope.

Reactive architectures are suitable for implementing simple agents to perform routine reaction tasks in dynamic environments that require fast action. Purely deliberative architectures are suitable for representing those agents that need to carry out extensive deliberation such as argumentation for plan selection, which sometimes involves search tasks of combinatorial complexity. Hybrid agent architectures, like Envelope, integrate both reaction and deliberation. Procedural Reasoning Systems (Ingrand et al., 1992) employ a single representation and control scheme for both reaction and deliberation, whereas Envelope-like-layered agent architectures, such as InteRRaP (Fischer et al., 1994), use different representations and algorithms (implemented in separate layers) to perform these functions.

A slightly more sophisticated approach than reaction is the Subsumption architecture (Brooks, 1986), which is built in horizontal layers of desired behaviors or levels of competence. A higher-level layer contains as a subset each lower level of competence. Thus, each layer of control can be built as a completely separate component and simply added to existing layers to achieve the overall level of competence. The layers of the Subsumption architecture are composed of networks of finite state machines augmented with timers. In the current implementation of Envelope, a BN model, implementing some kind of behavior, within the Situation Understanding component can be thought of as a composition of modules representing prewired lower-level behavior.

The BDI architecture (Rao and Georgeff, 1991) views the system as a rational agent having certain mental attitudes of Belief, Desire, and Intention (BDI), representing, respectively, the information, motivational, and deliberative states of the agent. These independent mental attitudes determine the system's behavior. Cohen and Levesque (1990) provided one of the first logical formalizations of intentions and the notion of commitment. The deliberative component of Envelope has been formalized via modal epistemic logics (see the implementation section) with constructs that are similar in semantics to the ones in the BDI architecture.

Other notable agent architectures are PRODIGY and ICARUS. PRODIGY (Minton, 1990) uses search through a problem space to achieve goals cast as first-order expressions. The search is based on means-ends analysis to solve problems by resolving goals. Learning consists of constructing control rules based on previous problem-solving experiences. ICARUS (Langley et al., 1991; Langley, 2005) is a reactive architecture for physical agents that provides separate memories for categories and skills, and a hierarchical organization on both memories. Charles River Analytics' SAMPLE model (Mulgund et al., 2000) of human behavior has been incorporated into the agent development environment GRADE (Harper et al., 2001).

The EPIC, ACT-R, and Soar models incorporate a detailed model of perception that faithfully reproduces the restrictions and timings of human perception via sensory systems. The motivation for the development of cognitive architectures within the robotics and software agent communities, as detailed above, is somewhat different in that they want robots to exhibit sophisticated behaviors, including use of natural language, speech recognition, visual understanding, problem solving, and learning, in complex analog environments and in real time. Robots are not limited to human sensors or effectors, e.g., robots can have lasers, wheels, nonhuman grippers, and work in hazardous environments without being affected by extreme heat or radiation. So the architectures found in the robotics community generalize some of the structures found in other cognitive architectures, as well as relaxing the adherence to human timing data for the performance of nonhuman sensors or effectors. This is in line with the development of Envelope, where we relaxed some constraints and embraced various algorithmic heuristics without solid psychological foundations.

11.3.3 C4I Architectures and Envelope

The three C4I architectures that are to be considered here to compare with Envelope have already been discussed in detail in Chapter 1. Table 11-1 (based on the table in Section 1.2 of Chapter 1) shows the overall relationship between these C4I architectures and Envelope.

JDL Data Fusion Model	Level 1 = Object Assessment	Level 2 = Situation Assessment	Level 3 = Threat Assessment	Action...Not Explicit in the JDL Model	Level 4 = Process Refinement
Knowledge Pyramid	Data	Information and Knowledge	Knowledge	Wisdom	n/a
Situation Awareness	Perception	Comprehension	Projection	Decision	n/a
OODA Loop	Observe	Orient	Decide	Act	n/a
Information Processing Hierarchy	Skill/Rule-Based Processing	Rule/Knowledge-Based Processing	Rule/Knowledge-Based Processing	n/a	n/a
Envelope	Environment Perception	Situation Understanding	Option Generation Plan Generation	Procedure Execution	Plan Generation

Table 11-1: C4I architectures and Envelope

Perception and fusion at JDL Levels 0 and 1 can be mapped into Envelope's Environment Perception component. Situation assessments can be thought of as part of the Situation Understanding component, and Envelope's option and plan generations support threat assessment. But Level 4 requires planning and schedule and thus is mapped to the deliberative portion of Envelope. Simple correspondence can be made between OODA and Envelope, where observation is carried out at the Environment Perception component, orientation is simply situation understanding, and decisions are made at the Option Evaluation component. Finally, actions are executed at the Procedure Execution component.

11.4 LEARNING, SYSTEMATICITY, AND LOGICAL OMNISCIENCE

Classical cognitive architectures support learning by creating new cognitive structures (e.g., production rules) based on the results of multistep problem solving. Cognitive architectures, such as ACT-R and Soar, tend to focus on skill learning. While important, we believe that it is equally important to understand how conceptual knowledge is learned, and organized for effective use. The Envelope model does both skill and structure learning depending on the kind of options we are evaluating. Soar's learning mechanism is known as chunking. It is an explanation-based mechanism that learns by finding the conditions for why a particular conclusion was drawn and then creating new rules, or chunks, that propose the conclusion given the conditions.

Learning within Envelope is defined in terms of establishing a direct connection between two components that are separated by two or more edges. So, for example, if perceiving an enemy unit frequently leads to the execution of a ground attack plan via the usual process cycle of goal formulation, option generation, and evaluation, then the whole process cycle can be replaced by a single step production rule with sufficient context information in the condition of the rule. The learned rule is then placed in the long-term memory.

If a repetitive sequence of processing steps in Envelope involves the generation of a goal from a situation, which in turn is based on certain perceptions of the environment, then the repetition is learned and encoded as another production rule. This learning step is essentially shortening the path from the Environment Perception component to the Goal Formulation component via the Situation Understanding component, bypassing the Situation Understanding component. A similar shortening of path occurs between the Environment Perception component and the Procedure Execution component. This phenomenon corresponds to human skill learning.

The basic idea behind systematicity (Fodor and Pylyshyn, 1988) is that any cognitive system able to do one thing of a given type is able to do other, structurally related, things of that type. Van Gelder and Niklasson (1994) challenge the validity of systematicity for classical architecture. Their challenge is based on an experimental investigation about scientists' understanding of propositional logic which found that around 30% of people do not perform identically on structurally identical Modus Tollens-based inferences, i.e., directly violate the hypothesis of systematicity of Modus Tollens. The systematicity hypothesis is true in case an agent is logically omniscient, that is, if the agent knows all valid formulae and its knowledge is closed under logical implication or Modus Ponens.

An agent based on the current implementation of Envelope via logical inferencing will be logically omniscient. In practice real agents are resource-bounded and therefore what is logically knowable is not realizable. The problem is defining knowledge and reasoning for agents in the presence of bounds. In the epistemic formalization, as in Envelope, one can restrict the depth of the modalities found in formulae. Fagin and Halpern (1988) introduce the syntactic notion of "awareness," where the set of formulae, which an agent is aware of, is just a list of formulae. Thus, it is possible to be aware of some formula F and not be aware of another formula G, even if F implies G. But the problem is that agents are still omniscient within the underlying logic used for modeling.

11.5 COMPUTATIONAL REALIZATION

An implementation of an intelligent agent based on Envelope involves updating its long- and short-term memories as the environment is perceived. The contents of the memories at any time point constitute its epistemic state (Gardenfors, 1988) at that time point. The Envelope approach provides various mental constructs for representing the epistemic state of an agent, which are necessary to represent and reason with the concepts in Envelope, such as belief, goal, and plan. The deliberative part of Envelope is embodied in a well-defined knowledge representation language, which explicitly supports the central concepts of decisions and plans, and associated constructs of goals, arguments, commitments, obligations, and constraints. The reactive part makes use of production rules and BNs.

The environment is perceived via various types of sensors, one of which is an image sensor. An image analyst's skill-based behavior will be able to detect engineering activities just by looking at the images. In the implementation of Envelope, the Environment Perception component generates image features from the perception via IMINT. The component then sends the features to each of the Goal Formulation, Situation Understanding, and Procedure Execution components. In the Situation Understanding component, the reactive or skill-based behavior of Envelope concludes engineering activity via a production rule of the form:

IF Image Features
THEN Engineering Activity

The engineering activity then in turn triggers the need for determining the presence of a possible missile launch site. Following the Rasmussen hierarchy, this task can be achieved in one of three ways (i.e., reactive, well-executed procedure, and deliberative) within the Situation Understanding component of Envelope depending on the agent's current state of skills and knowledge. We first illustrate a BN-based well-executed procedure within the Procedure Execution component for determining the presence of a missile launch site.

The BN in Figure 11-3 supports the classification of missile launch sites into sites for real (short- and long-range) missiles and dummy/no threat missiles. The strategic doctrine part of the BN is based on the distance of the missile launch site to potential targets, the vulnerability of the potential targets, the terrain adequacy, and the possibilities of cover and concealment. A site placed such that highly vulnerable potential targets are predominantly situated within the missile range is likely to be categorized as a real missile launch site. The BN illustrates a situation where a supporting hypothesis – the target vulnerability – is generated from more specific context knowledge – the target type and stage of the operations. If the target type is a command post and the context is placed in a postconflict stage, then the target vulnerability is low, which in turn reduces the chances of the presence of a long-range missile site. The physical doctrine evidence nodes for the BN includes the presence of camouflage, the existence of Transportable Erector Launchers (TELs), and the existence of engineering activities. Such activities can be detected by various types of sensors, including image intelligence sensors (IMINT). Enemy activity at a location with a sensor causes a report (TACREP) to be produced. The portion of the BN labeled "Blue Sensor" in the figure shows that IMINT produces reports upon detection of the enemy's engineering activities.

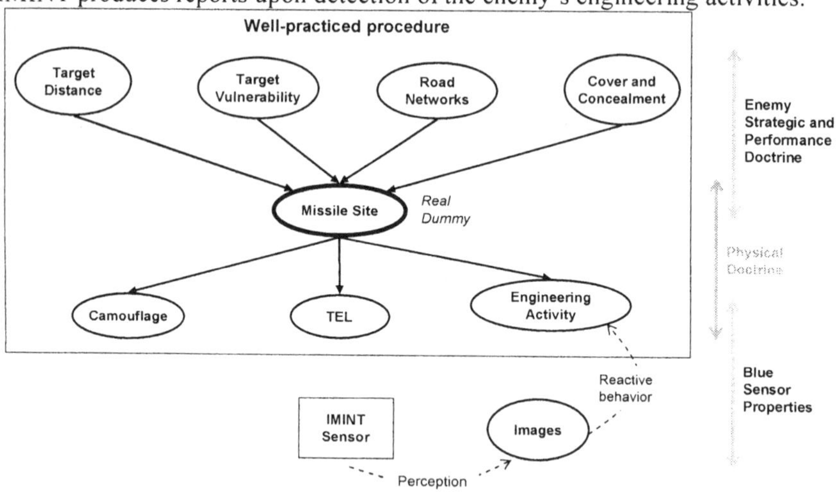

Figure 11-3: Example well-practiced procedure in BN implementing the missile-site determination process within the Situation Understanding component

The recognition of engineering activity has been achieved via a reactive production rule stored in the working memory. If the BN in Figure 11-3 is considered a well-practiced inference procedure then the human operator will immediately classify the site into one of the two possibilities based on the engineering activity evidence and other observations stored in the working memory. If no such procedure exists the engineering activity initiates knowledge-based problem solving via Symbolic Argumentation (see Chapter 9 for more details on Argumentation) (Das et al., 1997; Fox and Das, 2000; Das, 2005), implementing the missile-site determination process cycle originating and terminating at the Situation Understanding component of Envelope. Figure 11-4 shows how we represent such a problem-solving activity as a decision construct (Fox and Das, 2000).

```
decision:: identify-missile-site
  situation
    engineering-activity
  goal
    identify-missile-site
  options
    real;
    dummy
  arguments
    suitable-road-networks => support(real, 0.7);
    no cover-and-concealment => support(dummy, 0.8);
    camouflaged => support(not dummy, 6.0);
    ...
commits
    netsupport(X, U) & netsupport(Y, V) &
    netsupport(Z, W) & U > V & U > W => add(X).
```

Figure 11-4: Example knowledge-based problem solving via argumentation implementing the missile-site determination process cycle originating and terminating at the Situation Understanding component of Envelope

Such a decision construct has several component parts: an evoking situation, a goal, one or more candidate options, arguments for evaluating options, and one or more commitment rules. For the purpose of inference emulating mental processing, these and other constructs are translated to an executable modal logical syntax involving epistemic operators corresponding to beliefs and goals. A situation describes the situation or event that initiates problem-solving activity. In the example, a report about an engineering activity raises the goal of determining the site identification. Typically, a goal is represented by a state that the agent tries to bring about. Options are different possible explanations to satisfy the goal. In the example, a missile site could be one of the following three: long, short, or dummy.

Arguments are like production rules that define the arguments appropriate for choosing between candidates for the decision. In the sense of ACT-R, the arguments in Figure 11-4 are "declarative" in nature as the options are competing hypotheses (beliefs). But arguments are "procedural" when the options are competing plans or actions (example given below). A decision construct is typically concerned with evidence when the decision involves competing hypotheses, and additionally with preferences and values when the decision is concerned with actions or plans. Commitment rules define the conditions under which an option may be recommended, or taken autonomously, by an agent. It may include logical and/or numerical conditions on the argument and belief databases.

Once a missile has been categorized as long, short, or dummy, the operator wants to classify it as one with an N-B-C (Nuclear, Biological, Chemical) or conventional warhead. The new situation is that there is a missile of type long, short, or dummy. This new situation, which is generated via some internal process rather than perceiving the environment, triggers another rule or well-executed procedure for identifying the type of warhead mounted. The possibilities of missile warhead type are N-B-C or conventional. An example rule within the Situation Understanding component of Envelope in the form of a production rule is of the following form:

IF Real Missile
AND Country Possesses N-B-C Weapons
AND Big City within Range
THEN Likely Warhead is N-B-C

We can also build a BN as before as a well-executed procedure to determine the warhead type. In the absence of this kind of rule or procedure, the new situation initiates a deliberative process by generating a goal to identify the type. If such a goal is generated then a decision construct similar to the one in Figure 11-4 can be invoked to perform knowledge-based problem solving.

To illustrate decision options involving actions, consider a situation in which a missile launch has been detected. The time-critical decision to be made here is among the choice of interceptors. Three possible COAs are interceptions via three different types of missiles: PAC-3, PAC-2/GEM, and None. A PAC-3-type missile is much more expensive than a PAC-2/GEM type and thus its applications are usually limited to cases when warheads are suspected to be of N-B-C type. Figure 11-5 implements an agent behavior for interceptor selection based on a well-executed procedure in the form of an Influence Diagram (ID) (see Chapter 9 for more details on IDs). Such diagrams are extensions of BNs incorporating the notions of action and utilities (respectively, rectangles and diamonds in the diagram). The procedure takes into account the actual quantifiable cost and nonquantifiable cost (e.g., loss of lives) as well as other factors such as probability of kill. The main point to be noted here is that the agent will evaluate such a complex relationship as depicted in the ID using information from the working memory.

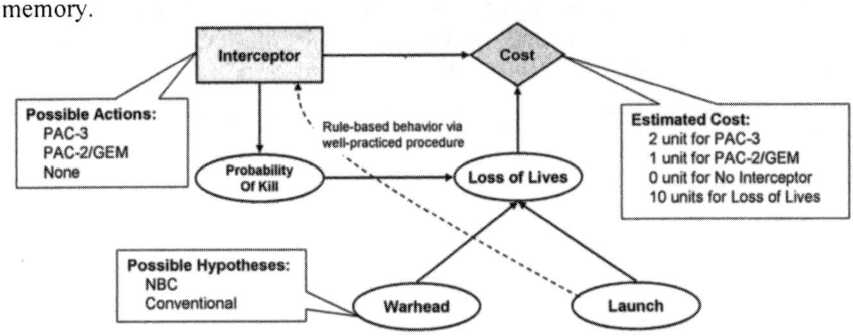

Figure 11-5: Example well-practiced procedure in ID implementing the missile-site determination process incorporating utility

In the absence of an ID as depicted above and the fact that some of the information may be stored in the long-term memory (e.g., missile costs and destruction capabilities), the interceptor selection process can trigger a problem-solving process. The generated goal in this case is to determine the interception option. An example decision construct that can be used to serve this purpose is shown in Figure 11-6.

```
decision:: alternative-interception
    situation
        launched-missile
    goal
        decide-alternative-interception
    candidates
        PAC-3;
        PAC-2/GEM;
        None
    arguments
        likely-warhead-is-NBC => support(PAC-3, 0.8);
        likely-warhead-is-conventional => support(PAC-2/GEM, 0.7);
        point-of-impact-is-non-residential &
            likely-warhead-is-conventional => support(PAC-2/GEM, 0.7);
        ...
    commits
        ... => plan(interception-by-PAC-3).
```

Figure 11-6: Example knowledge-based problem solving via argumentation implementing the interceptor selection process

One important distinction needs to be made between the construct in Figure 11-4 and the one in Figure 11-6. The commitment rule in Figure 11-4 updates the working memory with the agent's new belief, whereas the commitment rule in Figure 11-6, like the production rule in ACT-R, commits to a plan as shown in

Figure 11-7. The committed plan becomes an input to the Plan Generation component of Envelope. The component is currently built on Hierarchical Task Network (HTN)-based planning modules where each plan is hierarchically decomposed in subtasks and atomic tasks.

```
plan:: interception-by-PAC-3
    pre-conditions
        incoming-missile & plan(interception-by-PAC-3)
    subtasks
        track-and-engage;
        launch
        monitor
    scheduling-constraints
        first(track-and-engage);
        after(track-and-engage, launch);
        repeat(monitor)
    abort-conditions
        tracking-failed
         => abort(interception-by-PAC-3) & schedule(evacuation)
```

Figure 11-7: Example knowledge-based problem solving via argumentation implementing the planning process for an interception by PAC-3 in Envelope

As shown in Figure 11-7, a plan schema has four components: 1) preconditions, which must be true before the agent may commit to executing the plan; 2) subtasks, which specify the subplans and atomic actions from which this plan is built; 3) scheduling constraints, which specify the logical and temporal ordering relations that hold between the subtasks; and 4) abort conditions, which specify any conditions that, if true, will cause the plan to be aborted (and may schedule a new plan). The precondition in the example plan is satisfied when the use of PAC-3 is committed upon the observation of a launched missile. There are three subtasks: track and engage the missile, launch the selected interceptor, and monitor subsequent activities. The constraints are that the track and engage activity is first, then launch, and then continuous monitoring. The abort condition states that if tracking fails then abort the interception and schedule the evacuation around the predicted point of impact.

As in the case of HTN planning frameworks, a particular plan or sub-plan can sometimes be hierarchically decomposed in more than one way. For example, the monitoring sub-plan may be based only on space-based assets or may include both ground- and space-based assets, thus potentially generating more than one schedule. Each such schedule is a set of tasks to be carried out over time to achieve the desired plan. The selected schedule generates effective procedures and sends them sequentially to the Procedure Execution component. Note that a schedule will typically contain parallel tasks that will require simultaneous attention from motor and cognitive skills. In cases where there are two or more tasks that need to be run in parallel and both require motor skills, then delegation of one or both will be necessary by the agent. In such cases, the delegation task itself becomes a part of the procedure. Similar constraints are applicable for tasks requiring simultaneous cognitive processing.

Decision and plan constructs are translated and internally represented in the syntax of modal propositional logic. We provide here some example sentences that are translations of the decision construct shown in Figure 11-4. The situation and goal portion in the decision is translated to the following modal rule:

$$\langle bel \rangle engineering\_activity \rightarrow \langle goal \rangle identify\_missile\_site$$

The above sentence states that if it is believed that there is engineering activity, then a goal is generated to determine the missile site. A goal is considered to be achieved as soon as it becomes true. In the context of the decision, this is reflected in the following formulae:

$$\langle bel \rangle (real \wedge \neg dummy) \rightarrow \langle bel \rangle identify\_missile\_site$$
$$\langle bel \rangle (dummy \wedge \neg real) \rightarrow \langle bel \rangle identify\_missile\_site$$

The first of the above three sentences states that if the missile is believed to be of long-range type, but neither short nor dummy, then it is believed that the missile has been identified. In other words, the earlier goal is considered achieved upon believing that the missile is of long range. The representations for the arguments in the decision construct are

$$\langle bel \rangle suitable\_road\_networks \rightarrow \langle sup_{0.7} \rangle real$$
$$\langle bel \rangle \neg cover\_concealment \rightarrow \langle sup_{0.8} \rangle \neg dummy$$
$$\langle bel \rangle camouflaged \rightarrow \langle sup_{0.6} \rangle real$$

A detailed account of the current implementation of Envelope, along the formalization described in this section, can be found in Fox and Das (2000) and Das (2005).

11.6 SOME DISCUSSION

We have presented here an adaptive and symbolic cognitive architecture, Envelope, especially suitable for producing intelligent agents for a variety of goal-oriented information processing tasks. We have compared Envelope's foundational assumptions with that of well-known architectures in the field by taking examples from battlefield information processing tasks. The contribution of Envelope is not in superior cognitive components, but rather in integrating diverse aspects of intelligent behavior within a single practical framework.

The use of various specific AI algorithms (e.g., heuristics in task planning and scheduling, option evaluation via theorem proving, probabilistic evidence propagation) in the implementation of Envelope provides a mechanism for constraining the forms of models for improving efficiency, but has no substantial scientific psychological basis. Kieras (2005) termed them as speculative quasi-psychological constraints. One can use Envelope to determine the logical adequacy of a strategy for a task by relaxing psychological (and time) constraints to determine if the correct result is produced. If such a simulation shows the strategy is wrong, then its incompatibility with human thinking is revealed.

The generic nature of Envelope ensures its applicability across a variety of application domains. In this respect, we have found usefulness in the application of Envelope in its entirety, as well as in its various parts, via its current Prolog-based implementation. For example, the Domino model (Fox and Das, 2000), which is the precursor of Envelope, has been instantiated for a variety of domains, including clinical decision support. One such instantiation is the cognitive agent COGENT (Das and Grecu, 2000), where the Rasmussen interpretation of Envelope, as shown in Figure 11-2, provides guidelines for the design of the decision-aiding module, specifically, for defining functionality to support the missing operators in three major tasks: data preparation, situation assessment, and response recommendation.

Recent experience with Envelope is in several agent-based defense and space applications. While applying Envelope to defense applications we have found it to be a natural fit. As shown in the previous section, a typical defense application involves perception from the area of operations via distributed sensors, assessment of situations and threats based on the sensor information, and then generation of COAs from a set of options to counter threats. This cycle is of course central to Envelope. Moreover, breakdown of such a cycle in Envelope into various stages along its arrows helps subject matter experts to modularize their thinking, and, in turn, helps knowledge engineers to systematically elicit knowledge for situation assessment and action recommendation.

Envelope's symbolic reasoning is theoretically well founded in both classical and nonclassical logics. We have been engaged in enhancing Envelope's theoretical foundation and implementation towards a fully generic agent framework serving as a core for a variety of applications.

11.7 FURTHER READING

Many references have been provided throughout the chapter.

Chapter 12

Distributed Fusion

This chapter presents distributed fusion from the Situation Assessment (SA) and target tracking perspectives. The chapter also describes the relevance of distributed fusion to Network Centric Warfare (NCW) environments.

For SA in a distributed NCW environment, we distribute the cliques of a junction tree to be maintained by local fusion nodes within the environment. The underlying Bayesian Belief Network (BN) model of the junction tree represents the combined SA model of distributed warfighters in the context of a mission. Local fusion nodes communicate and coordinate with each other to improve their local estimates of the situations relevant to the mission, thereby contributing to the development of a Common Tactical Picture (CTP) of the mission via shared awareness. We make use of the message passing inference algorithm for junction trees that naturally fits within distributed NCW environments.

For target tracking in a distributed NCW environment, we produce an overall estimate of a target at a fusion center combining estimates from distributed sensors located at different fusion nodes that are tracking the same target. We make use of the Kalman Filter (KF) algorithm for estimating targets at local fusion nodes from sensor observations. Individual estimates from local fusion nodes are then combined at a fusion center, thereby generating evidence to be posted at appropriate cliques of BN fragments for SA.

12.1 CONCEPT AND APPROACH

Sensor networks consist of a large number of nodes of sensing devices densely distributed over the operational environment of interest. Nodes have wired or wireless connectivity that are tied to one or more backbone networks, such as the Internet, SIPRNET (Secret Internet Protocol Router Network), and NIPRNET (Nonclassified (unclassified but sensitive) Internet Protocol Router Network). Each sensor node has its own measurements collection and processing facility to estimate and understand the environment around it. Hence a sensor is situationally aware in terms of position and movements of targets and threats they pose. This awareness must be shared among all other nodes to generate an assessment of the situation of the environment as a whole (sometimes called the Common Tactical Picture (CTP)) for effective coordinated action.

Sensor nodes can be thought of as intelligent autonomous agents who communicate, coordinate, and cooperate with each other to improve their local situational awareness, and to assess the situation of the operational environment as a whole. The concept of distributed fusion refers to such decentralized processing environments consisting of autonomous sensor nodes and additional processing nodes without sensors, if necessary, to facilitate message communication, data storing, relaying, information aggregation, and assets scheduling. Some of the advantages of distributed fusion are reduced communication bandwidth, distribution of processing load, and improved survivability from a single point failure. The distributed fusion concept naturally fits within the upcoming NCW paradigm and its backbone command network, the Global Information Grid (GIG).

As a concrete example of distributed fusion, consider the decentralized processing environment as shown in Figure 12-1.

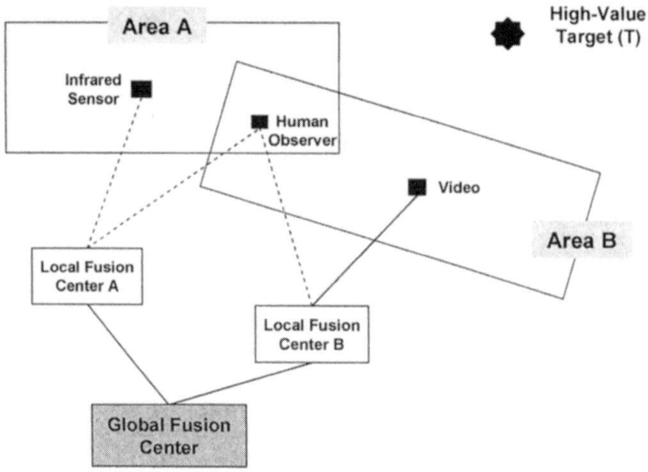

Figure 12-1: An example distributed fusion environment

We assume that there is a high-value target (top right of the figure) within a region of interest, and that the designated areas A and B surrounding the target are considered to be the most vulnerable. These two areas need to be under surveillance to detect any probing activities, which indicate a possible attack threat. The sensor coverage in areas A and B, shown in grey, are by an infrared sensor (SIGINT) and a video camera (IMINT), respectively. In addition, a human observer (HUMINT) is watching the area common between A and B. There are two local fusion centers for the two areas to detect any probing activity. The infrared sensor has a wireless connectivity with the local fusion center for area A, whereas the video camera has a wired connectivity to the local fusion center for area B for streaming video. Moreover, the human observer communicates wirelessly with the two local fusion centers. Each of the two centers fuses sensor data that it received to determine any possible probing activity. They then pass their assessments (i.e., higher-level abstraction, rather than raw sensor information, and thus saving bandwidth) to another fusion center that assesses the overall threat based on the reports of probing activities and other relevant prior contextual information.

In a centralized fusion environment, where observations from IMINT, HUMINT, and SIGINT are gathered at one place and fused, a BN model, such as the one in Figure 12-2, can be used for an overall SA. The model handles dependence among sensors and fusion centers via their representation in nodes and interrelationships.

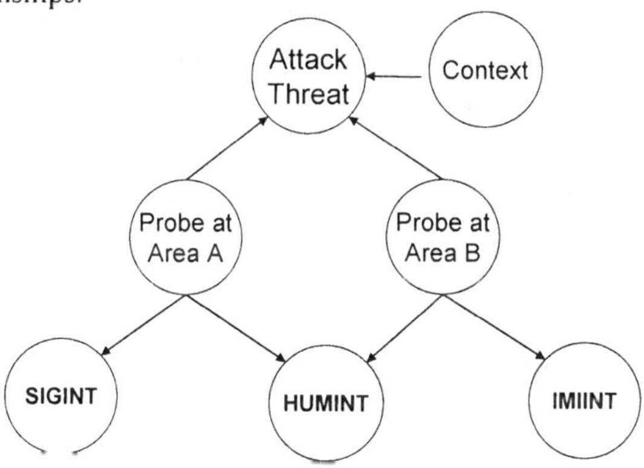

Figure 12-2: A centralized BN model for situation assessment

A probing activity at an area will be observed by those sensors that are covering the area, and the lower half of the BN models this. For example, SIGINT and HUMINT reports will be generated due to a probing activity at area A. Similarly, SIGINT and HUMINT reports will be generated due to a probing activity at area B. The upper half of the BN models the threat of an attack based on the probing activities at areas A and B, together with other contextual information.

In a decentralized environment, as illustrated in Figure 12-1, each of the three fusion centers contains only a fragment of the above BN model as shown in Figure 12-3.

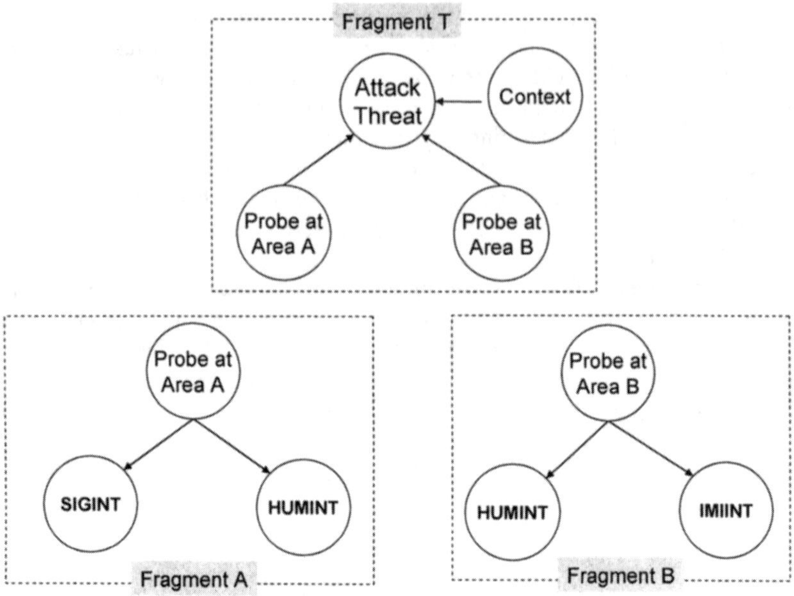

Figure 12-3: Distributed parts of the BN models

Local fusion centers A and B assess probing activities based on their local model fragments and send their assessments to the global fusion center, which then uses its model to determine the overall attack threat. But if the same HUMINT report is received by both the local fusion centers then the process has to make sure this common information is used only once; otherwise, there will be a higher than actual level of support for threat to be determined by the global fusion model. This is the *data incest problem* in a distributed fusion environment, which is the result of repeated use of identical information. *Pedigree* needs to be traced not only to identify common information, but also to assign appropriate trust and confidence on data sources. An information graph (Liggins et al., 1997), for example, allows common prior information to be found.

12.2 DISTRIBUTED FUSION ENVIRONMENTS

As shown in Figure 12-4, a typical distributed fusion environment is likely to contain a variety of fusion nodes to do a variety of tasks:

- Process observations generated from a cluster of heterogeneous sensors (e.g., the local fusion centers A and B in Figure 12-1, and nodes labeled 5 and 9 in Figure 12-4).
- Process observations generated from only a single sensor (e.g., nodes labeled 11, 12, and 13 in Figure 12-4).

- Perform a task (e.g., Situation Assessment (SA) and Threat Assessment (TA), Course of Action (COA) generation, planning and scheduling, Common Tactical Picture (CTP) generation, collection management) based on information received from other sensors in the environment and other information stored in databases (e.g., nodes labeled 1, 2, 3, 4, 6, 7, and 10 in Figure 12-4).
- Relay observations generated from sensors to other nodes (e.g., the node labeled 8 in Figure 12-4).

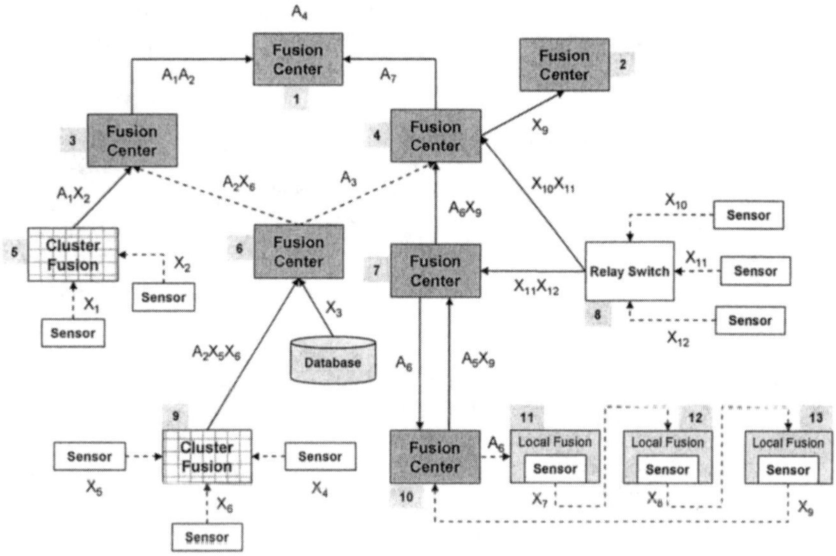

Figure 12-4: A typical distributed fusion environment

As shown in Figure 12-4, a fusion node receives values of some variables obtained either from sensor observations (X variables) or via information aggregation (A variables) by other nodes. Such values can also be obtained from databases. For example, the fusion center labeled 6 receives values of the variables A2, X5, and X6 from the cluster fusion node labeled 9 and values of the variable X3 from a database. Note that an arrow between two nodes indicates the flow of information in the direction of the arrow as opposed to a communication link. The existence of an arrow indicates the existence of at least a one-way communication link, not necessarily direct, via some communication network route. For example, there is a one-way communication link from the node labeled 3 to the node labeled 1. But a reverse communication link between these two nodes will also be necessary in implementing our message passing distributed fusion algorithm to be presented later.

Therefore, each node (fusion center, cluster fusion, relay switch, or local fusion) in a distributed fusion environment has knowledge of the states of some variables, called *local variables*, as shown in Figure 12-5. For example, the fusion node labeled 6 has knowledge of the X variables X3, X5, and X6, and A variables A2 and A3. The node receives values of the variables A2, X5, and X6 from the node labeled 9 and the variable X3 from a database. The node generates values of the variable A3 via some information aggregation operation. On the other hand, fusion node 9 receives measurements X3, X5, and X6 from a cluster of sensors, and generates A2; fusion node 8 relays values of the variables X10, X1, and X12 to other nodes; and fusion node 12 obtains measurements of X8 from a single sensor.

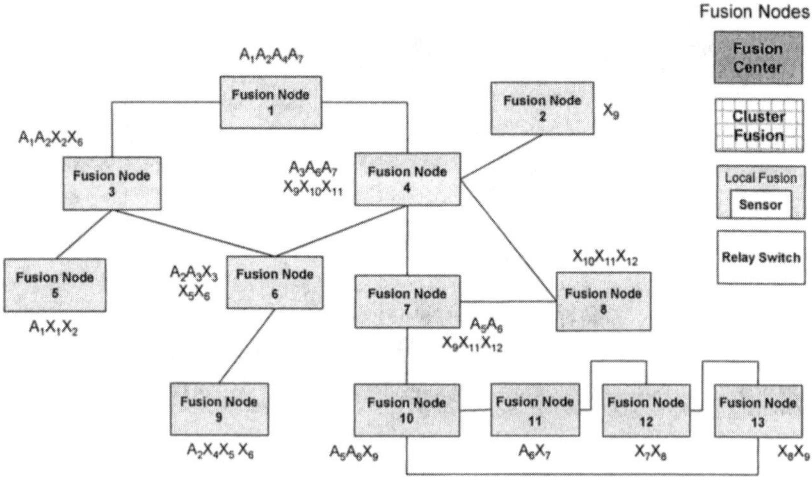

Figure 12-5: Network of distributed fusion nodes

Figure 12-6 shows four possible distributed fusion environments: centralized, hierarchical, peer-to-peer, and grid-based. Note that the direction of an arrow indicates information flow, and the existence of a communication link along the direction of an arrow.

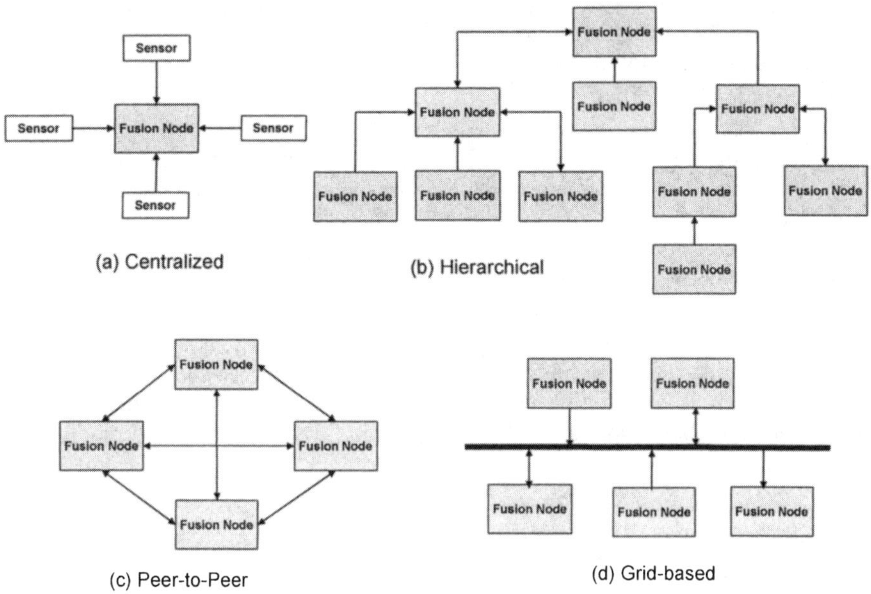

Figure 12-6: Possible distributed fusion environments

In a *centralized* environment, only the sensors are distributed and send their observations to a centralized fusion node. The centralized node then combines all the information to perform tracking or SA. In a *hierarchical* environment, the fusion nodes are arranged in a hierarchy with the higher-level nodes processing results from the lower-level nodes and possibly providing some feedback. The hierarchical architecture will be natural for applications where situations are assessed with an increasing level of abstractions along a command hierarchy, starting from tracking of targets at the bottom level. Considerable savings in communication can be achieved in a hierarchical fusion environment. In both *peer-to-peer* and *grid-based* distributed environments, every node is capable of communicating back and forth with every other node. This internode communication is direct in the case of a peer-to-peer environment, but some form of publish and subscribe communication mechanism is required for a grid-based environment.

12.3 ALGORITHM FOR DISTRIBUTED SITUATION ASSESSMENT

We start with a junction tree for a given BN model for SA as shown in Figure 12-2. The cliques of the tree are likely to be the clusters of nodes as indicated in Figure 12-3. Our distributed SA framework has four steps:

- Network formation
- Spanning tree formation
- Junction tree formation
- Message passing

The nodes of the sensor network first organize themselves into a network of fusion nodes similar to the one shown in Figure 12-5. Each fusion node has partial knowledge of the whole environment. This network is then transformed into a spanning tree so that neighbors have high-quality connections. In addition, the spanning tree formation algorithm optimizes the communication required by inference in junction trees. The algorithm can recover from communication and node failures by regenerating the spanning tree. Figure 12-7 is a spanning tree obtained from the network in Figure 12-5. The decision to sever the link between nodes 4 and 6, as opposed to between nodes 3 and 6, can be mitigated using the communication bandwidth and reliability information in the cycle of nodes 1, 3, 6, and 4.

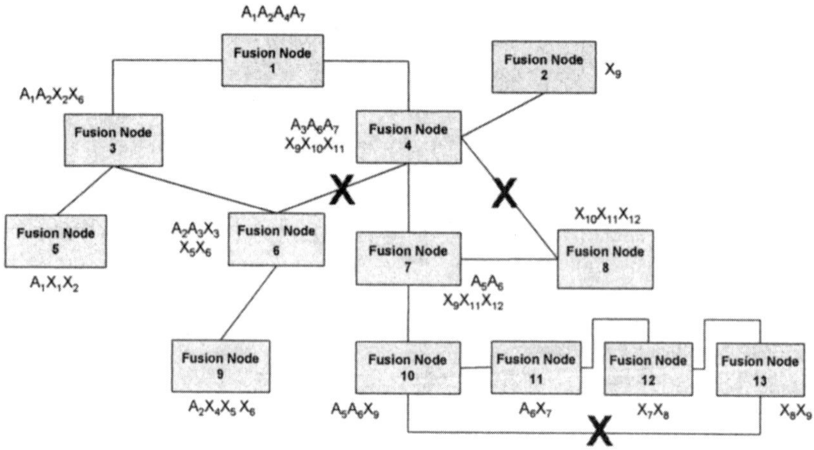

Figure 12-7: A spanning tree

Using pairwise communication link information between neighbors in a spanning tree, the nodes compute the information necessary to transform the spanning tree into a junction tree for the inference problem. Finally, the inference problem is solved via message passing on the junction tree.

During the formation of a spanning tree, each node chooses a set of neighbors so that the nodes form a spanning tree where adjacent nodes have high-quality communication links. Each node's clique is then determined as follows. If i is a node and j is a neighbor of i, then the variables reachable to j from i, R_{ij}, are defined recursively as

$$R_{ij} = D_i \bigcup_{k \in nbr(i)-\{j\}} R_{ki}$$

where D_i is the set of local variables of node *i*. A base case corresponds to a leaf node, which is simply that node's local variables. If a node has two sets of reachable variables to two of its neighbors that both include some variable *V*, then the node must also carry *V* to satisfy the running intersection property of a junction tree. Formally, node *i* computes its clique C_i as

$$C_i = D_i \bigcup_{\substack{j,k \in nbr(i) \\ j \neq k}} R_{ji} \cap R_{ki}$$

A node *i* can also compute its separator $S_{ij} = C_i \cap C_j$ with its neighbor *j* using reachable variables as

$$S_{ij} = C_i \cap R_{ji}$$

Figure 12-8 shows the junction tree obtained from the spanning tree in Figure 12-7.

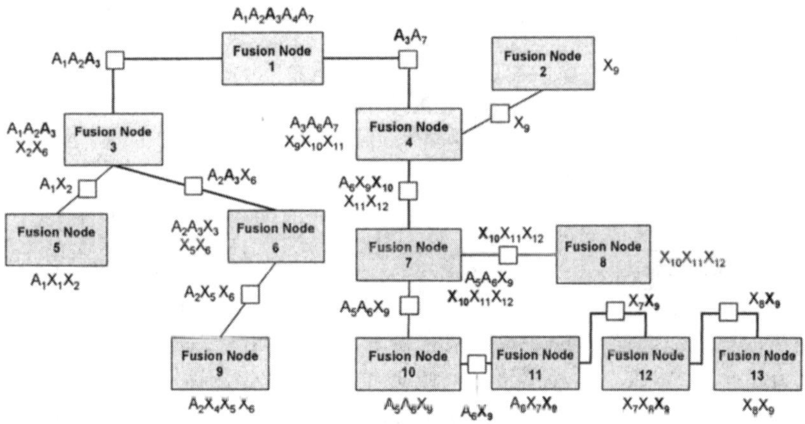

Figure 12-8: A junction tree from the distributed fusion environment

The variables reachable to a leaf node, for example, fusion node 9, are its local variables A_2, X_4, X_5, X_6. The variables reachable to an intermediate node, for example, fusion node 1, from its neighboring nodes 3 and 4 are

$$R_{31} = \{A_1, A_2, A_3, X_1, X_2, X_3, X_4, X_5, X_6\}$$

$$R_{41} = \{A_3, A_5, A_6, A_7, X_7, X_8, X_9, X_{10}, X_{11}, X_{12}\}$$

The local variable of the fusion node 1 is $D_1 = \{A_1, A_2, A_4, A_7\}$. Therefore, its clique is

$$C_1 = \{A_1, A_2, A_3, A_4, A_7\}$$

The formation of a suitable junction tree from a BN model for SA is the only part of our distributed fusion approach that is global in nature. Once we have formed a junction tree, the message passing algorithm can be run asynchronously on each node responding to changes in each others' states. Each time a node *i* receives a new separator variables message from a neighbor *j*, it recomputes its own clique and separator variables messages to all neighbors but j, and transmits them if they have changed from their previous values.

12.4 DISTRIBUTED KALMAN FILTER

As shown in Figure 12-9, we assume that a distributed KF environment consists of N Local Fusion nodes producing track estimates based on a single or multiple local sensor, and a Fusion Center combining these local estimates to a global one. An example distributed environment for tracking a ground vehicle on a road consists of: 1) a group of ground acoustic sensors laid on the road are coordinated by a local fusion node; 2) an aerial video feed provided by a UAV; and 3) GMTI data generated by JSTARS. A local fusion node requires feedback from the global estimate to achieve the best performance. Moreover, the global estimation has to cope with sensors running at different observation rates.

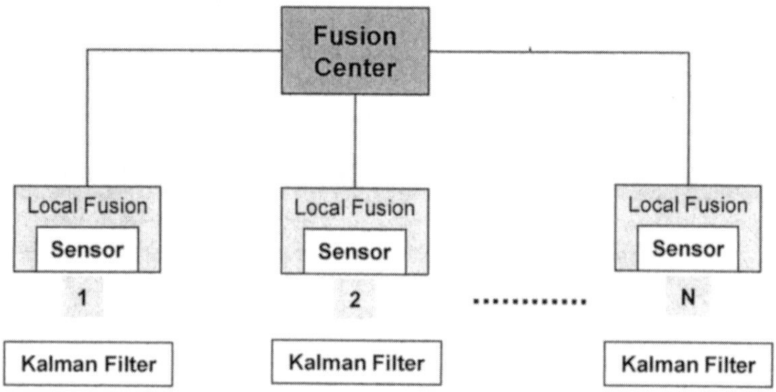

Figure 12-9: Distributed Kalman Filter

The target's dynamic is modeled in a transition model as

$$X_k = FX_{k-1} + W_{k-1}$$

where the state vector X_k is the estimate of the target at time instant k, and F is the transition mode matrix invariant of k, and W_k is the process white and Gaussian noise with zero-mean. The measurement models are given by

$$Z_{ik} = H_i X_k + V_{ik}$$

where Z_{ik} is the measurement or observed output state at the time step k from the i-th sensor ($i = 1, 2, ..., N$), H_i is the corresponding observation matrix invariant of k, and V_{ik} is the corresponding white and Gaussian noise with zero-mean. The centralized KF algorithm for estimating the target's state and error covariance matrix has the following two recursive steps:

Prediction

$$P_{k|k-1} = FP_{k-1|k-1}F^T + Q_{k-1}$$
$$\hat{X}_{k|k-1} = F\hat{X}_{k-1|k-1}$$

Filtering

$$P_{k|k}^{-1} = P_{k|k-1}^{-1} + \sum_{i=1}^{N} H_i^T R_{ik}^{-1} H_i$$

$$\hat{X}_{k|k} = P_{k|k}\left[P_{k|k-1}^{-1}\hat{X}_{k|k-1} + \sum_{i=1}^{N} H_i^T R_{ik}^{-1} Z_{ik}\right]$$

where $\hat{X}_{k|k}$ is the position estimate at time step k, $P_{k|k}$ is the error covariance matrix, and Q_k and R_{ik} are covariance matrices of the process and measurement noises, respectively. The inverse P^{-1} of the covariance matrix P is a measure of the information contained in the corresponding state estimate.

In a distributed KF environment, each local fusion node i produces its own estimate $\hat{X}_{i(k|k)}$ based on the information available from its sensor using the standard KF technique. These individual estimates are then fused together at the fusion center to produce the overall estimate $\hat{X}_{k|k}$.

As shown in Figure 12-10, there are two ways to carry out distributed KF (Liggins et al., 1997):

- Without feedback, meaning an individual fusion node performs target tracking based on its own local sensor measurements, and sends its estimation of the target state and error covariance matrix to the fusion center at every time step
- With feedback, meaning an individual fusion node sends its estimation to the fusion center as before, but obtains feedback from the fusion center in terms of the center's overall estimation combining results from individual local fusion nodes

Without Feedback

$$P_{k|k}^{-1} = P_{k|k-1}^{-1} + \sum_{i=1}^{N}\left[P_{i(k|k)}^{-1} - P_{i(k|k-1)}^{-1}\right]$$

$$\hat{X}_{k|k} = P_{k|k}\left[P_{k|k-1}^{-1}\hat{X}_{k|k-1} + \sum_{i=1}^{N}\left[P_{i(k|k)}^{-1}\hat{X}_{i(k|k)} - P_{i(k|k-1)}^{-1}\hat{X}_{i(k|k-1)}\right]\right]$$

With Feedback

$$P_{k|k}^{-1} = \sum_{i=1}^{N} P_{i(k|k)}^{-1} - (N-1) P_{k|k-1}^{-1}$$

$$\hat{X}_{k|k} = P_{k|k} \left[\sum_{i=1}^{N} P_{i(k|k)}^{-1} \hat{X}_{i(k|k)} - (N-1) P_{k|k-1}^{-1} \hat{X}_{k|k-1} \right]$$

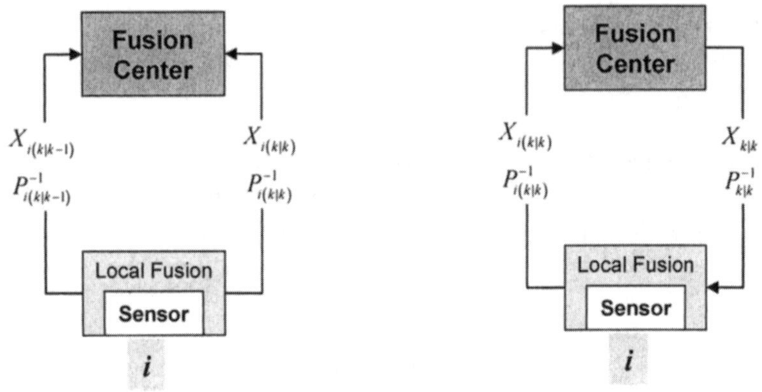

Figure 12-10: Distributed target tracking with and without feedback

Note in the above two cases of estimation that the fusion center fuses only the incremental information when there is no feedback. The new information is the difference between the current and previous estimates from the local fusion nodes. When there is feedback, the fusion node has to remove its own previously sent information before combining the local estimates, that is, the new information is the difference between the new estimate and the last feedback from the fusion center. The process of removing the common estimates is to make sure that the local estimates that are combined are independent.

12.5 RELEVANCE TO NETWORK CENTRIC WARFARE

The Network Centric Warfare (NCW) concept is a part of the DoD's effort to create a 21st century military by transforming its primarily platform centric force to a network centric force with the help of modern information technologies. NCW is predicated upon dramatically improved capabilities for information sharing via the Internet-like infrastructure. When paired with enhanced capabilities for sensing, information sharing can enable a force to realize the full potential of dominant maneuver, precision engagement, full-dimensional protection, and focused logistics.

As shown in Figure 12-11, NCW involves working in the intersection of three interconnected domains, namely physical, information, and cognitive.

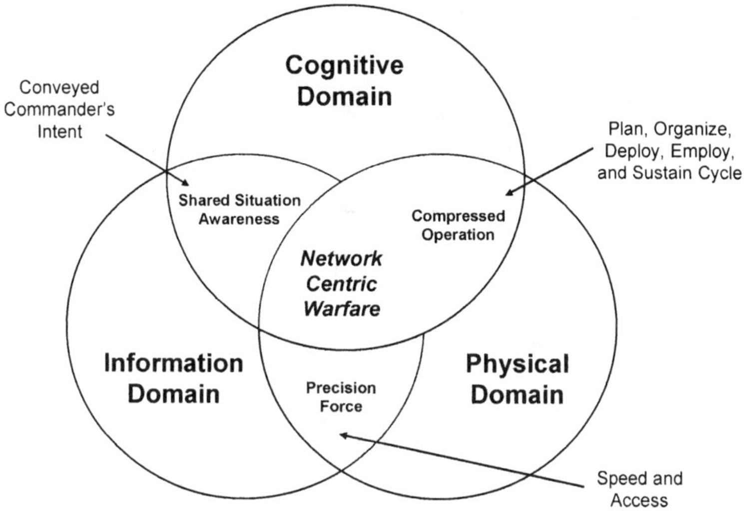

Figure 12-11: Conceptual vision of Network Centric Warfare (NCW, 2001)

The *physical domain* is the place where the situation the military seeks to influence exists. It is the domain where strikes, protections, and maneuverings take place across the environments of ground, sea, air, and space. It is the domain where physical platforms and the communications networks that connect them reside.

The *information domain* is where information is created, manipulated, and shared. It is the domain that facilitates the communication of information among warfighters. It is the domain where the command and control of modern military forces is communicated, and where the commander's intent is conveyed.

The *cognitive domain* is in the minds of the participants. It is the domain where perceptions, awareness, understanding, beliefs, and values reside and where, as a result of sense making, decisions are made.

From the perspective of distributed fusion presented in the earlier sections, it is the cognitive domain that provides warfighters with the capability to develop and share high-quality situational awareness. Fusion nodes representing warfighters communicate their assessments of situations via appropriate coordination and negotiation. As shown in Figure 12-12, autonomous intelligent agents can act on behalf of warfighters (Lichtblau, 2004) within an NCW environment to reduce their cognitive workload.

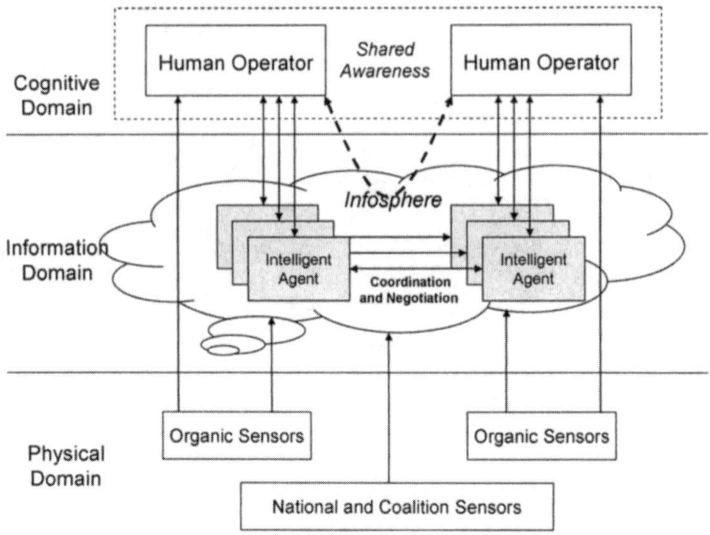

Figure 12-12: Role of intelligent agents for NCW

The GIG is a globally interconnected, end-to-end set of information capabilities, associated processes, and personnel for collecting, processing, storing, disseminating, and managing information on demand to warfighters, defense policymakers, and support personnel. The GIG is to operate within the information domain to enable the creation of a fusion network consisting of fusion nodes and their interconnections as needed.

12.6 FURTHER READING

Readers are recommended to consult Liggins et al. (1997) and Durrant-Whyte (2000) for an overall discussion on distributed fusion mostly from the target tracking perspective. Liggins et al. (1997) also discuss an approach to address the data incest problem via information graphs. There are alternative approaches to a distributed KF algorithm, for example, Rao and Durrant-Whyte (1991). But Schlosser and Kroschel (2004) present some experimental results from their study of the effect of communication rate between fusion nodes on the performance of a decentralized KF algorithm.

See Paskin and Guestrin (2004) for a more detailed account of a junction tree-based distributed fusion algorithm along the lines of the one presented here. The algorithm in the paper, in addition, optimizes the choice of junction tree to minimize the communication and computation required by inference. See Das et al. (2002) for distributing components of a BN for battlefield SA across a set of networked computers. The objective was to enhance inferencing efficiency and to allow computation at various levels of abstraction suitable for military hierarchical organizations.

There is an abundance of open source literature on NCW. A must read on NCW is Cebrowski and Garstka (1998), and also NCW (2001) and Cebrowski (2001). Further reading of topics related to NCW are on Effect-Based Operations (EBO) (Smith, 2002) and Sense and Respond Logistics (S&RL) (OFT, 2003). The NCW vision is being realized within the DoD branches, including in the Army via its FCS (Future Combat Systems) program, in the Navy (Antanitus, 2003), and in the Air Force (Sweet, 2004).

References

Aja-Fernández, S., Alberola-López, C., and Cybenko, G. V. (2002). "A fuzzy MHT algorithm applied to text-based information tracking," *IEEE Transactions on Fuzzy Systems*, 10(3).

Aliferis, C., and Cooper, G. (1994). "An evaluation of an algorithm for inductive learning of Bayesian Belief Networks using simulated datasets," *Proceedings of the 10th Conference on Uncertainty in Artificial Intelligence* (*UAI*), Seattle, WA.

Allen, J. F. (1984). "Toward a general theory of action and time," *Artificial Intelligence*, 23(2):123–154.

Anderson, J. R. (1996). "ACT: A simple theory of complex cognition," *American Psychologist*, 51:355–365.

Anderson, J. R., Bothell D., Byrne, M. D., Douglass, S., Lebiere, C., and Qin Y. (2004). "An integrated theory of the mind," *Psychological Review*, 111(4):1036–1060.

Antanitus, D. (2003). "FORCEnet Architecture," *Briefing to National Defense Industrial Association* (*NDIA*) *Conference*, San Diego, CA.

Anzano, E. S. (1999). "Fuzzy Clustering Means Algorithm for Track Fusion in U.S. Coast Guard Vessel Traffic Service Systems," Master's thesis, Naval Postgraduate School, Monterey, CA.

Arulampalam, S., Maskell, S., Gordon, N., and Clapp, T. (2002). "A tutorial on particle filters for on line nonlinear/non-Gaussian Bayesian tracking," *IEEE Transactions on Signal Processing*, 50:174–188.

Avitzour, D. (1995). "Stochastic simulation Bayesian approach to multitarget tracking," *IEE Proceedings on Radar and Sonar Navigation*, 142(2):41–44.

Bar-Shalom, Y. (1995). *Multitarget-Multisensor Tracking: Principles and Techniques*, YBS Publishing, Storrs, CT.

Bar-Shalom, Y., and Tse, E. (1975). "Tracking in a cluttered environment with Probabilistic Data Association," *Automatica*, 11:451–460.

Bar-Shalom, Y., and Fortmann, T. (1988). *Tracking and Data Association*, Mathematics in Science and Engineering Series, Vol. 179, Academic Press, NY.

Bar-Shalom, Y., and Li, X.-R. (1993). *Estimation and Tracking: Principles, Techniques, and Software*, Artech House, Norwood, MA.

Bar-Shalom, Y., and Li, X.-R. (1995). *Multitarget-Multisensor Tracking: Principles and Techniques*, YBS Publishing, Storrs, CT.

Bergman, N., and Doucet, A. (2000). "Markov chain Monte Carlo data association for target tracking," *Proceedings of the IEEE International Conference on Acoustics, Speech, and Signal Processing* (*ICASSP*), Istanbul, Turkey.

Bertsekas, D. (1988). "The auction algorithm: A distributed relaxation method for the assignment problem," *Annals of Operations Research, Special Issue on Parallel Optimization*, 14:105–123.

Bhowmik, S., and Sadhukhan, C. (2007). "Application of extended Kalman filter to tactical ballistic missile re-entry problem," *http://arxiv.org/ftp/arxiv/papers/0707/0707.1918.pdf.*

Binder, J., Koller, D., Russell, S., and Kanazawa, K. (1997). "Adaptive probabilistic networks with hidden variables," *Machine Learning*, 29:213–244.

Blackman, S. S., and Popoli, R. (1999). *Design and Analysis of Modern Tracking Systems*, Artech House, Norwood, MA.

Blasch, E., and Plano, S. (2003). "Level 5: User refinement to aid the fusion process," *Proceedings of the SPIE*, 5099, Multisensor, Multisource Information Fusion: Architectures, Algorithms, and Applications.

Blasch, E., and Plano, S. (2002). "JDL Level 5 fusion model: User refinement issues and applications in group tracking," *Proceedings of the SPIE*, 4729:270–279, Signal Processing, Sensor Fusion, and Target Recognition XI.

Bonissone, P., and Tong, R. (1985). "Reasoning with uncertainty in expert systems (editorial)," *International Journal of Man-Machine Studies*, 241–250.

Bourgeois, F., and Lassalle, J. C. (1971). "An extension of the Munkres algorithm for the assignment problem to rectangular matrices," *Communications of the Association for Computing Machinery*, 14(12):802–806.

Boyen, X., and Koller, D. (1998). "Tractable inference for complex stochastic processes," *Proceedings of the 14th Conference on Uncertainty in Artificial Intelligence* (*UAI*), Madison, WI.

Bratko, I. (2000). *Prolog Programming for Artificial Intelligence*, 3rd edition, Wokingham, Addison-Wesley, England.

Brooks, R. A. (1986). "A robust layered control system for a mobile robot," *IEEE Journal of Robotics and Automation*, RA-2:14–23.

Buntine, W. (1991). "Theory refinement on Bayesian networks," *Proceedings of the 7th Annual Conference on Uncertainty in Artificial Intelligence* (*UAI*), Los Angeles, CA.

Buntine, W. (1996). "A guide to literature on learning probabilistic network from data," *IEEE Transactions on Knowledge and Data Engineering*, 8:195–210.

Byrne, M. D. (2003). "Cognitive architecture," *The Human-Computer Interaction Handbook: Fundamentals, Evolving Technologies, and Emerging Applications*, 97–117, Erlbaum, Hillsdale, NJ.

Card, S., Moran, T., and Newell, A. (1983). *The Psychology of Human-Computer Interaction*, Lawrence Erlbaum, Hillsdale, NJ.

Cebrowski, A. K. (2001). *The Implementation of Network-Centric Warfare*, Office of the Force Transformation, Washington, DC.

Cebrowski, A. K., and Garstka, J. J. (1998). "Network Centric Warfare: Its origin and future," *Proceedings of the Naval Institute*, 124(1).

Chang, C. B. (1980). "Ballistic trajectory estimation with angle only measurements," *IEEE Transaction on Automatic Control*, AC-25:474–480.

Chang, K.-C. (2005). "Distributed estimation and multisensor tracking and fusion," Presentation downloaded from the Department of Systems Engineering, George Mason University, Fairfax, VA.

Chang, C. L., and Lee, R. C. T. (1973). *Symbolic Logic and Mechanical Theorem Proving*, Academic Press, NY.

Chellas, B. F. (1980). *Modal Logic: An Introduction*, Cambridge University Press, Cambridge, England.

Chickering, D. (2002). "Learning equivalence classes on Bayesian network structures," *Journal of Machine Learning Research*, **2**:445–498.

Chow, C. K., and Liu, C. N. (1968). "Approximating discrete probability distributions with dependence trees," *IEEE Transactions on Information Theory*, **14**:462–467.

Chung, K. L. (2000). *A Course in Probability Theory*, Academic Press, NY.

Clark, K. L. (1978). "Negation as failure," *Logic and Databases*, 293–322, Plenum Press, NY.

Clocksin, W. F., and Mellish, C. S. (2003). *Programming in Prolog*, 5th edition, Springer-Verlag, Berlin, Germany.

Cobb, B., and Shenoy, P. (2003). "Inference in hybrid Bayesian networks with mixtures of truncated exponentials," *Proceedings of the 6th Workshop on Uncertainty Processing*, 47–63, Czech Republic.

Cohen, P. R., and Levesque, H. J. (1990). "Intention is choice with commitment," *Artificial Intelligence*, 42(3):13–61.

Cook, S. A. (1971). "The complexity of theorem proving procedures," *Proceedings of the 3rd Annual ACM Symposium on the Theory of Computing*, ACM Press, NY, 151–158

Cooper, G. (1990). "The computational complexity of probabilistic inference using Bayesian Belief Networks," *Artificial Intelligence*, 42.

Cooper, G., and Herskovits, E. (1992). "A Bayesian method for the induction of probabilistic networks from data," *Machine Learning*, 9:309–347.

Copi, I. M. (1979). *Symbolic Logic*, Macmillan, New York.

Dang, H., Han, C., and Duan, Z. (2003). "A new data association approach for automotive radar tracking," *Proceedings of the 6th International Conference on Information Fusion*, Cairns, Australia.

Das, S. (1992). *Deductive Databases and Logic Programming*, Addison-Wesley, Wokingham, England.

Das, S., Fox, J., Elsdon, D., and Hammond, P. (1997). "A flexible architecture for autonomous agents," *Journal of Experimental and Theoretical Artificial Intelligence*, 9(4):407–440.

Das, S., and Grecu, D. (2000). "COGENT: Cognitive agent to amplify human perception and cognition," *Proceedings of the 4th International Conference on Autonomous Agents*, Barcelona.

Das, S., Grey, R., and Gonsalves, P. (2002). "Situation assessment via Bayesian Belief Networks," *Proceedings of the 5th International Conference on Information Fusion*, Annapolis, Maryland.

Das, S., Introne, J., Lawless, D., Hoyt, R., and Muza, S. (2004). "Probabilistic Unit Life Status Estimation (PULSE)," *Proceedings of the 7th International Conference on Information Fusion*, Stockholm, Sweden.

Das, S. (2005). "Symbolic argumentation for decision making under uncertainty," *Proceedings of the 8th International Conference on Information Fusion*, Philadelphia, PA.

Das, S., Lawless, D., Ng, B., and Pfeffer, A. (2005). "Factored particle filtering for situation assessment in urban environments," *Proceedings of the 8th International Conference on Information Fusion*, Philadelphia, PA.

Das, S., Kanjilal, P., and Lawless, D. (2006). "Spatiotemporal clustering for aggregating hostile units in cluttered environments," *Proceedings of the 9th International Conference on Information Fusion*, Florence, Italy (July).

Das, S. (2007). "Knowledge acquisition issue for model based Level 2/3 Fusion," Position Paper at the Invited Panel on Results from Levels 2/3 Fusion Implementations: Issues, Challenges, Retrospectives and Perspectives for the Future, *Proceedings of the 10th I nternational Conference on Information Fusion*, Quebec City, Canada.

Das, S. (2007). "Envelope of Human Cognition for Battlefield Information Processing Agents," *Proceedings of the 10th International Conference on Information Fusion*, Quebec, Canada.

Das, S. (2008). *Foundations of Decision-Making Agents: Logic, Probability, and Modality*, World Scientific/Imperial College Press, Singapore/London.

Dean, T., Firby, R. J., and Miller, D. (1988). "Hierarchical planning involving deadlines, travel time, and resources," *Computational Intelligence*, 4:381–398.

Deb, S., Pattipati, K., and Bar-Shalom, Y. (1977). "A generalized S-D assignment for multisensor multitarget state estimation," *IEEE Transaction on Aerospace and Electronic System*, 33(2):523–538.

Deerwester, S., Dumais, S. T., Furnas, G. W., Landauer, T. K., and Harshman R. (1990). "A. Indexing by latent semantic analysis," *Journal of the Society for Information Science*, 41(6):391–407.

Dempster, A. P. (1966). "New methods for reasoning towards posterior distributions based on sample data," *Annals of Mathematical Statistics*, **37**:355–374.

Dempster, A., Laird, N., and Rubin, D. (1977). "Maximum likelihood from incomplete data via the EM algorithm," *Journal of Royal Statistical Society B*, 39:1–38.

Deo, N. (1974). *Graph Theory with Applications to Engineering and Computer Science*, Prentice Hall, Englewood Cliffs, NJ.

Djuric, P., Kotecha, J., Zhang, J., Huang, Y., Ghirmai, T., Bugallo, M., and Miguez, J. (2003). "Particle filtering," *IEEE Signal Processing Magazine*, 20(5):19–38.

Doucet, A., de Freitas, N., Murphy, K., and Russel, S. (2000a). "Rao-Blackwellised particle filtering for dynamic Bayesian networks," *Proceedings of the 16th Conference on Uncertainty in Artificial Intelligence* (*UAI*), Stanford, CA.

Doucet, A., de Freitas, N., Murphy, K., and Russel, S. (2000b). "A simple tutorial on RBPF for DBNs," *http://www.cs.ubc.ca/~nando/software.html.*

Doucet, A., de Freitas, N., and Gordon, N. (eds.). (2001). *Sequential Monte Carlo Methods in Practice*, Springer-Verlag, NY.

Druzdzel, M. J., and van der Gaag, L. C. (1995). "Elicitation of probabilities for belief networks: Combining qualitative and quantitative information," *Proceedings of the 11th Conference on Uncertainty in Artificial Intelligence* (*UAI*), Montreal, Canada.

Dubois, D., and Prade, H. (1988). *Possibility Theory*, Plenum Press, NY.

Dubois, D., and Prade, H. (1993). "On possibility/probability transformations," *Fuzzy Logic*, 103–112.

Duda, R., and Hart, P. (1973). *Pattern Recognition and Scene Analysis*, Wiley, NY.

Duda, R. O., Hart, P. E., and Stork, D. G. (2001). *Pattern Classification*, 2nd edition, John Wiley & Sons, NY.

Dumais, S. T., Furnas, G. W., Landauer, T. K., and Deerwester, S. (1988). "Using latent semantic analysis to improve information retrieval," *Proceedings of CHI'88: Conference on Human Factors in Computing*, Washington, DC.

Durrant-Whyte, H. F. (2000). *A Beginners Guide to Decentralised Data Fusion*, Australian Centre for Field Robotics, The University of Sydney, Australia.

Endsley, M. R. (1988). "Design and evaluation for situation awareness enhancement," *Proceedings of the 32nd Annual Meeting of the Human Factors Society*, 97–101.

Endsley, M. (1995). "Towards a theory of situation awareness in dynamic systems," *Human Factor Journal*, 37(1):32–64.

Feller, W. (1968). *An Introduction to Probability Theory and Its Applications*, Vol. 1, John Wiley & Sons, NY.

Ernst, G., and Newell, A. (1969). *GPS: A Case Study in Generality and Problem Solving*, Academic Press, NY.

Fagin, R., and Halpern, J. Y. (1988). "Belief, awareness and limited reasoning," *Artificial Intelligence*, 34:39–76.

Ferguson, I. A. (1992). "Touring Machines: An Architecture for Dynamic, Rational, Mobile Agents," Ph.D. dissertation, Computer Laboratory, University of Cambridge.

Fine, S., Singer, Y., and Tishby, N. (1998). "The hierarchical hidden Markov model: Analysis and applications," *Machine Learning*, 32:42–62.

Fischer, K., Muller, J. P., and Pischel, M. (1994). "Unifying control in a layered agent architecture," *Technical Report TM 94-05*, DFKI GmbH.

Fodor, J. A., and Pylyshyn, Z. (1988). "Connectionism and cognitive architecture: A critical analysis," *Cognition*, 28:3–71.

Fox, J., and Das, S. (2000). *Safe and Sound: Artificial Intelligence in Hazardous Applications*, AAAI-MIT Press, Cambridge, MA.

Frank, O. (2003). *Multiple Target Tracking*, Dipl. El. Ing. Thesis, Swiss Federal Institute of Technology Zurich (ETH).

Friedman, N. (1998). "The Bayesian structural EM algorithm," *Proceedings of the 14th International Conference on Uncertainty in Artificial Intelligence* (*UAI*), Madison, WI.

Friedman, N., and Goldszmidt, M. (1996). "Learning Bayesian networks with local structure," *Proceedings of the 12th International Conference on Uncertainty in Artificial Intelligence* (*UAI*), Portland, OR.

Friedman, N., and Koller, D. (2003). "Being Bayesian about network structure: A Bayesian approach to structure discovery in Bayesian networks," *Machine Learning*, 50:95–126.

Fung, R. M., and Crawford, S. L. (1990). "Constructor: A system for the induction of probabilistic models," *Proceedings of the 8th Nation al Conference on Artificial Intelligence* (*AAAI*), Boston, MA.

Gallier, J. H. (2003). *Logic for Computer Science*, John Wiley & Sons, NY. Revised online version, *http://www.cis.upenn.edu/~jean/gbooks/ logic.html.*

Gardenfors, P. (1988). *Knowledge in Flux: Modeling the Dynamics and Epistemic States*, MIT Press, Cambridge, MA.

Geman, S., and Geman, D. (1984). "Stochastic relaxation, Gibbs distributions and the Bayesian restoration of images," *IEEE Transaction on Pattern Analysis & Machine Intelligence*, 6:721–742.

Ghahramani, Z. (2001). "An introduction to Hidden Markov Models and Bayesian networks," *International Journal of Pattern Recognition and Artificial Inteligence*, 15(1):9–42.

Gilks, W., Richerdson, S., and Spiegelhalter, D. (eds.). (1996). *Markov Chain Monte Carlo in Practice*, Chapman and Hall, London.

Ginsberg, M. L. (ed.). (1987). *Readings in Nonmonotonic Reasoning*, Morgan Kaufmann, Los Altos, CA.

Golub, G., and Van Loan, C. F. (1996). *Matrix Computations*, 3rd edition, The Johns Hopkins University Press, Baltimore.

Gordon, N. J. (1997). "A hybrid bootstrap filter for target tracking in clutter," *IEEE Transactions on Aerospace and Electronic Systems*, 33(1):353–358.

Gordon, N., Salmond, D., and Smith, A. (1993). "Novel approach to nonlinear/non-Gaussian Bayesian state estimation," *IEE Proceedings on Radar and Signal Processing*, 140(2):107–113.

Handschin, J., and Mayne, D. (1969). "Monte Carlo technique to estimate the conditional expectation in multi-stage nonlinear filtering," *International Journal of Control*, 9(5):547–559.

Harper, K., Ton, N., Jacobs, K., Hess, J., and Zacharias, G. (2001). "GRADE: Graphical Agent Development Environment," *Proceedings of the 10th Conference on Computer Generated Forces and Behavior Representation*, Norfolk, VA.

Hall, D., and Llinas, J. (2001). *Handbook of Multisensor Data Fusion*, CRC Press, Boca Raton, FL.

Heckerman, D. E. (1996). "A tutorial on learning Bayesian networks," *Technical Report: MSR-TR-95-06*, Microsoft Corporation, Redmond, WA.

Heckerman, D. E., and Shortliffe, E. H. (1992). "From certainty factors to belief networks," *Artificial Intelligence in Medicine*, **4**:35–52.

Heckerman, D. E., Geiger, D., and Chickering, D. M. (1995). "Learning Bayesian networks: The combination of knowledge and statistical data," *Machine Learning*, 20(3).

Herskovits, E., and Cooper, G. (1990). "Kutato: An entropy-driven system for construction of probabilistic expert systems from databases," *Proceedings of the 6th International Conference on Uncertainty in Artificial Intelligence* (*UAI*), Cambridge, MA.

Hintikka, J. (1962). *Knowledge and Belief: An Introduction to the Logic of the Two Notions*, Cornell University Press, NY.

Hojsgaard, S., and Thiesson, B. (1995). "BIFROST: Block recursive models induced from relevant knowledge, observations, and statistical techniques," *Computational Statistics and Data Analysis*, 19:155–175.

Howard, R. A., and Matheson, J. E. (1981). "Influence diagrams," *Readings on the Principles and Applications of Decision Analysis*, 721–762, Strategic Decisions Group, Menlo Park, CA.

Huang, C., and Darwiche, A. (1996). "Inference in Belief Networks: A procedural guide," *International Journal of Approximate Reasoning*, 15(3):225–263.

Hudlicka, E. (2004). "Beyond cognition: Modeling emotion in cognitive architectures," *Proceedings of the 6th International Conference on Cognitive Modeling*, 118–123, Lawrence Earlbaum, Mahwah, NJ.

Hue, C., Le Cadre, J.-P., and Perez, P. (2002). "Sequential Monte Carlo methods for multiple target tracking and data fusion," *IEEE Transaction on Signal Processing*, 50(2):309–325.

Hughes, G. E., and Cresswell, M. J. (1996). *A New Introduction to Modal Logic*, Routledge, London.

Ingrand, F. F., Georgeff, M. P., and Rao. A. S. (1992). "An architecture for real-time reasoning and system control," *IEEE Expert*, 7(6).

Isard, M., and Blake, A. (1998). "Condensation – Conditional density propagation for visual tracking," *International Journal of Computer Vision*, 29:5–28.

Jackson, P. (1998). *Introduction to Expert Systems*, Addison-Wesley, Wokingham, England.

Jain, A. K., Murthy, M. N., and Flynn, P. J. (1999). "Data clustering: A review," *ACM Computing Surveys*, 31(3):264–323.

Jensen, F. V., Lauritzen, S. L., and Olesen. K. G. (1990). "Bayesian updating in causal probabilistic networks by local computations," *Computational Statistics*, Q. 4, 269–282.

Jensen, F., Jensen, F. V., and Dittmer, S. L. (1994). "From influence diagrams to junction trees," *Proceedings of the 10th Conference on Uncertainty in Artificial Intelligence* (*UAI*), Seattle, WA.

Jensen, F. V. (1996). *An Introduction to Bayesian Networks*, Springer-Verlag, NY.

Jensen, F. V. (2002). *Bayesian Networks and Decision Graphs*, Springer-Verlag, NY.

Julier, S. J., and J. K. Uhlmann. (1997). "A new extension of the Kalman filter to nonlinear systems," *P roceedings of the 11th International Symposium on Aerospace/Defense Sensing*, Simulation and Controls, Multi Sensor Fusion, Tracking and Resource Management II, SPIE.

Just, M. A., and Carpenter, P. N. (1992). "A capacity theory of comprehension: Individual differences in working memory," *Psychological Review*, 99, 122–149.

Kalata, P. (1992). "$\alpha - \beta$ target tracking systems: A survey," *Proceedings of the American Control Conference*, 832–836.

Kalman, R. (1960). "A new approach to linear filtering and prediction problems," *Journal of Basic Engineering*, 82:35–45.

Kanazawa, K., Koller, D., and Russell, S. (1995). "Stochastic simulation algorithms for dynamic probabilistic networks," *Proceedings of the 11th Conference on Uncertainty in Artificial Intelligence* (*UAI*), Montreal, Canada.

Kanjilal, P., Das, S., and Lawless, D. (2007). "Spatiotemporal clustering from noisy data," *Proceedings of the IJCAI Workshop on Spatial and Temporal Reasoning*, Hyderabad, India.

Karlsson, R., Gustafsson, F., and Karlsson, T. (2003). "Particle filtering and Cramer-Rao lower bound for underwater navigation," *Proceedings of the IEEE Conference on Acoustics, Speech, and Signal Processing* (ICASSP), Hong Kong.

Karp, R. (1972). "Reducibility among combinatorial problems," *Complexity of Computer Computations*, 85–103, Plenum Press, NY.

Kieras, D. (2005). "A survey of cognitive architectures," *http://www.cs.cmu.edu /~bej/CognitiveModelingForUIDesign/*.

Kieras, D., and Meyer, D. E. (1997). "An overview of the EPIC architecture for cognition and performance with application to human-computer interaction," *Human-Computer Interaction*, 12:391–438.

Klein, G., Ross, K. G., Moon, B. M., Klein, D. E., Hoffman, R. R., and Hollnagel, E. (2003). "Macrocognition," *Human-Centered Computing*, IEEE Computer Society Press, Washington, DC.

Klir, G. J. (1993). "Information-preserving probability-possibility transformations: Recent developments," *Fuzzy Logic*, 417–428.

Klir, G. J., and Yuan, B. (1995). *Fuzzy Sets and Fuzzy Logic: Theory and Applications*, Prentice Hall, Saddle River, NJ.

Kong, A., Liu, J. S., and Wong, W. H. (1994). "Sequential imputation method and Bayesian missing data problems," *Journal of the American Statistical Association*, 89:278–288.

Kowalski, R. A. (1979). "Algorithm = Logic + Control," *Communications of the ACM*, 22(7):424–435.

Krause, P. (1998). "Learning probabilistic networks," *The Knowledge Engineering Review*, 13:321–351.

Kripke, S. A. (1963). "Semantical analysis of modal logic I: Normal modal propositional calculi," *Zeitschrift fur Mathematische Logik und Grundladen der Mathematik* (*ZMLGM*), 9:67–96.

Laird, J. E., Newell, A., and Rosenbloom, P. S. (1987). "Soar: An architecture for general intelligence," *Artificial Intelligence*, 33:1–64.

Lam, W., and Bacchus, F. (1994). "Learning Bayesian Belief Networks: An approach based on the MDL principle," *Computational Intelligence,* 10(4).

Langley, P., McKusick, K. B., Allen, J. A., Iba, W. F., and Thompson, K. (1991). "A design for the ICARUS architecture," *SIGART Bulletin*, 2:104–109.

Langley, P. (2005). "An adaptive architecture for physical agents," *Proceedings of the 2005 IEEE/WIC/ACM International Conference on Intelligent Agent Technology*, IEEE Computer Society Press, Washington, DC.

Lauritzen, S. (1995). "The EM algorithm for graphical association models with missing data," *Computational Statistics and Data Analysis*, 19:191–201.

Lauritzen, S., and Spiegelhalter, D. (1988). "Local computations with probabilities on graphical structures and their applications to experts systems," *Journal of Royal Statistical Society B*, 50(2):154–227.

Leake, D. B. (ed.). (1996). *Case-Based Reasoning: Experiences, Lessons, and Future Directions*, AAAI Press/MIT Press, Cambridge, MA.

Lehman, J. F., Laird, J., and Rosenbloom, P. (2005). "A gentle introduction to Soar, an architecture for human cognition," *An Invitation to Cognitive Science (2nd edition), Volume 4: Methods, Models and Conceptual Issues*, MIT Press, Cambridge, MA.

Lemmon, E. J. (1977). "*An introduction to modal logic*," American Philosophical Quarterly, Monograph No. 11, University of Illinois Press, Champaign, IL.

Lerner, U., Segal, E., and Koller, D. (2001). "Exact inference in networks with discrete children of continuous parents," *Proceedings of the 17th Annual Conference on Uncertainty in Artificial Intelligence* (*UAI*), Seattle, WA.

Leung, H., and Wu, J. (2000). "Bayesian and Dempster-Shafer target identification for radar surveillance," *IEEE Transactions on Aerospace and Electronic Systems*, 36(2):432–447.

Lewis, R. L. (1999). "Cognitive modeling, symbolic," *The MIT Encyclopedia of the Cognitive Sciences*, MIT Press, Cambridge, MA.

Lichtblau, D. E. (2004). "The critical role of intelligent software agents in enabling net-centric command and control," *Command and Control Research and Technology Symposium, The Power of Information Age Concepts and Technologies*, San Diego, CA.

Liggins, M. E., Chong, C.-Y., Kadar, I., Alford, M. G., Vannicola, V., and Thomopoulos, S. (1997). "Distributed fusion architectures and algorithms for target tracking," *Proceedings of the IEEE*, 85(1):95–107.

Llinas, J., Bowman, C., Rogova, G., Steinberg, A., Waltz, E., and White, F. (2004). "Revisiting the JDL data fusion model II," *Proceedings of the 7th International Conference on Information Fusion*, Sweden.

Lloyd, J. W. (1987). *Foundations of Logic Programming*, 2nd edition, Springer-Verlag, Berlin, Germany.

Łukasiewicz, J. (1970). *Selected Works* (ed.: L. Borkowski), North-Holland Publication Company, Amsterdam and PWN, Warsaw.

Magee, D., and Boyle, R. (2002). "Detecting lameness using 're-sampling condensation' and 'multi-stream cyclic Hidden Markov Models,'" *Image and Vision Computing*, 20:581–594.

McCarthy, J. (1980). "Circumscription: A form of nonmonotonic reasoning," *Artificial Intelligence*, 13:27–39.

Meila, M., and Jordan, M. I. (1998). "Estimating dependency structure as a hidden variable," *Advances in Neural Information Processing Systems* (*NIPS*), 10, MIT Press, Cambridge, MA.

Mendelson, E. (1987). *Introduction to Mathematical Logic*, Wadsworth & Brooks/Cole Advanced Books and Software, Pacific Grove, CA.

Meyer, F. G., and Chinrungrueng, J. (2005). "Spatiotemporal clustering of fMRI time series in the spectral domain," *Medical Image Analysis*, 9:51–68.

Meyer, J.-J. Ch., and van der Hoek, W. (1995). "Epistemic logic for AI and computer science," *Cambridge Tracts in Theoretical Computer Science*, 41, Cambridge University Press, Cambridge, England.

Minton, S. (1990). "Quantitative results concerning the utility of explanation-based learning," *Artificial Intelligence*, 42:363–391.

Mitchell, T. (1997). *Machine Learning*. McGraw-Hill, NY.

Morefield, C. (1977). "Application of 0-1 integer programming to multitarget tracking problems," *IEEE Transaction on Automatic Control*, 22(3):302–312.

Mulgund, S., Harper, K., Zacharias, G., and Menke, T. (2000). "SAMPLE: Situation Awareness Model for Pilot-in-the-Loop Evaluation," *Proceedings of the 9th Conference on Computer Generated Forces and Behavior Representation*, Orlando, FL.

Muller, J. P., and Pischel, M. (1994). "Modeling interacting agents in dynamic environments," *Proceedings of the 11th European Conference on Artificial Intelligence* (*ECAI*), 709–713.

Munkres, J. (1957). "Algorithms for assignment and transportation problems," *Journal of the Society of Industrial and Applied Mathematics*, 5(1):32–38.

Murphy, K. (2002). "Dynamic Bayesian Networks: Representation, Inference and Learning," Ph.D. thesis, University of California, Berkeley.

NCW. (2001). *Network Centric Warfare*, Department of Defense, Report to Congress.

Neapolitan, R. E. (2003). *Learning Bayesian Networks*, Prentice Hall, Upper Saddle River, NJ.

Newell, A. (1990). *Unified Theories of Cognition*, Harvard University Press, Cambridge, MA.

Newell, A., and Simon, H. (1972). *Human Problem Solving*, Englewood Cliffs, Prentice Hall, NJ.

O'Keefe, R. A. (1990). *Craft of Prolog*, MIT Press, Cambridge, MA.

ODNI. (2008). "How do we collect intelligence?" Office of the Director of National Intelligence, *http://www.dni.gov/who_what/what_collection.htm*.

OFT. (2003). *Operational Sense and Respond Logistics: Co-Evolution of an Adaptive Enterprise*, Concept Document, Office of Force Transformation.

OODA Loop. (2001). *http://www.d-n-i.net/fcs/ppt/boyds_ooda_loop.ppt*.

Papadimitriou, C. H. (1993). *Computational Complexity*, Addison Wesley, Reading, MA.

Paskin, M., and Guestrin, C. (2004). "Robust probabilistic inference in distributed systems," *Proceedings of the 20th Conference on Uncertainty in Artificial Intelligence (UAI)*, Banff, Canada.

Pearl, J. (1986). "A constraint-propagation approach to probabilistic reasoning," *Proceedings of the 2nd Conference on Uncertainty in Artificial Intelligence (UAI)*, Philadelphia, PA.

Pearl, J. (1988). *Probabilistic Reasoning in Intelligent Systems: Networks of Plausible Inference*, Morgan Kaufmann, San Mateo, CA.

Pearl, J. (1990). "Reasoning under uncertainty," *Annual Review of Computer Science*, 4:37–72.

Pfeffer, A., Das, S., Lawless, D., and Ng, B. (2007). "Global/local dynamic models," *Proceedings of the 20th International Joint Conference on Artificial Intelligence (IJCAI)*, Hyderabad, India.

Pincus, S. M. (1991). "Approximate entropy as a measure of system complexity," *Proceedings of the National Academy of Science*, 88:2297–2301.

Pitt, M., and Shephard, N. (1999). "Filtering via simulation: Auxiliary particle filters," *Journal of the American Statistical Association*, 94(446).

Quinlan, J. R. (1986). "Induction of decision trees," *Machine Learning*, 1(1):81–106.

Quinlan, J. R. (1993). *C4.5: Programs for Machine Learning*, Morgan Kaufmann, San Francisco, CA.

Rabiner, L. (1989). "A tutorial on Hidden Markov Models and selected applications in speech recognition," *Proceedings of the IEEE* 77, 257–286.

Ramoni, M., and Sebastiani, P. (1997). "Discovering Bayesian Networks in incomplete databases," *Technical Report KMI-TR-46*, Knowledge Media Institute, The Open University, Milton Keynes, UK.

Rao, A. S., and Georgeff, M. P. (1991). "Modeling rational agents within a BDI architecture," *Proceedings of the 2nd Conference on Knowledge Representation and Reasoning*, Cambridge, MA.

Rao, B. S., and Durrant-Whyte, H. F. (1991). "Fully decentralised algorithm for multisensor Kalman filtering," *IEE Proceedings-Control Theory and Applications*, 138(5):413–420.

Rasmussen, J. (1983). "Skills, rules and knowledge: Signals, signs and symbolism, and other distinctions in human performance models," *IEEE Transactions on Systems, Man, and Cybernetics*, 12:257–266.

Rasmussen, J. (1986). *Information Processing and Human Machine Interaction: An Approach to Cognitive Engineering*, North Holland, NY, 1986.

Reason, J. (1990). *Human Error*, Cambridge University Press, Cambridge, UK.

Rebane, G., and Pearl, J. (1987). "The recovery of causal poly-trees from statistical data," *Proceedings of the Uncertainty in Artificial Intelligence (UAI)*, Seattle, WA.

Reid, D. B. (1979). "An algorithm for tracking multiple targets," *IEEE Transactions on Automatic Control (AC)*, 24(6):843–854.

Reiter, R. (1978). "On closed world databases," *Logic and Databases*, 55–76, Plenum Press, NY.

Reiter, R. (1980). "A logic for default reasoning," *Artificial Intelligence*, 13:81–132.

Robinson, J. A. (1965). "A machine-oriented logic based on the resolution principle," *Journal of the Association for Computing Machinery (ACM)*, 12:23–41.

Rumelhart, D. E. (1984). "The emergence of cognitive phenomena from sub-symbolic processes," *Proceedings of the 6th Annual Conference of the Cognitive Science Society*, Boulder, Colorado, Erlbaum, Hillsdale, NJ.

Russel, J., Binder, J. Koller, D., and Kanazawa, K. (1995). "Local learning in probabilistic networks with hidden variables," *Proceedings of the 14th International Joint Conference on Artificial Intelligence (IJCAI)*, Montreal, Canada.

Russell, S. J., and Norvig, P. (2002). *Artificial Intelligence: A Modern Approach*, Prentice Hall, Upper Saddle River, NJ.

Sahami, M. (1996). "Learning limited dependence Bayesian classifiers," *Proceedings of the 2nd International Conference on Knowledge Discovery and Data Mining*, Portland, OR.

Salerno, J., Hinman, M., and Boulware, D. (2004). "Building a framework for situation awareness," *Proceedings of the 7th International Conference on Information Fusion*, Stockholm, Sweden.

Schlosser, M. S., and Kroschel, K. (2004). "Communication issues in decentralized Kalman filters," *Proceedings of the 7th International Conference on Information Fusion*, Stockholm, Sweden.

Schulz, D., Fox, D., and Hightower, J. (2003). "People tracking with anonymous and ID-sensors using Rao-Blackwellised particle filters," *Proceedings of the*

International Joint Conference on Artificial Intelligence (*IJCAI*), Acapulco, Mexico.

Shachter, R. D. (1986). "Evaluating influence diagrams," *Operations Research*, 34:871–882.

Shachter, R. D., and Heckerman, D. (1987). "Thinking backward for knowledge acquisition," *AI Magazine*, **8**(3):55–61.

Shachter, R. D., and Peot, M. (1990). "Simulation approaches to general probabilistic inference on Belief Networks," *Proceedings of the 6th Conference on Uncertainty in Artificial Intelligence* (*UAI*), Cambridge, MA.

Shafer, G. (1976). *Mathematical Theory of Evidence*, Princeton University Press, Princeton, NJ.

Shenoy, P. P. (1992). "Valuation-based systems for Bayesian decision analysis," *Operations Research*, 40(3):463–484.

Shortliffe, E. H. (1976). *Computer-based Medical Consultations: MYCIN*, Elsevier, NY.

Shortliffe, E. H., and Buchanan, B. G. (1975). "A Model of Inexact Reasoning in Medicine," *Mathematical Biosciences*, **23**(3/4):351–379.

Singh, M. (1997). "Learning Bayesian networks from incomplete data," *Proceedings of the 14th National Conference on Artificial Intelligence* (*AAAI*), Providence, RI.

Singh, M., and Valtorta, M. (1995). "Construction of Bayesian network structures from data: A brief survey and an efficient algorithm," *International Journal of Approximate Reasoning*, 12:111–131.

Smets Ph. (1988) Belief Functions.

Smets, P. (1988). "Belief functions," In *Non Standard Logics for Automated Reasoning*, 253–286, Academic Press, London.

Smets, P. (1991). "Varieties of Ignorance and the need for well founded theories," *Information Sciences*, 57–58:135–144.

Smets, P. (1994). "Non standard probabilistic and non probabilistic representations of uncertainty," *Proceedings of the 5th International Conference on Information Processing and Management of Uncertainty in Knowledge-Based Systems* (IPMU), Paris, France.

Smets, P., and Kennes, R. (1994). "The transferable belief model," *Artificial Intelligence*, 66:191–234.

Smith, E. A. (2002). *Effects Based Operations – Applying Network Centric Warfare in Peace, Crisis, and War*, DoD Command and Control Research Program, CCRP.

Sowa, J. F., and Majumdar, A. K. (2003). "Analogical reasoning," *Conceptual Structures for Knowledge Creation and Communication*, *LNAI*, 2746, Springer-Verlag.

Spiegelhalter, D., and Lauritzen, S. (1990). "Sequential updating of conditional probabilities on directed graphical structures," *Networks*, 20:579–605.

Spirtes, P., Glymour, C., and Scheines, R. (1990). "Causality from probability," *Proceedings of Advanced Computing for the Social Sciences*, Williamsburg, VA.

Srinivas, S., Russell, S., and Agogino, A. M. (1990). "Automated construction of sparse Bayesian networks from unstructured probabilistic models and domain information," *Proceedings of the 5th Conference on Uncertainty in Artificial Intelligence* (*UAI*), New York, NY.

Steinberg, A. N., Bowman, C. L., and White, Jr., F. E. (1998). "Revisions to the JDL data fusions models," *Proceedings of the 3rd NATO/IRIS Conference*, Quebec City, Canada.

Stephanou, H. E., and Lu, S.-Y. (1988). "Measuring consensus effectiveness by a generalized entropy criterion," *IEEE Transactions on Pattern Analysis and Machine Intelligence*, 10(4):544–554.

Sterling, L., and Shapiro, E. (1994). *The Art of Prolog*, MIT Press, Cambridge, MA.

Stevens, M. R., and Snorrason, M. (2003). "Automatic target segmentation using PMMW imagery," *Image and Vision Computing*, 21(7), Elsevier Publishing.

Stockmeyer, L. J., and Meyer, A. R. (1973). "Word problems requiring exponential time," *Proceedings of the 5th ACM Symposium on Theory of Computing*, 1–9.

Stone, L. D., Barlow, C. A., and Crowin, T. L. (1999). *Bayesian Multiple Traget Tracking*, Artech House, Norwood, MA.

Streit, R., and Lugiubuhl, T. (1994). "Maximum likelihood method for probabilistic multi-hypothesis tracking," *Proceedings of SPIE*, 2235:394–405.

Stoll, R. R. (1963). *Set Theory and Logic*, W. H. Freeman and Company, NY (Reprint edition (1979) by Dover Publications).

Sun, R. (2004). "Design for cognitive architectures," *Philosophical Psychology*, 17:3.

Suzuki, J. (1996). "Learning Bayesian Belief Networks based on the MDL principle: An efficient algorithm using the branch and bound technique," *Proceedings of the International Conference on Machine Learning*, Italy.

Sweet, N. (2004). "The C2 Constellation: A US Air Force Network Centric Warfare program," *Command and Control Research and Technology Symposium*, San Diego, CA.

Thiesson, B. (1995). "Accelerated quantification of Bayesian networks with incomplete data," *Proceedings of First International Conference on Knowledge Discovery and Data Mining*, Montreal, QU.

Thiesson, B., Meek, C., Chickering, D. M., and Heckerman, D. (1998). "Learning mixtures of Bayesian networks," *Proceedings of the 14th Conference on Uncertainty in Artificial Intelligence* (*UAI*), Madison, WI.

Thomas, A., Spiegelhalter, D., and Gilks, W. (1992). "Bugs: A program to perform Bayesian inference using Gibbs' sampling," *Bayesian Statistics* 4, 837–842, Oxford University Press, UK.

Toulmin, S. (1956). *The Uses of Argument*, Cambridge University Press, Cambridge, England.

USAF. (1998). "Air Force Pamphlet 14-210," *Intelligence, USAF Intelligence Targeting Guide*, Department of Defense.

Van Emden, M. H., and Kowalski, R. A. (1976). "The semantics of predicate logic as a programming language," *Journal of the Association for Computing Machinery*, 23(4):733–742.

Van Gelder, T., and Niklasson, L. (1994). "Classicalism and cognitive architecture," *Proceedings of the 16th Annual Conference of the Cognitive Science Society*, Erlbaum, Hillsdale, NJ.

Verma, T. S., and Pearl, J. (1992). "An algorithm for deciding if a set of observed independencies has a causal explanation," *Proceedings of the 8th International Conference on Uncertainty in Artificial Intelligence (UAI)*, Stanford, CA.

Wen, W. X. (1990). "Optimal decomposition of Belief Networks," *Proceedings of the 6th Conference on Uncertainty in Artificial Intelligence (UAI)*, Cambridge, MA.

White, Jr., F. E. (1987). "Data fusion lexicon," Joint Directors of Laboratories, Technical Panel for C3, Data Fusion Sub-Panel, Naval Ocean Systems Center, San Diego.

White, Jr., F. E. (1988). "A model for data fusion," *Proceedings of the 1st National Symposium on Sensor Fusion*, Vol. 2.

Wilson, R. (1996). *Introduction to Graph Theory*, Addison-Wesley, Reading, MA.

Yager, R. R., Kacprzyk, J., and Fedrizzi, M. (eds.) (1994). *Advances in the Dempster-Shafer Theory of Evidence*, John Wiley & Sons.

Yager, R. R. (2004). "On the determination of strength of belief for decision support under uncertainty Part II: Fusing strengths of belief," *Fuzzy Sets and Systems*, 142:129–142.

Zadeh, L. A. (1965). "Fuzzy sets," *Information and Control*, 8:338–353.

Zadeh, L. A. (1978). "Fuzzy sets as the basis for a theory of possibility," *Fuzzy Sets and Systems*, 1:3–28.

Zadeh, L. A. (2002). "What is fuzzy logic and what are its applications?," PowerPoint Presentation.

Zweig, G. (1998). "Speech Recognition with Dynamic Bayesian Networks," Ph.D. thesis, University of California, Berkeley.

About the Author

Subrata Das is the chief scientist at Charles River Analytics, Inc., (www.cra.com) in Cambridge, Massachusetts. Dr. Das leads R&D projects funded by DoD, DARPA, and NASA in the areas of high-level and distributed information fusion, decision-making under uncertainty, intelligent agents, planning and scheduling, and machine learning. His technical expertise includes mathematical logics, probabilistic reasoning including Bayesian belief networks, symbolic argumentation, particle filtering, and a broad range of computational artificial intelligence (AI) techniques. Dr. Das held research positions at Imperial College and Queen Mary and Westfield College, both part of the University of London. He received his Ph.D in computer science from Heriot-Watt University in Scotland and an M.Tech from Indian Statistical Institute.

Dr. Das has published many journal and conference articles. He is the author of the books *Deductive Databases and Logic Programming* (Addison-Wesley, 1992) and *Foundations of Decision Making Agents: Logic, Modality, and Probability* (World Scientific/Imperial College Press, 2008). He has also coauthored the book *Safe and Sound: Artificial Intelligence in Hazardous Applications* (MIT Press, 2000).

Dr. Das is a member of the editorial board of the journal *Information Fusion*. He is in the process of editing a special issue on agent-based information fusion for the journal. He has been a regular contributor, a technical committee member, a panel member, and a tutorial lecturer at the International Conferences on Information Fusion. He has also given several series of tutorials on multisensor data fusion on behalf of the Technology Training Corporation.

Index

The Artech House Electronic Warfare Library

Electronic Intelligence: The Analysis of Radar Signals, Second Edition, Richard G. Wiley

Electronic Warfare for the Digitized Battlefield, Michael R. Frater and Michael Ryan

Electronic Warfare in the Information Age, D. Curtis Schleher

Electronic Warfare Target Location Methods, Richard A. Poisel

EW 101: A First Course in Electronic Warfare, David Adamy

Foundations of Communications Electronic Warfare, Richard A. Poisel

High-Level Data Fusion, Subrata Das.

Information Warfare and Organizational Decision-Making, Alexander Kott, editor

Information Warfare Principles and Operations, Edward Waltz

Introduction to Communication Electronic Warfare Systems, Richard A. Poisel

Knowledge Management in the Intelligence Enterprise, Edward Waltz

Mathematical Techniques in Multisensor Data Fusion, Second Edition, David L. Hall and Sonya A. H. McMullen

Modern Communications Jamming Principles and Techniques, Richard A. Poisel

Principles of Data Fusion Automation, Richard T. Antony

Stratagem: Deception and Surprise in War, Barton Whaley

Statistical Multisource-Multitarget Information Fusion, Ronald P. S. Mahler

Tactical Communications for the Digitized Battlefield, Michael Ryan and Michael R. Frater

Target Acquisition in Communication Electronic Warfare Systems, Richard A. Poisel

For further information on these and other Artech House titles, including previously considered out-of-print books now available through our In-Print-Forever® (IPF®) program, contact:

Artech House
685 Canton Street
Norwood, MA 02062
Phone: 781-769-9750
Fax: 781-769-6334
e-mail: artech@artechhouse.com

Artech House
46 Gillingham Street
London SW1V 1AH UK
Phone: +44 (0)20-7596-8750
Fax: +44 (0)20-7630-0166
e-mail: artech-uk@artechhouse.com

Find us on the World Wide Web at: www.artechhouse.com